AF785200

RHS

Wildlife of a garden

'It is, I find, in zoology as it is in botany: all nature is so full, that that district produces the greatest variety which is the most examined.'

Gilbert White,
Letter XX to Thomas Pennant
8 October 1768

RHS

Wildlife of a garden

A THIRTY-YEAR STUDY

Jennifer Owen

CABI

CABI is a trading name of CAB International

CABI
Nosworthy Way
Wallingford
Oxfordshire OX10 8DE
UK

Tel: +44 (0)1491 832111
E-mail: info@cabi.org
Website: www.cabi.org

CABI
200 Portland Street
Boston
MA 02114
USA

Tel: +1 (0)617 682 9015
E-mail: cabi-nao@cabi.org

First published in 2010 by the Royal Horticultural Society.
This reprint published in 2026 by CABI in association with the Royal Horticultural Society.

© Royal Horticultural Society 2010, 2026. All rights, including for text and data mining, AI training, and similar technologies, are reserved. No part of this publication may be reproduced in any form or by any means, electronically, mechanically, by photocopying, recording or otherwise, without the prior permission of the copyright owners.

The views expressed in this publication are those of the author(s) and do not necessarily represent those of, and should not be attributed to, CAB International (CABI). Any images, figures and tables not otherwise attributed are the author(s)' own. References to internet websites (URLs) were accurate at the time of writing. CAB International and, where different, the copyright owner shall not be liable for technical or other errors or omissions contained herein. The information is supplied without obligation and on the understanding that any person who acts upon it, or otherwise changes their position in reliance thereon, does so entirely at their own risk. Information supplied is neither intended nor implied to be a substitute for professional advice. The reader/user accepts all risks and responsibility for losses, damages, costs and other consequences resulting directly or indirectly from using this information.

CABI's Terms and Conditions, including its full disclaimer, may be found at https://www.cabidigitallibrary.org/terms-and-conditions.

A catalogue record for this book is available from the British Library, London, UK.

ISBN-13: 9781836995531(hardback)
9781836995548 (paperback)
9781836995555 (ePDF)
9781836995562 (ePub)

DOI: 10.1079/9781836995562.0000

RHS editor: Mike Grant
Designer: Simon Garbutt
CABI Commissioning editor: Rebecca Stubbs
CABI Production editor: Tracy Head

Printed and bound in the UK by Halstan & Co. Ltd, Amersham

RHS Publisher: Helen Griffin
RHS Consultant editor: Simon Maughan
RHS Head of editorial: Tom Howard

The Royal Horticultural Society is the UK's leading gardening charity dedicated to advancing horticulture and promoting good gardening. Its charitable work includes providing expert advice and information in print, online and at its five major gardens and annual shows, training gardeners of every age, creating hands-on opportunities for children to grow plants, and sharing research into plants, wildlife, wellbeing and environmental issues affecting gardeners.

For more information, visit: www.rhs.org.uk or call 020 3176 5800.

Contents

Figures, tables & lists

Figures

Tables

Lists

Acknowledgements

IDENTIFICATION OF THE MANY INSECTS and other invertebrates trapped in the garden from 1972 to 2001 would not have been possible were it not for the unstinting cooperation of many people, who regularly identified large samples. I am indebted to Michael Archer (wasps and solitary bees), L.R. Cole (sawflies), Jon Daws (spiders), J.P. Dear (Calliphoridae and Sarcophagidae), the late Don Goddard (beetles and lacewings), Peter Kirby (Hemiptera), Derek Lott (beetles), John Mouseley (large Brachycera – Diptera), the late Denis Owen (moths and ladybirds), John E. Owen (moths), Colin Plant (lacewings), D.T. Richardson (myriapods), the late John Sankey (harvest-spiders), John Sellick (Psocoptera and psociform Neuroptera), David Sheppard (sawflies), the late C.J. Smith (spiders), the late Henry and Marjorie Townes (Ichneumonidae etc.) and Phil Withers (Psychodidae).

Additional identifications were made by R. Belshaw, U. Bowen, J.F. Burton, Ian Evans, Nona Finch, the late Don Hall-Smith, L. Jones-Walters, Richard Owen, Keith Porter, Denis Ratcliffe, Adrian Rundle, K.G. Smith, Mark Sterling, Alan Stubbs, S.L. Sutton and the late Derek Whiteley. Ken Thompson kindly supplied copies of all publications arising from the BUGS project. Janet Dawson of Leicester Museums, Arts and Records Services has always been willing to help with enquiries and checklists. Eric Bonser of the Newtown Linford Meteorological Station kindly supplied weather data for the period 1987–2001. Technical staff in the Department of Applied Biology at De Montfort University, Leicester, provided chemicals and glassware. Cambridge University Press have been very helpful in dealing with various copyright issues.

To all these I tender my sincere thanks. Without their endeavours, there would have been no book to write, although any errors in the text are mine.

Jennifer Owen, Leicester, September 2010

Foreword

In my modest library I have a shelf reserved for what I regard as 'important' books. And here will reside this extraordinary treatise.

Within the burgeoning list of wildlife gardening books to which I have contributed there is a plethora of fanciful advice on how to improve your patch for wildlife. But this book is different: it's a superb account of exactly what that wildlife is. It's a giant of a book, the best of British, the handiwork of an endangered species – a naturalist who has found untold riches on her doorstep and who has cultured the ability to identify, record and represent it in an eminently readable way.

In 1972 Jennifer Owen began a precise audit of her Leicester garden and these are the results, the analysis and conclusions. It describes the most intensively studied 741m^2 of very typical suburbia on record. Indeed, forget gardens, this catalogue of skilful toil has few rivals anywhere. What makes it so important is not that this able naturalist has so diligently snooped at her space, but that she has kept it up for more than thirty years. Continuous, accurate standardised data …scientific gold. And how it glistens! Indeed, it pertinently warns, with justified examples, of the caution that should be applied to short-term studies. It also highlights trends that are sympathetic with changes in our climate, not in Antarctica but in the English Midlands – in her own back yard, and ours.

And such revelations are not rare in this book. For example, our pre-occupation with the value of native species as invertebrate food plants clearly needs a rethink. In this garden 46 species of moth larvae used 40 native plants, while 38 species used 75 alien plants. And the caterpillars of just one species fed on 12 native but a staggering 42 alien plants.

But it's not all joy to behold. Predictably, the trends for the vast majority of species from all taxa have been steadily or steeply downward over the study period. Most species use the garden primarily as a feeding station. Thus it is a remarkably rich environment but it's no utopia either. Like the rest of us, Jennifer is short of house sparrows and hedgehogs. The sheer abundance of wildlife in the 1970s has become a sadly accurate statistical memory: bars on a chart, fatter figures further up the page.

It concludes with some stimulating ideas concerning the value of gardens as a wildlife resource in the UK, both now and in the immediate future. It leaves those of us who have 'garden lists' awestruck, although Jennifer does admit that she didn't record housefly in her study, and hardcore urban birders will think the bird list is short of a few 'certs' such as nocturnal migrant waders.

But, all in all, and with apologies to William Blake,

'To see a world in a scrap of urban land
And heaven in a non-native flower,
Turn the pages in the palm of your hand
And read this book for an hour.'

Chris Packham, New Forest, October 2010

New introduction to the reprint

Jennifer Owen's 30-year study remains one of the most remarkable pieces of ecological research ever undertaken in Britain and is on a par with Darwin's (1881) best-selling work on earthworms. It is a work of record whose value has only deepened with time, and whose meticulous observations continue to shape how we understand gardens, biodiversity, and the ecological potential of our cultivated spaces. The RHS, and the broader biodiversity community, regard this volume not simply as a classic, but as a definitive document, one that secures a uniquely invaluable body of research for future generations.

What Jennifer Owen achieved in her suburban Leicestershire garden from 1972 to 2001 is unprecedented. No garden, indeed, arguably no single habitat or location, in Britain has been studied with such sustained intensity across so many taxa. The scale of the research has not been replicated since, nor is it likely to be. In an era before digital photography, smartphones, or automated sensors, Owen assembled a unique dataset from daily observations to seasonal analyses, from malaise trap catches to plant lists. Her archive stands as a monument to ecological dedication, the product of patience, skill, and a naturalist's eye sharpened through decades of close watching.

The context of its creation is nearly as compelling as the results. Owen transformed an 'ordinary' suburban garden into one of the most intensively studied ecological sites in Europe. This work demonstrates that gardens are not simply a refuge for wildlife but a functioning habitat and a dynamic ecosystem. The concept that gardens could be habitats *in their own right*, rather than artificial backdrops, or merely refuges, was radical when her work began. Today, gardens are increasingly recognised for their value as biodiversity hotspots, with more than half of UK butterfly, reptile and amphibian species, and over 40% of birds and mammals being found in the UK's gardens (The Royal Horticultural Society, 2025).

Indeed, *RHS Wildlife of a Garden* is more than a wildlife gardening book; it is an ecological study of a habitat type that was barely recognised at the time. As Owen herself argued, gardens comprise a contrived plant diversity of native and non-native plants and structural heterogeneity, maintaining an "unstable community" through continuous management, yet this very dynamism creates niches and food webs of remarkable complexity. Her insights inspired projects including the Sheffield BUGS Project (https://wlgf.org/garden-science/garden-ecology/sheffield-bugs-project), which investigated the biodiversity supported by gardens across whole cities, and influenced research such as RHS 'Plants for Bugs' (https://www.rhs.org.uk/science/conservation-biodiversity/plants-for-bugs), which explored the role of non-native plants and demonstrated how garden plant choice supports biodiversity.

The influence of this book on the gardening world has been profound: it catalysed national conversations about native versus alien plants; influenced guidance on gardening for pollinators and other wildlife; and helped shift cultural perceptions of the garden from ornament to ecosystem.

Its impact extended far beyond horticulture. From ladybirds to psocids, from bumblebees to moths, Owen describes patterns and seasonal rhythms of hoverflies, annual fluctuations in moth populations, and the staggering diversity of beetles in a small garden. The study even includes six species of parasitoid wasp new to science, a testament to the intensity and consistency of Owen's observations, and the expertise of those who helped identify what she found.

One of Owen's most important contributions has been to dismantle the myth that gardens, especially those rich in exotic plants, were "biological deserts". She quoted C.S. Elton's 1966 claim to that effect before demonstrating, with evidence, just how wrong it was. Her garden provided a counterexample so compelling that it helped begin to shift scientific thinking on exotic plants, garden ecology, and the conservation potential of urban landscapes.

The research also positioned gardens within national ecological debates. Comparing her garden's fauna with UK fauna as a whole, Owen revealed the astonishing proportions of the UK's national biodiversity recorded in a single back garden. This comparison remains one of the most striking arguments ever made for the ecological significance of gardens, and one that subsequent research and reports have reinforced.

Owen recognised that gardens, unlike many declining traditional 'semi-natural' habitats, could be increasing with urban expansion. With growing public interest in wildlife-friendly gardening and concern over pesticides, she noted that gardens were becoming ever more hospitable as habitats—perhaps even more so now than when she first posed the question in her 1975 paper: *Suburban Gardens: England's Most Important Nature Reserve?* (Owen & Owen, 1975).

Her work continues to resonate not only scientifically but culturally. Many readers first encountered her ideas in unexpected places. For me it was with delight in the question on the BBC's QI program (https://www.youtube.com/watch?v=4tJ3KFVKq6I), "Where is the best place to discover a new species?". "Not the Amazon rainforest," was the answer, "but your back garden!"

Importantly, Owen was not a wildlife gardener. She did not set out to create wildflower meadows or woodland walks; instead, she demonstrated that an ordinary, productive garden, used for recreation, for daily life, is also an extraordinary wildlife habitat. That message is more relevant today than ever.

If there is a profound message to draw from *RHS Wildlife of a Garden*, it is that meaningful ecological work does not require wilderness. It requires attention. It requires curiosity. It requires the courage to look closely at what is already around us.

This reprint ensures that Owen's legacy remains alive. It safeguards a unique body of research, makes an important work available to a new generation, and reminds us that the most remarkable ecological discoveries can be found just outside the back door.

Andy Salisbury, RHS Head of Plant Health, March 2026

References

Darwin, C.R. (1881) *The formation of vegetable mould, through the action of worms, with observations on their habits.* John Murray, London.

Elton, C.S. (1966) *The Pattern of Animal Communities.* Methuen, London.

Owen, J. & Owen, D.F. (1975) Suburban gardens: England's most important nature reserve? *Environmental Conservation* 2: 53–59. Available at: https://www.jstor.org/stable/44516384.

The Royal Horticultural Society (2025) *The RHS State of Gardening Report 2025* (eds Griffiths, A., Gush, M.B., Salisbury, A., Sutcliffe, C., Redmond, H., Sachs, A., & Könyves, K.). The Royal Horticultural Society, London. Available at: rhs.org.uk/stateofgardening (accessed 31 March 2026).

Jennifer Owen (1936–2015)

Jennifer Owen (née Bak) was born in Leicester on 9 November 1936, the daughter of F. Alan Bak, a hosiery firm owner and amateur ornithologist, and Kate Bak. She studied zoology at the University of Oxford where she met ecologist Denis Owen (1931–1996) in 1955. They married in 1958 and both took up positions at the University of Michigan where Jennifer did a PhD on a social wasp species. They subsequently worked in academic roles at universities in Africa and Sweden. In 1971 she moved back to the UK and took up a research position at the University of Leicester. In 1972 she started recording all the wildlife in her garden in the Humberstone suburb of Leicester. After 15 years, her initial results were published in *The Ecology of a Garden: The First Fifteen Years* (1991). After 30 years of recording (1972–2001), a more comprehensive account was presented in this updated book, *RHS Wildlife of a Garden: A Thirty-Year Study* (first published in 2010). Jennifer and Denis had two children, Sue and Richard. In 2010 she was awarded the RHS Veitch Memorial Medal and the British Ecological Society's Ecological Engagement Award. As well as numerous scientific papers, she wrote several books, including *Feeding Strategy* (University of Chicago Press, 1980), *Garden Life* (Chatto & Windus, 1983) and *The Ecology of a Garden* (Cambridge University Press, 1991). She lived with multiple sclerosis for the latter part of her life and died in 2015, aged 78.

Mike Grant, original RHS commissioning editor, March 2026

1 : Setting the scene

Many birds such as wood pigeon, *Columba palumbus* (opposite), find good feeding opportunities in gardens. This species often nests in tall trees, as it does in the author's Leicester study garden

On returning to Leicester in 1971, after 13 years of living abroad, and having bought a house in the suburbs with a medium-sized garden, my late husband Denis and I were keen to investigate the garden fauna. Our gardens in Kampala, Uganda, and in Freetown, Sierra Leone, had proved to be rich in wildlife and full of interest. The Freetown garden had more species of butterflies than nearby rainforest because, as well as forest butterflies, it also harboured much of the savanna butterfly fauna. This made us look more closely at what it is that makes gardens so rich. It quickly became clear that one crucial factor is the mosaic of open spaces and shady areas that the gardener creates, and this is why the garden butterfly fauna in Freetown included both savanna and forest species. Another crucial factor is the high plant diversity the gardener contrives, providing a habitat rich in feeding and breeding opportunities for insects, probably richer than any natural habitat of comparable area.

With this tropical experience, we were impatient to find out whether an English garden would also prove to be a rich wildlife habitat, and so immediately started investigating our garden fauna, not only by observation but, more critically, by a range of trapping methods. The early results, discussed in an article for *Environmental Conservation* (Owen, J. & Owen, D.F. 1975), were most encouraging, leading us to speculate that suburban gardens might be England's most important nature reserve. This suspicion has been strengthened by more recent studies, such as the University of Sheffield's BUGS project which began in the late1990s (Smith *et al.* 2006).

There is nothing special about the Leicester study garden: most gardens must harbour as diverse a fauna. Facts are essential, however, to make the case for the significance of gardens, and to that end I continued the trapping programme and record-keeping from 1972 until 2001, giving a total period of 30 years. The first 15 years of the investigation were detailed in *The Ecology of a Garden* (Owen, J. 1991). The present book summarises and completes the story, and tries to identify any changes and trends during the 30-year period.

Wildlife gardens

There is a fashion, dating from about 1980, for 'wildlife gardening', although this is usually interpreted as emulating the countryside. It is far from being a case of laissez faire, for the wildlife gardener puts a lot of time and labour into creating wildflower meadows, marshy ponds or even woodland walks. Few people attempt to improve, as wildlife habitats, ordinary gardens that are productive and attractive in the conventional sense; instead, an effort is made to create outside the back door a most ungarden-like habitat. This is admirable in its own way, but overlooks the possibilities of having both the productive garden that most people want, and a wildlife habitat at the same time.

Although many books describe in a generalized way the rich fauna of gardens and advise on enhancing them as wildlife habitats (see *e.g.* Baines 1984, 1985; Chinery 1977, 1986; Gibbons & Gibbons 1988), there is remarkably little reliable and documented information available on what actually occurs. The journals of learned societies contain many short notes, each on one particular record of, say, an unusual beetle or a scarce moth taken in a garden, but there is no main body of knowledge to put such records of garden insects into context. Local natural history societies often have a great deal of information on record, but it is not generally available.

Other studies

As long ago as 1945, Richard Fitter published a fascinating account of the natural history of London (Fitter 1945), in which he recognised the richness of gardens as habitats, particularly for birds. Davis (1978) reviewed some of the changes in the insect fauna of London with urbanization, and summarized the detailed investigations of A.A. Allen, who collected beetles in his southeast London garden from 1926 to 1973, recording more than 700 species.

A partial survey of the natural history of Buckingham Palace garden was undertaken in the 1960s by the South London Entomological and Natural History Society (Bradley & Mere 1966, McClintock *et al.* 1964). This cannot be regarded as a typical garden because of its size (16.03ha) and its seclusion, but it is in a densely built-up area of London. In the 1990s, the Head Gardener at Buckingham Palace invited what is now the London Natural History Society to coordinate a second, more detailed survey.

In the 30 years since the first survey, not only had there been some changes in garden management, but urban development had encroached on the garden and pollution from traffic had increased (although levels of sulphur dioxide had decreased). As well as 322 species of flowering plants growing wild and 58 species of birds, 30 of them breeding in the garden, the 1990s survey identified more than 2,160 species of insects and 207 of other invertebrates (Plant, 1999 and 2001). There are no comparable lists for gardens elsewhere in Britain and only selective reports from continental Europe.

This situation is changing, however, for in the late 1990s a group of ecologists at the University of Sheffield, headed by Ken Thompson, started an exhaustive investigation of 61 gardens in Sheffield. The Biodiversity in Urban Gardens in Sheffield (BUGS) Project is an investigation of the resource that domestic gardens provide for biodiversity and ecosystem functioning, the factors that influence the levels of plant biodiversity associated with different gardens, and ways in which features of gardens can be manipulated to enhance biodiversity (Smith, R.M. *et al.* 2006). Sixty gardens are being studied by pitfall trapping, 61 by litter sampling, and in 16 of them Malaise traps are being operated.

In 2004 to 2007, the BUGS project was extended to gardens in Leicester, Oxford, Cardiff and Edinburgh (see e.g. Loram *et al.* 2007). At least 12 scientific papers have been published on the findings and Ken Thompson has written a popular book (Thompson 2006).

Since 1968, Rothamsted Research has had a nationwide network of moth light traps in a variety of habitats, including gardens, manned by volunteers, and the charity Butterfly Conservation produced a report on the findings (Fox *et al.* 2006). Nevertheless, Frankie & Ehler (1978), in a review of the ecology of insects in urban environments, commented that "our ignorance in this area can be attributed, in large part, to the general dearth of relevant ecological studies", and evidently found little published information on gardens.

One detailed study of a garden insect fauna is Frank Lutz's celebrated description of the insect fauna of his 'yard' in suburban New York (Lutz 1941). He did not trap systematically, but caught 1,402 species of insect in his 23m x 61m lot (or garden). It is noteworthy that his garden was managed in the way that most people look after their suburban plot, and in four consecutive years in which he entered for a garden contest, he won awards. Evidently, in the United States, as in Britain, a conventional suburban garden can be rich in insects, given sympathetic management.

House sparrows, *Passer domesticus*, have recently declined in numbers nationally, partly because of tidying up of roof crevices in which they nest, and partly because of a decline in the insects essential for feeding their young

URBAN WILDLIFE

Massive urbanization, in developed and developing countries alike, has dramatically altered the physical environment, and hence the flora and fauna. Natural ecosystems are replaced by man-made ecosystems, which are organized for the well-being of humans, and in which they are the dominant species. To a large extent, plants and animals are physically excluded by the way in which the land is used, but two categories of animal persist in urban environments. First, there are those characterized by their tolerance of and capacity to exploit man and his activities. Typical of this category are the opportunistic rats, mice, cockroaches, house sparrows (*Passer domesticus*) and feral pigeons (*Columba livia*) found in the heart of cities the world over, functioning mainly as scavengers. Secondly, there are local species that find food and living space in the green areas of cities, particularly on the suburban fringe. Exactly what these species are in a city varies geographically, but such animals as blackbirds (*Turdus merula*) and foxes (*Vulpes vulpes*) do particularly well in English cities, often living at greater densities than in the surrounding countryside.

Climate

The climate of a city differs considerably from that of the surrounding countryside. Cities are always warmer than their environs, and this 'urban heat island' effect extends to the suburbs, where flowering and other seasonal events are earlier than in nearby countryside. Leicester city temperatures show a marked heat island effect: at 23.15hrs on 19 August 1966, when the temperature at the edge of the city was 17.8°C, that in the centre was 21.1°C (Pye 1972b).

As a consequence of higher temperatures there is continuous water loss from a city, the relative humidity is less, and soils quickly dry out, particularly in the summer. There is less incident radiation because of dust particles in the air, and cities are more cloudy and may be twice as foggy (Gill & Bonnet 1973). Cities are less windy than their surroundings, but grid-like street layouts cause gusts and eddies (Miess 1979). Since the 1960s, air pollution, which had become of major concern in cities, has decreased with less use of solid fuels. Pollution-sensitive lichens have returned to cities, although there are still major problems with gases from vehicle exhausts. These environmental effects undoubtedly affect a suburban garden such as mine.

Cities as wildlife habitats

The range and availability of spatial habitats and breeding sites for animals varies with the age of a city and the degree of urbanization; the more modern and urbanized a city, the fewer animals find a home there. In general, animal diversity, whether of birds and mammals or of invertebrates, tends to decrease towards the built-up city centre, where there is less space and there are fewer feeding opportunities. Since the late 1960s, however, an increasing number of European and North American cities have had an element of wildlife conservation in their planning, which means that habitats are created, nature reserves are set aside, and there is increasing tolerance (and, often, active encouragement) of urban wildlife.

Green spaces are found in all cities, and are important refuges for wildlife. Wasteland, railway embankments and roadside verges are soon colonized by a variety of plants. The foliage, flowers and seeds of these plants provide food for insects and birds, particularly Lepidoptera and their larvae, and finches, such as linnets (*Carduelis cannabina*), greenfinches (*C. chloris*) and goldfinches (*C. carduelis*). The typical wild plants of urban areas, that grow wherever there is cleared land, or even from crannies in walls and cracks in pavements, are species that produce large quantities of small seed, are fast-growing, and are tolerant of a wide range of physical conditions and pollution. Examples include *Tridax procumbens*, a weed in the daisy family, found throughout the tropics but originating in Central and tropical South America, and the ubiquitous *Buddleja davidii* of European cities, which originates from China. The green corridor along a railway or stream may be an important route to and from the centre of a city for foxes and other large animals, and railways have aided the dispersal of plants such as Oxford ragwort (*Senecio squalidus*).

Self-sown Oxford ragwort, *Senecio squalidus*, in the author's garden

Parks, gardens and ornamental plantings incorporate a range of spatial habitats and feeding opportunities for animals, particularly for birds and insects. The predilection of planners for

monocultures, however, has often been the cause of herbivorous insects getting out of hand, such as the millions of aphids that drip honeydew from lime trees lining English streets, and gypsy moth (*Lymantria dispar*) caterpillars that denude North American street trees.

Rivers, canals, reservoirs and ornamental lakes accommodate a range of fish and other aquatic life, together with many birds, even in the heart of a city. The lake in St. James's Park, in the heart of London, supports introduced, and pinioned, pelicans, cormorants and ducks; more significantly, large numbers of pochard (*Aythya farina*) and tufted duck (*A. fuligula*) visit to feed, and both have bred in the park. Coot (*Fulica atra*) and moorhen (*Gallinula chloropus*) are resident, and herons (*Ardea cinerea*) not only feed there but have established a breeding colony.

Since the inter-war years, foxes, *Vulpes vulpes*, have become a familiar sight in many towns and villages, including Leicester. A survey by Leicester Wildlife Group in 1984 estimated that there were 19 fox family units (each averaging one dog fox, one vixen and 4.8 cubs, often accompanied by a barren female) in the city as a whole

Other sorts of animal are typical of suburban and greener areas of towns and cities, and may occasionally stray into the densely built-up centre. These include opportunists, such as foxes and collared doves (*Streptopelia decaocto*) in Britain, and also species abundant in the environs of a city, such as hedgehogs (*Erinaceus europaeus*) and robins (*Erithacus rubecula*). These creatures are either enclosed within city boundaries, by development around their habitat, or are able to extend their range into the peripheral, greener areas of a city. Deer are maintained in semi-wild condition in large parks in some European cities, such as the fallow deer (*Dama dama*) and red deer (*Cervus elephas*) of Richmond Park, London. Exotic species sometimes become naturalized from introductions or escapes, so that flocks of rose-ringed parakeets (*Psittacula krameri*) are a familiar suburban sight in southeastern England.

GARDEN WILDLIFE

A look at urban wildlife leads naturally on to consideration of gardens, because gardens are particularly important for the animals of city and suburb. Most of the animals found in parks and green spaces in cities will also be found in gardens, especially the larger ones. In gardens the world over, birds and insects exploit the feeding and breeding opportunities offered by dense growth, plant diversity and an abundance of flowers, fruit and seeds.

Local bird species, particularly opportunists, may be abundant in town gardens, whether the bulbuls (Pycnonotidae), wax finches (Estrildidae), robin chats (Muscicapidae), barbets (Capitonidae) and sunbirds (Nectariniidae) of Africa; the blue jays (*Cyanocitta cristata*), cardinals (*Richmondena cardinalis*), chickadees (Paridae), leaf warblers (Parulidae) and robins (*Turdus migratorius*) of North America; or the thrushes (Turdidae), tits (Paridae), robins (*Erithacus rubecula*), greenfinches and warblers (Sylviidae) of an English garden.

Small, ground-feeding doves find a niche around human settlements and are found in gardens and parks in cities all over the world: laughing doves (*Streptopelia senegalensis*) in Africa, mourning doves (*Zenaidura macroura*) in North America, introduced spotted doves (*Streptopelia chinensis*) and laughing doves in Australia, and collared doves in Europe.

Many gardens in European and North American cities have bird-feeding tables, which are visited by a wide variety of birds, and the provision of food undoubtedly attracts birds to gardens and may improve their chances of survival in bleak winters.

Exactly what wildlife frequents a garden depends on its locality and size (and on where it is situated geographically, of course), but birds and insects in particular, because they fly, are present in variety and abundance, even in inner-city gardens. Much has been published on gardens and their wildlife, but for the most part it is generalized or anecdotal, and, as already mentioned, there have been surprisingly few detailed studies. To most people, garden wildlife seems to mean vertebrates (particularly mammals and birds), butterflies, and only a few of the more spectacular among insects and other invertebrates. There is even a tendency to exclude plants from consideration.

Most professional ecologists, moreover, have tended to dismiss gardens as man-made and artificial and of no relevance to wildlife. Elton (1966) considered them 'in the direction of biological deserts', as he assumed that 'the multitude of exotic plant species ... have so few animals attached to them.' How wrong he was, as investigation of my garden has demonstrated.

Gardens in the UK

Largely because the average householder in Britain wants to own and work a plot of land around or behind their house, the total area of garden is astonishingly large, and a greater proportion of the land than in any other country (Owen, J. 1991). It is estimated that approximately 8,000km^2 of the UK is covered by domestic gardens (Loram *et al.* 2007). This is a significant proportion, particularly now that many people, concerned about loss of biodiversity in the wider countryside, are managing their private plots in a way that is sympathetic to wildlife. On the basis of a 1986 survey by the City Wildlife Project, 27.6% of the area of Leicester was then under gardens, although the BUGS2 Project suggested that by the end of the 20th century, this had diminished somewhat, to 26.8% (18.0km^2).

THE TOPOLOGY OF LEICESTERSHIRE & RUTLAND

Leicestershire has much in common topologically with its smaller neighbour, the county of Rutland, and they were for a time combined for administrative purposes. Although they are, once again, separate counties, it is geographically logical to consider them as one entity. Together they form a compact area of 2,548km^2, with the conurbation of greater Leicester in the centre. The region is traditionally thought of as a land of rolling pastures and wooded valleys, the grass fields enclosed by well-cared-for hawthorn hedges, but in the last 50 years or so, the aspect of the countryside has changed considerably, as hedges have been removed to enlarge fields given over to arable crops.

Leicestershire is divided by the River Soar, which runs from south to north

through the centre, into two distinct regions. These differ in their physical geography, their historical development, and their modern character (Pye 1972a). The city of Leicester, which began as a pre-Roman settlement and Roman military centre at an important river crossing, is situated in the wide, flat river valley. The city was originally established on river gravel terraces, but in the last hundred years or so has extended out onto the alluvial soils of the flood plain and up onto the higher land beyond.

Map showing Leicestershire and Rutland (above) and position in Britain (below). The author's study garden is situated in the city of Leicester (see map p40 for detail of garden location)

My garden is in the eastern part of the city, at an altitude of about 72m above sea level, on the slopes above the older parts of the city, which lie at an average altitude is 45.7m. The underlying rock of the city, and much of the county west of the Soar, is Triassic Keuper Marl, but from about 15km east of the river the base rock is Jurassic. Rounded ironstone hills form the higher land (maximum elevation 230m) in the east of the county, and Precambrian rocks of the Charnian series, thrusting through the Keuper Marl, form the craggy outcrops of Charnwood Forest (maximum elevation 278m) in the west.

Most of the south of the county, the land to the east of the Soar, and that between southeastern Charnwood and the river, is overlain by Pleistocene boulder clay, in places 21m deep, within which are patches of sand and of gravel. In most of eastern Leicestershire, and Rutland, the boulder clay forms deep soils that support rich grassland. The soil immediately to the east of the city, and

An undulating arable agricultural landscape, typical of Leicestershire and Rutland, seen from the A47 on the eastern outskirts of Leicester. In the middle distance is the village of Scraptoft, now a suburb of the city. Beyond that, and closer to the centre of the city, is Humberstone, the area of the city where the author's study garden is situated. The high ground on the skyline beyond is Charnwood Forest

including my garden, is rather poorly drained clay soil, which is graded Class 3 agricultural land, and has a tendency to waterlog.

What is now Leicestershire and Rutland was probably once densely covered with mixed deciduous woodland, which, it is assumed, was dominated by small-leaved lime (*Tilia cordata*) and ash (*Fraxinus excelsior*), with possibly some pedunculate oak (*Quercus robur*). Between 4,000 and 5,000 years ago, agriculture was introduced to Britain, substantial forest clearance began, and by the end of the Roman era the wildwood was only a dim memory in England. The Saxons, who settled from about 400AD, found England as a whole only about 30% wooded, and Late Saxon farming reduced woodland cover to perhaps 15%, about twice what it is now (Colebourn & Gibbons 1987).

Agricultural development

An invasion from Denmark in the 9th century led to Leicestershire's becoming part of the Danelaw. By the time of Domesday (1086), Leicestershire was one of the most densely peopled and extensively cultivated counties in medieval England (Hoskins 1957). In the 12th century, much of the county was declared 'forest' (in other words a hunting reserve, which might be pasture woodland or 'waste', but rarely woodland as such). From then on there was further clearance of woodland and extension of farming. The first enclosures took place in the 14th century, when sheep and cattle grazing was extended, the open fields were reorganized, arable land was converted to pasture and there was a further reduction of 'waste' (woodland, scrub and other common land).

More than a thousand years of cultivation, clearing, draining and establishment of open field systems meant that by the late 16th century the land had reached its ultimate development as wide-open countryside (Hickling 1978), despite some enclosures. (It is, however, possible that agricultural developments of recent decades, towards large arable 'prairies', are returning the countryside to that position). There was further enclosure with the introduction of permanent grassland in the 17th century, and in the 18th century, ploughed land was left to grass over and become pasture; by 1870 the area was once again noted for its grassland. In summary, apart from the loss of Leicester Forest East and the Forest of Rutland, and the contraction of some woods, there was little change in the general aspect of the counties in the seven centuries from Domesday to the time of the parliamentary enclosures.

A considerable proportion of the county had been fenced by landowners before the onset of parliamentary enclosures in the second half of the 18th and first quarter of the 19th centuries, when the rest was enclosed. Because Georgian county-planners only made roads of the most heavily used tracks, the parliamentary enclosures produced a landscape with fewer roads and more field paths than in counties such as Devon that had been enclosed in medieval times. The planners also created the new straight roads with wide verges that are now so characteristic of Leicestershire and Rutland. Despite the effects of the Enclosure Awards, the county's characteristic small fields and dense hedges owe much to Robert Bakewell, an influential 18th century agriculturist. He farmed in the county and advocated the partitioning of large tracts of grazing land into fields of 10 acres (4.05ha). Organized fox-hunting developed in Leicestershire during the 1770s, just in time to enjoy the exhilaration of galloping across miles of unfenced country. An effect of fox-hunting on the countryside, subsequently beneficial to wildlife, was the establishment of fox coverts – spinneys of roughly 1–8ha with dense understoreys.

Intensive agriculture

In 1900, 80% of Leicestershire and Rutland's cultivated land was under permanent grass. There was, however, a temporary change from pasture to arable in the First World War (1914–1918) and then much more dramatically in the Second World War (1939–1945). Since then, the county's landscape has changed again, with the grubbing out of hedges, felling of woods, drainage of marshes, the disappearance of the farm horse, ploughing up of grassland, and increasing use of pesticides and fertilizers on the large, open fields. In the 1930s, only 1% of the county's farmland was under arable crops, but by 1992, 51.85% of farmland was arable (pers. comm., Michael Jeeves, Head of Conservation, Leicestershire and Rutland Wildlife Trust).

There are, however, still many areas of grassland, although now it is mainly uniformly green 'improved' pasture (*i.e.* improved for grazing, not wildlife) rather than the flower-spangled meadows and pastures that once flourished; the prevalence of ridge-and-furrow contours shows that much of the remaining grass was once under the plough, long before the enclosures. However, despite modern developments, the landscape is still one of villages, with comparatively small fields (particularly in the west), large hedges and rough corners; farming is the major land use in the two counties, occupying at least 80% of the land.

Grazing land (above left) on the borders of Leicestershire and Rutland

The city of Leicester seen from the high ground of Charnwood Forest, northwest of the city

HABITATS & WILDLIFE OF LEICESTERSHIRE & RUTLAND

Leicestershire and Rutland have never been renowned for their wildlife; there are richer, more spectacular woodlands, grasslands, moorlands, wetlands and aquatic habitats elsewhere in the British Isles. They are, however, counties of remarkably varied terrain and soils, despite the uniformity imposed by modern agricultural techniques. Their apparent lack of interest to the naturalist may be largely a consequence of lack of knowledge. Increased investigation recently by local naturalists is leading to more and more discoveries of species (particularly animals), new to the county, and rediscoveries of species that were thought to have been lost.

In terms of distribution of plants in particular, but also of animals, centrally placed Leicestershire is in an interesting position in Britain. It lies at a midpoint in relation to the decline in rainfall from northwest to southeast, the fall in mean summer temperature from south to north, and the fall from west to east in mean winter temperature. This is superimposed on the geological contrast between the Precambrian Charnwood rocks with their acid soils in the north-west, and the ironstones and oolitic limestones in the east. Consequently, western and northern species are restricted to Charnwood, and Continental species to the limestone.

Until recently, Leicestershire (excluding Rutland) was one of the most poorly forested counties in lowland Britain, with only 2.3% of its area covered by woodlands of over 5 acres (2.02ha), only four other counties having less wood-

Bluebells, *Hyacinthoides non-scripta* (above left), are found in many areas of ancient woodland in the two counties

Rutland Water (above right) has become a popular recreational amenity for sailing and fishing, but it has great value for water birds; the successful introduction of breeding ospreys has been a triumph for the Leicestershire & Rutland Wildlife Trust and Anglian Water

land, proportionately, and one the same. There was, however, a total area of just over 4,850ha made up of 1,200 woods exceeding 0.4ha, and at least 50 of these are thought to be primary woodlands directly continuous with the wildwood of prehistoric times, although they have been much utilized and modified by man (Primavesi & Evans 1988).

Thanks to the recent development of the National Forest, which extends into northwest Leicestershire, 4% of the county's area is now wooded (Michael Jeeves, pers. comm.), a welcome change. Most of the ancient woods are classified as ash/field maple (*Fraxinus excelsior/Acer campestre*), although in the 18th and 19th centuries pedunculate oak was planted in many. They have a rich ground flora dominated by dog's mercury (*Mercurialis perennis*) and bluebell (*Hyacinthoides non-scripta*). The bird fauna of the ash/maple woods, of the sessile oak/birch (*Quercus petraea/Betula pendula*) woods on the acid soils of Charnwood Forest, and of many recent woods, is rich and varied.

In the last 60 years, the area of permanent pasture has shrunk. What remains has been 'improved' so that species-rich grassland is now rare in the county. The Leicestershire and Rutland Wildlife Trust regards the preservation of the few remaining unimproved grassland areas as a high priority and has been able to acquire a total of 400ha as nature reserves. As fields have been enlarged for arable farming, grubbing-up of hedges and tidying up of rough corners and headlands has had a deleterious effect on insects, birds and other animals, although this has been countered by implementation of national environmental remediation schemes, such as setaside, by many farmers.

The area of fresh water in the two counties is greater than ever before in

The author's reinstated garden pond in 1984. Garden ponds are invaluable breeding sites for frogs and newts, as well as attracting many insects

historic times, not least because of the creation of Rutland Water reservoir, completed in 1975, which covers 1,417ha and has a perimeter of 27.3km, alongside which are marshy wetland habitats. This has had a marked effect on birdlife, as well as other aquatic species.

Charnwood Forest, with its sandy soil, heath and moorland vegetation, and sessile oak/birch woodland, was the last part of the county to be enclosed. It still has an extensive deer park, much used by Leicester people as a recreational area. Like much of the Forest, this park, from which over 500 species of beetle have been recorded, has a rich fauna. Some of the more notable species, however, have disappeared in the last century, probably because of human pressure. Man has dominated the Leicestershire landscape throughout historic times, but does so particularly now.

Although the wildlife of Leicestershire and Rutland was previously thought to be impoverished by comparison with many other parts of the UK, recent research, particularly on the animals, is inconclusive. Many good habitats and wildlife sites are gone for ever, but the work of the Leicestershire and Rutland Wildlife Trust, directly as landowner, and also through the local authorities and by influencing the public at large, is helping to safeguard what is left.

The city of Leicester

A great deal is known about Leicester city's wildlife habitats, as a consequence of the Leicester Habitat Survey (Lomax 1987), conducted by the City Wildlife Project (now called Environ), which regularly updates the survey. Of the city's 7,337ha, 64.9% is built up (this figure includes gardens), 18.8% is managed as formal open space, such as public parks, 11.3% is agricultural land, including allotments, and 5% is land left to nature, either intentionally or by neglect. Out of the 35.1% of open space, only 1.9% consists of A* land (sites of city-wide ecological importance), the best examples of habitat types in the city, with a great diversity of flora and fauna, and locally rare or uncommon species.

Although less than a third of the formal open space (parks, etc) has high ecological value, the roughs on the 250ha of golf courses harbour voles, which support stoats (*Mustela erminea*), weasels (*M. nivalis*), kestrels (*Falco tinnunculus*) and tawny owls (*Strix aluco*). There is an increasing trend towards planting groups of native trees and dense shrubbery, all of which improves the formal open space as bird habitat.

The city's birds are well known, thanks to the active and thriving Leicestershire and Rutland Ornithological Society, and the city appears to be rich in birds, mainly because it is well supplied with parks and other green areas. Apart from parks, Leicester has more than 24,000 street trees, and there are tree-lined walks and squares in the heart of the city.

Grey squirrels, *Sciurus carolinensis*, are present in Leicester's parks and squares and they regularly visit gardens to feed from bird tables

Woodland is regarded as one of the most important habitat types in the city, mainly because there is little of it, only 78.42ha. Apart from two ancient remnants of the 13th-century Leicester Forest, most of the woodland is plantations. Some woods are rich in birds, including woodpeckers (*Picus viridis*, *Dendrocopus major* and *D. minor*), nuthatches (*Sitta europea*), treecreepers (*Certhia familiaris*) and warblers, such as willow warblers (*Phylloscopus trochilus*) and chiffchaffs (*P. collybita*), and these woods are also home to small mammals and foxes (*Vulpes vulpes*).

South of the city one can find rich, unmanaged grasslands alongside the River Soar and the canal, and also some areas of wetland, which city-wide is a scarce habitat. Despite modifications to reduce flooding and aid navigation, many stretches of canal and river are diverse, rich habitats, although the same cannot be said for streams, which have been straightened and concrete-lined. Apart from garden ponds, the number of ponds in the city has declined dramatically in the last century, which makes those in gardens all the more important.

The buildings of the city furnish nesting sites for swallows (*Hirundo rustica*), swifts (*Apus apus*), house martins (*Delichon urbica*), spotted flycatchers (*Muscicapa striata*) and kestrels, as well as the ubiquitous feral pigeons, house sparrows (*Passer domesticus*) and starlings (*Sturnus vulgaris*). Bats of three species, pipistrelle (*Pipistrellus pipistrellus*), long-eared (*Plecotus auritus*) and noctule (*Nyctalus noctua*), hawk for insects over the city, and Daubenton's bats (*Myotis daubentoni*) hunt over the almost rural expanse of meadows by the river. There are grey squirrels (*Sciurus carolinensis*) in all the parks and squares.

At least 2,025ha (about 27.6% of the city as a whole) are under garden of one sort or another, and more than 230ha (11.3% of city gardens) are judged to be good wildlife habitats. One large garden, about 3km from the centre of Leicester on a busy main road, was investigated from 1953 to 1962, and found to have 22 species of bird breeding there (Hickling 1978). The wildlife value of inner-city gardens is limited, but those of suburbia are rich in insects, birds and amphibians. Gardens are also a regular haunt of the one mammal for which some information is available – the fox. They are seen regularly, even in the city centre, and in most large gardens there is a den under a shed or outbuilding. Many people put out food for foxes, from fish and chips to chicken joints. With increased general awareness of, and interest in, the environment, the outlook is good for wildlife in the city of Leicester, despite its inauspicious geographical position.

2 : Studying the Leicester garden

THE SUBURB OF WHICH MY GARDEN is a part began to develop in the early twentieth century as a mixture of out-of-town residences for the middle class, and planned and enlightened working-class housing. These were grafted onto the ancient, once-rural village of Humberstone. Since then, the surrounding and intervening land has been built on, mainly in the inter-war years. It now forms a typical suburb, varied in housing type and age. The percentage of vegetation cover is high, as most of the houses have gardens, often quite large ones.

My garden is well-established, and the pattern of the surrounding suburb is so settled that it takes a considerable imaginative leap to picture what it was like a hundred years ago, let alone two centuries ago or even more. The land was once part of the Humberstone Hall estate, which was sold and broken up in 1919. My house was built in the mid-1920s and the garden laid out in 1927; its basic design remained unchanged from then until after the completion of the study in 2001.

Before 1923, the site of the garden was a mixed plantation, possibly from 1846. After 1923, the surrounding area gradually became built up, although, at the time of building, my house and its immediate neighbours represented the most easterly extension of the city. The garden is now nearly 2km within the city boundary, surrounded by suburbia; it is a corner plot on a busy road, only 3.8km from the city centre. In 1972 there were open fields and a mixed farm 800m to the north of the garden. Development of the area began in 1989, the farmhouse was demolished and the fields disappeared beneath a supermarket and housing. A small brook flanked by tall trees that passes 125m south of the garden was channelled in concrete and most of the trees felled in the late 1980s. There are two mature parks with old trees within a radius of 450m, the surrounding gardens are well-established, and many are densely planted.

Herbaceous border in the author's garden (opposite) with peonies (*Paeonia officinalis*), red valerian (*Centranthus ruber*) and perennial cornflower (*Centaurea montana*)

Fig. 2.1 Leicester garden plan

For more detail, see Figs 2.2 and 2.3 overleaf

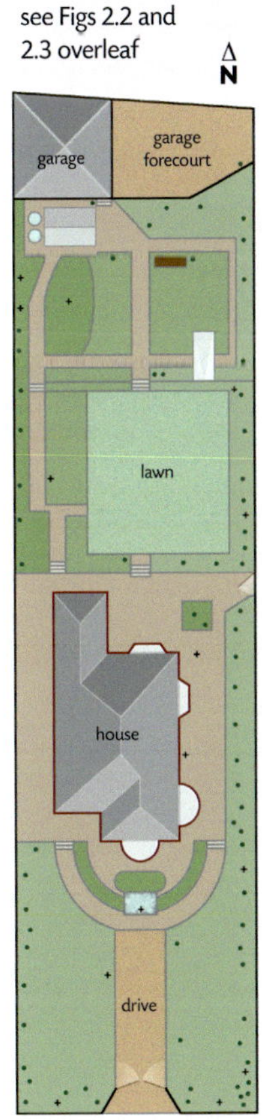

The garden layout

The Leicester plot (grid reference SK 624054) is not quite rectangular, being 15.24m on the south side, 14.25m on the north, 61.57m on the west and 60.55m on the east side, along which runs another road, constructed some time after 1923. The design of the garden, and its dense and varied planting, make it look bigger than it is, even though the house is in the centre of the 930m^2 plot, and part is taken up by the garage and its forecourt (see Figs. 2.1, 2.2 & 2.3).

The area of the entire plot, including the house and garage, is 930m^2, of which 741m^2 is garden. It incorporates all the elements of a typical suburban garden (as described in Owen, J. 1983). It has a lawn, flower beds, herbaceous borders,

Fig. 2.2 Detailed plan of the front garden

paths
steps
● trees and large shrubs
+ former trees and large shrubs
f annual flowers
p perennials
Δ N

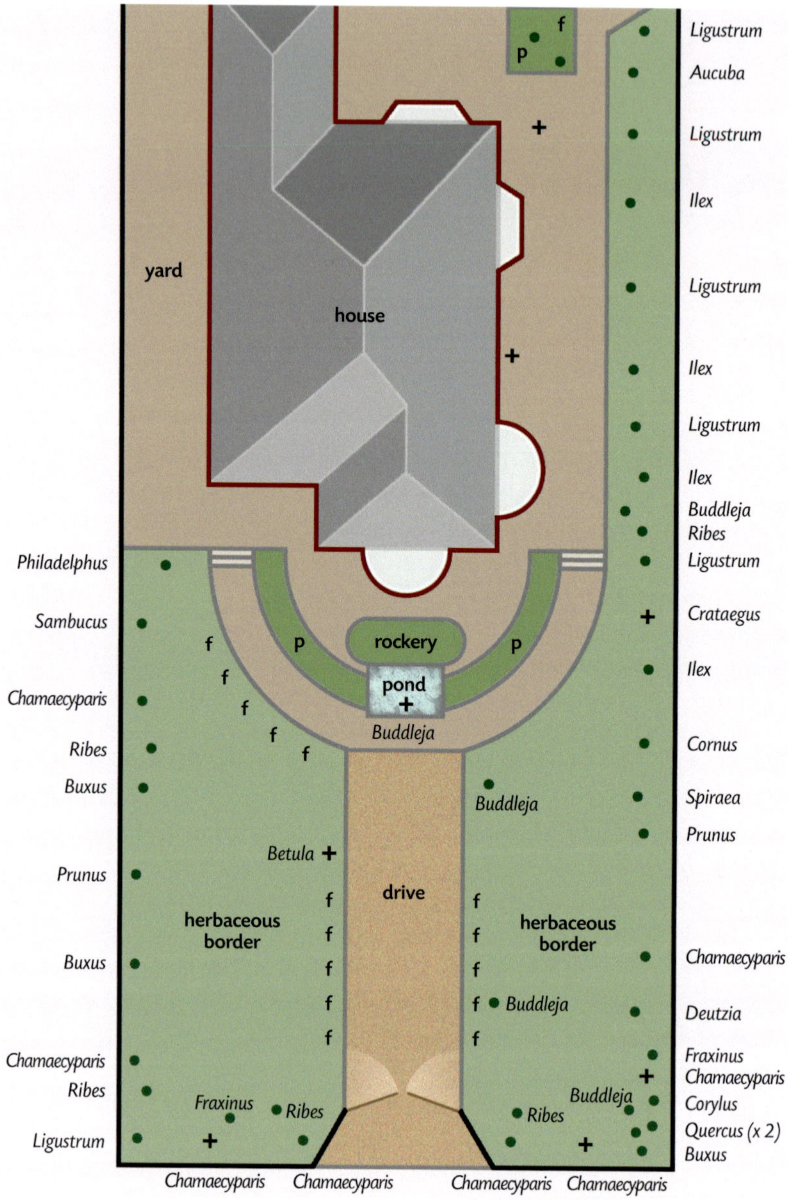

rockeries, vegetable patches, fruit bushes, an apple tree, flowering and evergreen shrubs, deciduous trees, conifers, a pond, a compost heap, a glasshouse, paths and paved areas. The vegetation tends to be denser and more luxuriant than in many gardens, but it is nonetheless neat, attractive and productive, and does not differ markedly from neighbouring gardens.

There have been two significant changes since the garden layout was described in detail in Owen, J. 1991. By the early 1990s, the large silver birch tree (*Betula pendula*) by the south side of the lawn behind the house had started to shade the lawn completely in summer. In December 1994 an attempt was made to pollard it, but it did not survive. The stump was removed and replaced by a buddleia (*Buddleja davidii*). A seedling from the birch was planted elsewhere.

In late 1999 much of the mixed-planting bed to the west of the lawn was

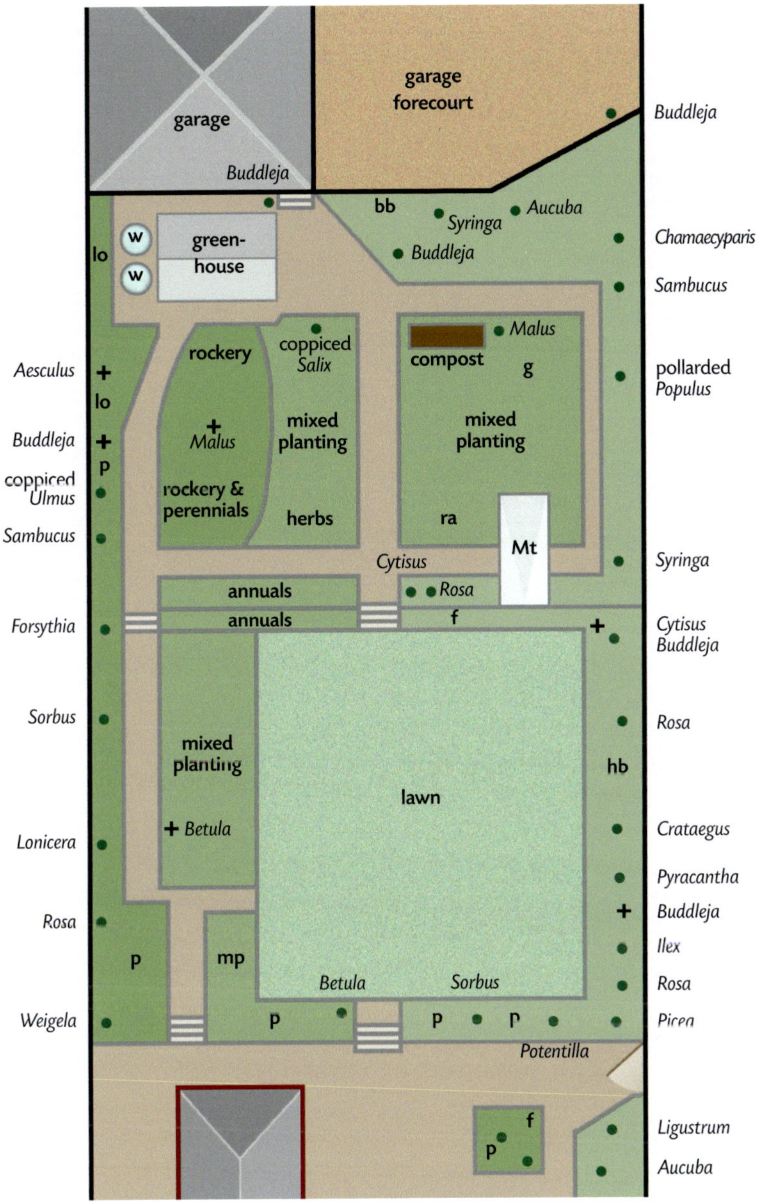

Fig. 2.3 Detailed plan of the back garden

turfed over to increase the area on which my grandsons can play football. I do not feel that extending the lawn area has greatly affected the character of the garden, but the loss of the big birch tree has made a significant difference, both visually and in terms of habitat diversity and biomass. Apart from these alterations, the skeleton of the garden and its basic layout did not change during the 30-year period of the study (1972–2001), although, inevitably, the vegetation became denser and taller.

Every effort was made over those 30 years to enhance the garden as a wildlife habitat. To this end, efforts were made to maintain, for as much of the year as possible, an abundance of flowers at which insects can feed, and flowering periods were extended by dead-heading. Plant diversity was kept high, increasing the feeding opportunities for insects such as caterpillars, and structural diversity

View from back of house towards garage in May 1980, with *Forsythia* x *intermedia* in flower. This was a standard viewpoint for photographs showing lawn and large bed for mixed planting, the so-called 'lawn bed'

was maintained to maximize different microhabitats, creating many-layered vegetation and a network of open spaces and shade. Berry-bearing shrubs were planted to attract birds; perennials were left to go to seed to feed seed-eating birds. Dead vegetation was left standing until after Christmas, to let insects and other invertebrates complete their life cycles.

I also felt it was important to maintain good ground cover, even by weeds, to provide shelter for ground beetles, spiders, centipedes and the like. This not only diversifies the garden flora but also helps in pest management, as these ground predators climb plants at night to feed on soft-bodied insects such as caterpillars. No pesticides were used, other than the occasional use of slug pellets. Fortunately, the garden has never had a serious pest problem, not only because of the ground cover, but also because all vegetables have been grown mixed with flowers and other aromatic plants. These deterred potential pests or, in the case of aphids, attracted hoverflies whose larvae feed on aphids.

The density and variety of the garden vegetation are direct consequences of the way in which the garden has been managed since 1971. To some extent this represents a departure from normal gardening practice, although the garden remains neat, attractive and productive. Pruning and clearing of woody vegetation and weeding are always kept to the minimum necessary for healthy growth and avoidance of competition. Interplanting aromatic plants such as French marigolds (*Tagetes patula*) and herbs with cabbages and the like (*Brassica* cultivars) is known to deter potential pests such as large white and small white ('cabbage white') butterflies (*Pieris brassicae* and *P. rapae*) and flea beetles such as *Phyllotreta cruciferae*, because it confuses the egg-laying females that locate the host plant by its smell (Root 1973, Tahvanainen & Root 1972). Since gravid female hoverflies feed before they lay eggs (Pollard 1971), they can be attracted to such plants as beans by planting flowers nearby at which they can feed before laying eggs among clusters of aphids. This is why I mix flowers into my vegetable planting. It not only minimises pest damage, but also looks attractive.

A similar view, but including the silver birch tree, later in the year, showing the bean poles and flowers of tufted phlox, *Phlox douglasii*, beyond the lawn

A similar view, later in the year, with denser vegetation, including hogweed (*Heracleum sphondylium*) and Egyptian onion (*Allium cepa* Proliferum Group) in the foreground

THE GARDEN'S FLORA

The occurrence in the garden of different species of flowering plants, whether intentionally grown or growing of their own accord, was recorded annually from 1975 to 2001. A total of 91 plant families, excluding grasses, was recorded; some were well represented, whereas 39 were represented by a single species. The best-represented families were *Asteraceae* with 64 species, *Lamiaceae* (26), *Rosaceae* (22), *Brassicaceae* (21), *Papilionaceae* (17), *Solanaceae* (15), *Caryophyllaceae* (15), *Scrophulariaceae* (14) and *Ranunculaceae* (13). The plant list was cumulative, and the species did not all occur at the same time.

E.J. Salisbury's remarks about gardens (Salisbury 1935) are particularly appropriate here. He stated that gardens 'comprise an artificial assemblage of many kinds of plants, the nature and number of which are mainly dependent on our arbitrary whims and fancies. Some that grow are only too patently

unfitted for the conditions in which they live, and many more fail to produce offspring and so when they perish must perforce be replaced by artificial propagation or by importation. The plants of a garden grow not where they will but where they must.' He went on to describe the need to combat competition from weeds, plants that come in of their own accord, pointing out that 'The garden is, in fact, an unstable community which only continual labour maintains in seeming stability.'

Categorising the garden's flora

During the 12-year period of 1975–1986, 360 plant species grew in the garden. Of these, 214 were alien, and 146 natives; 266 of those species were cultivated and 94 came in of their own accord. In the 15-year period of 1987–2001, however, only 330 plant species grew in the garden; of these, 190 were aliens and 140 natives; 245 species were cultivated and 85 came in of their own accord. I am not clear why there should be this reduction in the later period, as I was not aware of gardening with less fervour. Overall, from 1975 to 2001, the number of alien species was 240; of natives, 170; of cultivated species, 307; and of those that came in of their own accord, 103. In total, 410 species grew in the garden between 1975 and 2001, the annual total varying from 216 (1996) to 268 (1985).

Black-eyed Susan (*Rudbeckia hirta*) (above) and sweet tobacco (*Nicotiana alata*) growing beside young brassica plants

Canadian fleabane (*Conyza canadensis*), an occasional 'weed' in the Leicester garden (below)

Although many aliens are cultivated and most weeds are native species, it is important to remember that not all aliens are intentionally grown, neither do all native plants come in of their own accord as 'weeds'. For instance, weeds include such aliens as leptinella (*Cotula squalida*) from New Zealand, Canadian fleabane (*Conyza canadensis*) from North America, spotted dead-nettle (*Lamium maculatum*) from continental Europe and large-flowered pink-sorrel (*Oxalis debilis*) from South America. The plants that I cultivated include such natives as rowan (*Sorbus aucuparia*), Jacob's ladder (*Polemonium caeruleum*), shrubby cinquefoil (*Potentilla fruticosa*), and cabbage (*Brassica oleracea*). Naturalized species, even such well-established ancient introductions as sycamore (*Acer pseudoplatanus*), are listed here as aliens.

In listing the garden's flora, a species was scored for every year that it was present, but was scored in terms of the manner of its first appearance, either deliberately cultivated or spontaneously appearing of its own accord. Many cultivated plants are perennial and so persist for many years; others, such as borage (*Borago officinalis*) and pot marigold (*Calendula officinalis*), continue because they seed themselves in the garden. Similarly, some plants that have come in of their own accord, such as ash (*Fraxinus excelsior*) and elder (*Sambucus nigra*), persist indefinitely, and others, such as goat's-beard (*Tragopogon pratensis*) and wood avens (*Geum urbanum*), are short-lived and seed themselves year after year. All such plants were scored in terms of the manner of their first arrival

Lawn grasses allowed to flower for identification, with yellow iris (*Iris pseudacorus*), rosebay willowherb (*Chamerion angustifolium*) and ivy (*Hedera helix*) in the foreground

every year when they were present. For instance, borage was first cultivated from seed before 1975, and has since self-seeded every year; it was scored as a cultivated plant every year.

On the other hand, some plants that came in of their own accord, such as creeping-Jenny (*Lysimachia nummularia*), spotted dead-nettle (*Lamium maculatum*), feverfew (*Tanacetum parthenium*) and hogweed (*Heracleum sphondylium*), were encouraged for their value as ground cover or their attractiveness, and persist; they were scored every year as spontaneous introductions.

The majority of cultivated plants were identified using Gault (1976), Hay & Synge (1971), Nicholson *et al.* (1972), Brickell (1989) and Mitchell (1974). Native and naturalized plants were identified using Fitter, Fitter & Blamey (1974), Rose (1981) and Keble Martin (1967), but nomenclature, both scientific and vernacular, follows Stace (1997).

The deliberately cultivated plants are listed on pp22–26 and the spontaneous introductions are listed on pp26–28. Although these would be called weeds by most gardeners, in the context of my garden it is a misleading word. A plant is a weed only if I decide I do not want it; if I am tolerant, there are, by definition, no weeds in my garden. Hence my distinction between cultivated plants (those intentionally introduced to the garden) and plants that arrived spontaneously. Manifestly part of the garden's natural history, they are listed by family.

In 1975, 1976 and 1977, I attempted to identify all the grasses in the garden. To this end, those growing in flower and vegetable beds were left to flower, and areas of lawn were left uncut. The majority of the species recorded were probably introduced to the garden as lawn grasses. I did not try to distinguish between deliberately cultivated and spontaneous introductions. In all, 24 species were identified between 1975 and 1977 (see list opposite). Four species were aliens: hard fescue, from continental Europe, is a common lawn grass; oats, barley and wheat, derived from species found in southwest Asia, grew from discarded guinea-pig food. Hubbard (1974) was used for identification, and nomenclature follows Stace (1997). Many of these grasses persisted in the garden.

The flora of my garden is rich and varied, with up to 268 species of flowering plants, other than grasses, recorded in any one year, and an average of 226 ►*p36*

CULTIVATED PLANTS

Species of flowering plants, excluding grasses, that have been cultivated or intentionally grown in the garden are listed by family below. There is great variation between species in growth form, in duration of cultivation, and in their contribution to the total quantity of plant material, i.e. plant biomass, in the garden.

KEY:
T trees; **S** shrubs and woody climbers;
P herbaceous perennials; **L** low-growing perennials;
A annuals and biennials; **B** bulbs, corms and other monocotyledons; **W** plants growing in pond.
1 present for thirty years or more;
2 regularly present;
3 occasionally present.
* plants making a significant contribution to the biomass

Pinaceae

1 Norway spruce *Picea abies* (L.) Karst. **T1**

Cupressaceae

2 *Lawson's cypress *Chamaecyparis lawsoniana* (A. Murr.) Parl. **T1**

Taxaceae

3 Yew *Taxus baccata* L. **T2**

Nymphaeaceae

4 Water-lily *Nymphaea* cultivar **W3**

Ranunculaceae

5 Winter aconite *Eranthis hyemalis* (L.) Salisb. **P3**
6 Love-in-a-mist *Nigella damascena* L. **A2**
7 Larkspur *Consolida ajacis* (L.) Schur. **A3**
8 Delphinium **Delphinium* cultivar **P2**
9 Japanese anemone *Anemone hupehensis* Lemoine **P3**
10 Clematis *Clematis* cultivar **S3**
11 Summer pheasant's-eye *Adonis aestivalis* L. **A3**
12 *Columbine *Aquilegia vulgaris* L. **P1**
13 *Columbine *Aquilegia* cultivar **P1**
14 Chinese meadow-rue *Thalictrum delavayi* Franchet **P1**

Berberidaceae

15 Darwin's barberry *Berberis darwinii* Hook. **S2**
16 Japanese barberry *Berberis thunbergii* DC. **S3**
17 *Oregon grape *Mahonia aquifolium* (Pursh) Nutt. **S1**

Papaveraceae

18 *Opium poppy *Papaver somniferum* L. **A1**
19 Alpine poppy *Papaver alpinum* L. **A2**
20 Ladybird poppy *Papaver commutatum* Fisch. & Mey **A3**
21 Oriental poppy *Papaver orientale* L. **P2**
22 Californian poppy *Eschscholzia californica* Cham. **A2**
23 Mexican tulip poppy *Hunnemannia fumariifolia* Sweet **A3**

Fumariaceae

24 Bleeding-heart *Dicentra formosa* (Haw.) Walp. **P2**
25 Yellow corydalis *Pseudofumaria lutea* (L.) Borkh. **P3**

Ulmaceae

26 *Wych elm *Ulmus glabra* Huds. **T2**

Juglandaceae

27 Walnut *Juglans regia* L. **T3**

Fagaceae

28 Sessile oak *Quercus petraea* (Matt.) Liebl. **T3**
29 Pedunculate oak *Quercus robur* L. **T2**

Betulaceae

30 *Hazel *Corylus avellana* L. **T1**

Nyctaginaceae

31 Marvel of Peru *Mirabilis jalapa* L. **A3**

Chenopodiaceae

32 *Spinach *Spinacia oleracea* L. **A3**
33 *Beet *Beta vulgaris* L. **A2**

Caryophyllaceae

34 Corncockle *Agrostemma githago* L. **A2**
35 Rose-of-heaven *Silene coeli-rosa* (L.) Godr. **A3**
36 Rock soapwort *Saponaria ocymoides* L. **P3**
37 Annual baby's-breath *Gypsophila elegans* M. Bieb. **A3**
38 Baby's-breath *Gypsophila paniculata* L. **A3**
39 Cheddar pink *Dianthus gratianopolitanus* Vill. **L3**
40 Pink *Dianthus plumarius* L. **L1**
41 Maiden pink *Dianthus deltoides* L. **L3**
42 *Sweet William *Dianthus barbatus* L. **A2**
43 *Border pink *Dianthus* cultivar **L1**

Polygonaceae

44 *Russian vine *Fallopia baldschuanica* (Regel) Holub **S2**
45 *Rhubarb *Rheum × hybridum* Murray **P1**
46 *Sorrel *Rumex acetosa* L. **P2**

Paeoniaceae

47 *Peony *Paeonia officinalis* L. **P1**

Clusiaceae

48 Rose-of-Sharon *Hypericum calycinum* L. **P2**
49 Tall tutsan *Hypericum × inodorum* Mill. **S2**

Malvaceae

50 *Common mallow *Malva sylvestris* L. **P2**
51 *Royal mallow *Lavatera trimestris* L. **A2**
52 *Hollyhock *Alcea rosea* L. **P1**

Violaceae

53 Wild pansy *Viola tricolor* L. **P1**
54 Pansy *Viola* cultivar **P1**

Passifloraceae

55 Passion flower *Passiflora caerulea* L. **S3**

Examples of annual, biennial and perennial plants in the Leicester garden (left–right): meadow-foam, *Limnanthes douglasii* (annual); evening primrose, *Oenothera biennis* (biennial); and mock orange, *Philadelphus coronarius* (perennial shrub)

Tamaricaceae

56	Tamarisk	*Tamarix ramosissima* Ledeb. **S3**

Cucurbitaceae

57	Melon	*Cucumis melo* L. **A3**
58	Cucumber	*Cucumis sativus* L. **A2**
59	*Marrow	*Cucurbita pepo* L. **A2**

Begoniaceae

60	Begonia	*Begonia* Semperflorens Cultorum Group **A3**

Salicaceae

61	*Lombardy poplar	*Populus nigra* L. 'Italica' **T1**
62	*Crack willow	*Salix fragilis* L. **T1**

Brassicaceae

63	Wallflower	*Erysimum cheiri* L. **P1**
64	*Dame's-violet	*Hesperis matronalis* L. **P2**
65	Virginia stock	*Malcolmia maritima* W.T. Aiton **A3**
66	*Brompton stock	*Matthiola incana* (L.) R. Br. **A2**
67	Water-cress	*Rorippa nasturtium-aquaticum* (L.) Hayek **W3**
68	*Garden arabis	*Arabis alpina* L. subsp. *caucasica* (Willd. ex Schltdl.) Briq. **L1**
69	*Aubretia	*Aubrieta deltoidea* DC. **L1**
70	*Honesty	*Lunaria annua* L. **A2**
71	Golden Alison	*Aurinia saxatilis* (L.) Desv. **L2**
72	*Perennial candytuft	*Iberis sempervirens* L. **L1**
73	Garden candytuft	*Iberis umbellata* L. **A3**
74	*Cabbage etc.	*Brassica oleracea* L. **A1**
75	Sea-kale	*Crambe maritima* L. **P3**
76	*Radish	*Raphanus sativus* L. **A2**

Ericaceae

77	*Cross-leaved heath	*Erica tetralix* L. **L1**

Primulaceae

78	Primrose	*Primula vulgaris* Hudson **L2**
79	Primula	*Primula* × *polyantha* Mill. **L1**

Hydrangeaceae

80	*Mock orange	*Philadelphus coronarius* L. **S1**
81	*Deutzia	*Deutzia scabra* Thunb. **S1**
82	Hydrangea	*Hydrangea macrophylla* (Thunb.) Ser. **S1**

Grossulariaceae

83	Redcurrant	*Ribes rubrum* L. **S3**
84	Blackcurrant	*Ribes nigrum* L. **S2**
85	*Flowering currant	*Ribes sanguineum* Pursh. **S1**
86	*Gooseberry	*Ribes uva-crispa* L. **S1**

Crassulaceae

87	Butterfly stonecrop	*Sedum spectabile* Boreau **P1**
88	Caucasian stonecrop	*Sedum spurium* M. Bieb. **L1**
89	Reflexed stonecrop	*Sedum rupestre* L. **L1**
90	English stonecrop	*Sedum anglicum* Huds. **L3**

Saxifragaceae

91	Astilbe	*Astilbe chinensis* **B3**
92	Elephant-ear	*Bergenia cordifolia* (Haw.) Sternb. **P2**
93	Pink saxifrage	*Saxifraga* cultivar **L1**
94	*White saxifrage	*Saxifraga* cultivar **L1**
95	Coral flower	*Heuchera sanguinea* **P3**

Rosaceae

96	Van Houtte's spiraea	**Spiraea* × *vanhouttei* (Briot) Carrière **S1**
97	*Buck's beard	*Aruncus dioicus* (Walter) Fernald **P2**
98	*Raspberry	*Rubus idaeus* L. **S1**
99	*Loganberry	*Rubus* × *loganobaccus* L.H. Bailey **S1**
100	*Blackberry	*Rubus fruticosus* agg. L. **S1**
101	*Shrubby cinquefoil	*Potentilla fruticosa* L. **S1**
102	*Strawberry	*Fragaria* × *ananassa* Duchesne **P2**
103	Lady's mantle	*Alchemilla mollis* (Buser) Rothm. **P2**
104	Rose	**Rosa* cultivar **S1**

105	Almond	*Prunus dulcis* (Miller) D.A. Webb **T3**
106	*Cherry plum	*Prunus cerasifera* Ehrh. **T1**
107	Wild cherry	*Prunus avium* (L.) L. **T3**
108	*Japanese cherry	*Prunus serrulata* Lindl. **T2**
109	Japanese quince	*Chaenomeles japonica* (Thunb.) Spach **S2**
110	*Apple	*Malus domestica* Borkh. **T1**
111	*Rowan	*Sorbus aucuparia* L. **T1**
112	*Firethorn	*Pyracantha coccinea* Roem. **S2**
113	*Hawthorn	*Crataegus monogyna* Jacq. **T1**

Papilionaceae

114	French bean	*Phaseolus vulgaris* L. **A3**
115	*Runner bean	*Phaseolus coccineus* L. **A1**
116	Asparagus pea	*Lotus tetragonolobus* L. **A2**
117	Broad bean	*Vicia faba* L. **A3**
118	*Sweet pea	*Lathyrus odoratus* L. **A1**
119	*Pea	*Pisum sativum* L. **A3**
120	Russell lupin	*Lupinus* × *regalis* Bergmans **P2**
121	Kew broom	*Cytisus* × *kewensis* Bean **S2**
122	*Warminster broom	*Cytisus* × *praecox* Wheeler **P3**
123	Chinese wisteria	*Wisteria sinensis* (Sims) Sweet **S3**

Haloragaceae

124	Spiked water-milfoil	*Myriophyllum spicatum* L. **W3**

Thymelaeaceae

125	Mezereon	*Daphne mezereum* L. **S1**

Myrtaceae

126	Omeo gum	*Eucalyptus neglecta* Maiden **T3**
127	Snow gum	*Eucalyptus pauciflora* Sieber ex. Spreng. subsp. *niphophila* (Maiden & Blakely) L.A.S. Johnson & Blaxell **T3**

Onagraceae

128	*Evening primrose	*Oenothera biennis* L. **A1**
129	Clarkia	*Clarkia unguiculata* Lindl. **A2**
130	Godetia	*Clarkia amoena* (Lehm.) Nels. & Macbr. **A3**
131	Fuchsia	*Fuchsia magellanica* Lam. **S1**

Cornaceae

132	Dogwood	*Cornus alba* L. **S1**
133	Spotted laurel	*Aucuba japonica Thunb. **S1**

Celastraceae

134	Spindle	*Euonymus europaeus* L. **S2**
135	Evergreen spindle	*Euonymus japonicus* Thunb. **S3**

Aquifoliaceae

136	*Holly	*Ilex aquifolium* L. **T1**

Buxaceae

137	*Box	*Buxus sempervirens* L. **S1**

Euphorbiaceae

138	Castor oil plant	*Ricinus communis* L. **A3**
139	Caper spurge	*Euphorbia lathyris* L. **A2**
140	Mediterranean spurge	*Euphorbia characias* L. **P3**

Rhamnaceae

141	Alder buckthorn	*Frangula alnus* Mill. **S2**
142	Ceanothus	*Ceanothus* cultivar **S3**

Vitaceae

143	Grape vine	*Vitis vinifera* L. **S3**

Hippocastanaceae

144	*Horse-chestnut	*Aesculus hippocastanum* L. **T3**

Rutaceae

145	Mexican orange	*Choisya ternata* Kunth. **S3**
146	Rue	*Ruta graveolens* L. **P3**

Geraniaceae

147	*Meadow cranesbill	*Geranium pratense* L. **L2**
148	*Ashy cranesbill	*Geranium cinereum* Cav. **L1**
149	Geranium	*Pelargonium* cultivar **A3**

Limnanthaceae

150	*Meadow-foam	*Limnanthes douglasii* R. Br. **A2**

Tropaeolaceae

151	Nasturtium	*Tropaeolum majus* L. **A2**

Balsaminaceae

152	Busy Lizzie	*Impatiens* cultivar **A3**

Araliaceae

153	*Ivy	*Hedera helix* L. **L1**

Apiaceae

154	Sea holly	*Eryngium maritimum* L. **P2**
155	Chervil	*Anthriscus cerefolium* (L.) Hoffm. **A3**
156	*Fennel	*Foeniculum vulgare* Mill. **P1**
157	Celery	*Apium graveolens* L. var. *dulce* (Mill.) DC. **A3**
158	Parsley	*Petroselinum crispum* (Mill.) Nyman ex A.W. Hill **A1**
159	*Lovage	*Levisticum officinale* W.D.J. Koch. **P2**
160	Parsnip	*Pastinaca sativa* L. var. *hortensis* Gaudin **A3**
161	Carrot	*Daucus carota* L. subsp. *sativus* (Hoffm.) Arcang. **A3**

Apocynaceae

162	*Greater periwinkle	*Vinca major* L. **L1**

Asclepiadaceae

163	Cruel plant	*Araujia sericifera* Brot. **A3**
164	Bloodflower	*Asclepias curassavica* L. **A3**
165	Common milkweed	*Asclepias syriaca* L. **A3**

Nolanaceae

166	Chilean bellflower	*Nolana acuminata* L. Miers ex Dun. **A3**

Solanaceae

167	Deadly nightshade	*Atropa bella-donna* L. **P3**
168	Henbane	*Hyoscyamus niger* L. **A3**

169 Sweet pepper *Capsicum annuum* L. **A3**
170 *Tomato *Lycopersicon esculentum* Mill. **A1**
171 Chilean potato tree *Solanum crispum* Ruiz & Pav. **S3**
172 Eggplant *Solanum melongena* L. **A3**
173 Potato *Solanum tuberosum* L. **P2**
174 *Thorn apple *Datura stramonium* L. **A2**
175 *Sweet tobacco *Nicotiana alata* Link & Otto **A1**
176 Tree tobacco *Nicotiana glauca* R.C. Graham **A3**
177 Petunia *Petunia* cultivar (Hook.) Vilm. **A2**
178 Painted tongue *Salpiglossis sinuata* Ruiz & Pav. **A3**
179 Poor man's orchid *Schizanthus pinnatus* Ruiz & Pav. **A3**

Convolvulaceae

180 Dwarf convolvulus *Convolvulus tricolor* L. **A2**

Polemoniaceae

181 Jacob's-ladder *Polemonium caeruleum* L. **P3**
182 Phlox **Phlox paniculata* L. **P2**
183 Tufted phlox *Phlox douglasii* Hook. **L1**
184 Annual phlox *Phlox drummondii* **A3**

Hydrophyllaceae

185 Baby blue eyes *Nemophila menziesii* Hook. & Arn. **A3**

Boraginaceae

186 *Purple viper's bugloss *Echium plantagineum* L. **A1**
187 Lungwort *Pulmonaria officinalis* L. **P3**
188 *Russian comfrey *Symphytum × uplandicum* Nyman **P2**
189 Garden anchusa *Anchusa azurea* Mill. **P3**
190 Cape bugloss *Anchusa capensis* Thunb. **A3**
191 *Borage *Borago officinalis* L. **A1**
192 *Forget-me-not *Myosotis arvensis* (L.) Hill **A1**

Verbenaceae

193 Verbena *Verbena* cultivar **P3**

Lamiaceae

194 Large-flowered hemp-nettle *Galeopsis speciosa* Mill. **A3**
195 Garden catmint *Nepeta × faassenii* Bergmans ex Stearn **P1**
196 *Balm *Melissa officinalis* L. **P2**
197 *Hyssop *Hyssopus officinalis* L. **P2**
198 *Pot marjoram *Origanum majorana* L. **P2**
199 Garden thyme *Thymus vulgaris* L. **P3**
200 *Wild thyme *Thymus polytrichus* A. Kern ex Borbás **L1**
201 Lemon thyme *Thymus × citriodorus* Pers. **P3**
202 *Peppermint *Mentha × piperita* L. **P1**
203 *Spearmint *Mentha spicata* L. **P1**
204 *Lavender *Lavandula angustifolia* Mill. **P1**
205 Rosemary *Rosmarinus officinalis* L. **P2**
206 Clary *Salvia sclarea* L. **P3**
207 *Sage *Salvia officinalis* L. **P1**
208 Scarlet sage *Salvia splendens* Sello **A3**
209 Bergamot *Monarda didyma* L. **P2**
210 Sweet basil *Ocimum basilicum* L. **A3**
211 Bush basil *Ocimum minimum* L. **A3**

Callitrichaceae

212 Common water-starwort *Callitriche stagnalis* Scop. **W3**

Buddlejaceae

213 *Butterfly bush *Buddleja davidii* Franch. **S1**

Oleaceae

214 *Forsythia *Forsythia × intermedia* Zabel **S1**
215 *Lilac *Syringa vulgaris* L. **S1**
216 *Garden privet *Ligustrum ovalifolium* Hassk. **S1**

Scrophulariaceae

217 Blood-drop-emlets *Mimulus luteus* L. **P3**
218 Snapdragon *Antirrhinum majus* L. **A1**
219 Annual toadflax *Linaria maroccana* Hook. **A2**
220 *Foxglove *Digitalis purpurea* L. **A1**
221 Brooklime *Veronica beccabunga* L. **W3**
222 Spiked speedwell *Veronica spicata* L. **P1**
223 Veronica *Veronica* cultivar **P3**
224 Mask flower *Alonsoa warscewiczii* Regel. **A3**
225 Cape jewels *Nemesia strumosa* Benth. **A3**

Campanulaceae

226 Canterbury bell *Campanula medium* L. **A3**
227 Clustered bellflower *Campanula glomerata* L. **P3**
228 *Creeping bellflower *Campanula rapunculoides* L. **P1**
229 Garden lobelia *Lobelia erinus* L. **A2**

Rubiaceae

230 Caucasian crosswort *Phuopsis stylosa* Benth. & Hook. **P3**

Caprifoliaceae

231 Himalayan honeysuckle *Leycesteria formosa* Wall. **S2**
232 *Weigela *Weigela florida* (Bunge) A. DC. **S1**
233 *Honeysuckle *Lonicera periclymenum* L. **S1**

Valerianaceae

234 *Red valerian *Centranthus ruber* (L.) DC. **P1**

Dipsacaceae

235 Sweet scabious *Scabiosa atropurpurea* **A3**

Asteraceae

236 Globe thistle *Echinops ritro* L. **P2**
237 *Perennial cornflower *Centaurea montana* L. **P1**
238 Cornflower *Centaurea cyanus* L. **A2**
239 *Lettuce *Lactuca sativa* L. **A1**
240 Treasureflower *Gazania rigens* (L.) Gaertn. **A3**
241 *Canadian goldenrod *Solidago canadensis* L. **P1**
242 *Confused Michaelmas daisy *Aster novi-belgii* L. **P1**
243 Italian aster *Aster amellus* L. **P1**
244 Eastern showy aster *Aster spectabilis* Hook. & Arn. **P1**
245 *Fleabane *Erigeron* cultivar **P1**
246 China aster *Callistephus chinensis* (L.) Nees **A3**
247 *Tansy *Tanacetum vulgare* L. **P2**
248 *Pyrethrum *Tanacetum coccineum* (Willd.) Grierson **P1**

249 Tarragon *Artemesia dracunculus* L. **P3**
250 *Lavender cotton *Santolina chamaecyparissus* L. **P1**
251 *Fernleaf yarrow *Achillea filipendulina* Lam. **P2**
252 Annual chrysanthemum *Ismelia carinata* (Schousb.) Sch. Bip. **A3**
253 Chrysanthemum *Chrysanthemum* cultivar **P2**
254 Silver ragwort *Senecio cineraria* DC. **P3**
255 Oxford ragwort *Senecio squalidus* L. **A2**
256 Shrub ragwort *Brachyglottis* 'Sunshine' **P3**
257 Plantain-leaved leopard's bane *Doronicum plantagineum* L. **P2**
258 *Pot marigold *Calendula officinalis* L. **A1**
259 *Black-eyed Susan *Rudbeckia hirta* L. **A2**
260 Sunflower *Helianthus annuus* L. **A2**
261 Fern-leaved beggar-ticks *Bidens ferulifolia* (Jacq.) DC. **A3**
262 Large-flowered tickseed *Coreopsis grandiflora* Hogg ex Sweet. **P3**
263 Lanceleaf tickseed *Coreopsis lanceolata* L. **A3**
264 *Mexican aster *Cosmos bipinnatus* Cav. **A2**
265 *Dahlia *Dahlia* cultivar **P2**
266 African marigold *Tagetes erecta* L. **A2**
267 *French marigold *Tagetes patula* L. **A1**
268 Signet marigold *Tagetes tenuifolia* Cav. **A3**
269 Sneezeweed *Helenium autumnale* L. **P2**
270 Flossflower *Ageratum houstonianum* **A3**
271 African daisy *Arctotis* × *hybrida* hort. **A3**
272 Swan River daisy *Brachyscome iberidifolia* Benth. **A3**
273 *Shasta daisy *Leucanthemum* × *superbum* L. **P1**
274 Purple coneflower *Echinacea purpurea* (L.) Moench **A3**
275 Smooth oxeye *Heliopsis helianthoides* (L.) Sweet **P2**
276 Gayfeather *Liatris spicata* (L.) Willd. **P3**
277 Sea rosemary *Ozothamnus rosmarinifolius* (Labill.) Sweet **P3**
278 Umbrella milkwort *Tolpis barbata* (L.) Gaertn. **A3**
279 Youth-and-age *Zinnia elegans* Jacq. **A3**

Lemnaceae

280 Common duckweed *Lemna minor* L. **W2**

Asphodelaceae

281 Red-hot poker *Kniphofia uvaria* (L.) Oken. **B3**

Liliaceae

282 Tulip *Tulipa* cultivar **B1**
283 Madonna lily *Lilium candidum* L. **B1**

Convallariaceae

284 *Lily-of-the-valley *Convallaria majalis* L. **B1**
285 Solomon's-seal *Polygonatum multiflorum* (L.) All. **B1**

Hyacinthaceae

286 Squill *Scilla* species **B2**
287 Bluebell *Hyacinthoides non-scripta* (L.) Rothm. **B1**
288 Hyacinth *Hyacinthus orientalis* L. **B1**
289 *Grape hyacinth *Muscari armeniacum* Leichtlin ex Baker **B1**

Alliaceae

290 Chives *Allium schoenoprasum* L. **B2**
291 Onion *Allium cepa* L. **B2**
292 Egyptian onion *Allium cepa* L. Proliferum Group **B1**
293 Few-flowered leek *Allium paradoxum* (M. Bieb.) G. Don **B2**
294 Leek *Allium porrum* L. **B2**
295 Round-headed leek *Allium sphaerocephalon* L. **B3**
296 African lily *Agapanthus praecox* Willd. subsp. *orientalis* (F.M. Leight) F.M. Leight **B3**

Amaryllidaceae

297 Snowdrop *Galanthus nivalis* L. **B1**
298 *Daffodil *Narcissus* cultivar **B1**

Ruscaceae

299 Butcher's broom *Ruscus aculeatus* L. **S2**

Alstroemeriaceae

300 Peruvian lily *Alstroemeria aurea* Graham **B3**

Iridaceae

301 *Yellow iris *Iris pseudacorus* L. **B1**
302 Bearded iris *Iris germanica* L. **B2**
303 Dutch iris *Iris hollandica* hort. **B3**
304 Golden crocus **Crocus chrysanthus* (Herb.) Herb. **B1**
305 Spring crocus *Crocus vernus* (L.) Hill **B1**
306 Yellow crocus *Crocus* × *stellaris* Haw. **B1**
307 *Montbretia *Crocosmia* × *crocosmiiflora* (Lemoine) N.E. Br. **B1**

SPONTANEOUS INTRODUCTIONS

Ranunculaceae

308 Creeping buttercup *Ranunculus repens* L.
309 Lesser celandine *Ranunculus ficaria* L.

Papaveraceae

310 Common poppy *Papaver rhoeas* L.
311 Long-headed poppy *Papaver dubium* L.
312 Welsh poppy *Meconopsis cambrica* (L.) Vig.

Cannabaceae

313 Hop *Humulus lupulus* L.

Urticaceae

314 Common nettle *Urtica dioica* L.
315 Mind-your-own-business *Soleirolia soleirolii* (Req.) Dandy

Betulaceae

316 Silver birch *Betula pendula* Roth.

Chenopodiaceae

317 Fig-leaved goosefoot *Chenopodium ficifolium* Sm.
318 Fat-hen *Chenopodium album* L.
319 Common orache *Atriplex patula* L.

Caryophyllaceae

320	Common chickweed	*Stellaria media* (L.) Vill.
321	Common mouse-ear	*Cerastium fontanum* Baumg.
322	Little mouse-ear	*Cerastium semidecandrum* L.
323	Procumbent pearlwort	*Sagina procumbens* L.

Polygonaceae

324	Redshank	*Persicaria maculosa* Gray
325	Knotgrass	*Polygonum aviculare* L.
326	Black-bindweed	*Fallopia convolvulus* (L.) Á. Löve
327	Broad-leaved dock	*Rumex obtusifolius* L.

Clusiaceae

328	Tutsan	*Hypericum androsaemum* L.

Violaceae

329	Sweet violet	*Viola odorata* L.
330	Common dog-violet	*Viola riviniana* Rchb.

Salicaceae

331	Sallow	*Salix* species

Brassicaceae

332	Hedge mustard	*Sisymbrium officinale* (L.) Scop.
333	Garlic mustard	*Alliaria petiolata* (M. Bieb.) Cavara & Grande
334	Hairy bitter-cress	*Cardamine hirsuta* L.
335	Sweet Alison	*Lobularia maritima* (L.) Desv.
336	Shepherd's-purse	*Capsella bursa-pastoris* (L.) Medik.
337	Charlock	*Sinapis arvensis* L.

Primulaceae

338	Creeping-Jenny	*Lysimachia nummularia* L.
339	Scarlet pimpernel	*Anagallis arvensis* L. subsp *arvensis*

Crassulaceae

340	Biting stonecrop	*Sedum acre* L.
341	White stonecrop	*Sedum album* L.

Rosaceae

342	Cut-leaved blackberry	*Rubus laciniatus* Willd.
343	Wild strawberry	*Fragaria vesca* L.
344	Wood avens	*Geum urbanum* L.
345	Dog rose	*Rosa canina* L.
346	Cotoneaster	*Cotoneaster* species

Papilionaceae

347	Common bird's-foot-trefoil	*Lotus corniculatus* L.
348	Hairy tare	*Vicia hirsuta* (L.) Gray
349	Common vetch	*Vicia sativa* L.
350	Black medick	*Medicago lupulina* L.
351	White clover	*Trifolium repens* L.
352	Hop trefoil	*Trifolium campestre* Schreb.
353	Broom	*Cytisus scoparius* (L.) Link

Onagraceae

354	Great willowherb	*Epilobium hirsutum* L.
355	American willowherb	*Epilobium ciliatum* Raf.
356	Rosebay willowherb	*Chamerion angustifolium* (L.) Holub

Euphorbiaceae

357	Petty spurge	*Euphorbia peplus* L.

Aceraceae

358	Sycamore	*Acer pseudoplatanus* L.

Oxalidaceae

359	Procumbent yellow-sorrel	*Oxalis corniculata* L.
360	Large-flowered pink-sorrel	*Oxalis debilis* Kunth

Geraniaceae

361	Herb-Robert	*Geranium robertianum* L.

Apiaceae

362	Hogweed	*Heracleum sphondylium* L.

Solanaceae

363	Black nightshade	*Solanum nigrum* L.
364	Bittersweet	*Solanum dulcamara* L.

Convolvulaceae

365	Field bindweed	*Convolvulus arvensis* L.
366	Hedge bindweed	*Calystegia sepium* (L.) R. Br.

Lamiaceae

367	Hedge woundwort	*Stachys sylvatica* L.
368	Spotted dead-nettle	*Lamium maculatum* (L.) L.
369	Red dead-nettle	*Lamium purpureum* L.
370	Henbit dead-nettle	*Lamium amplexicaule* L.
371	Common hemp-nettle	*Galeopsis tetrahit* L.
372	Bugle	*Ajuga reptans* L.
373	Selfheal	*Prunella vulgaris* L.

Plantaginaceae

374	Greater plantain	*Plantago major* L.

Oleaceae

375	Ash	*Fraxinus excelsior* L.
376	Wild privet	*Ligustrum vulgare* L.

Scrophulariaceae

377	Great mullein	*Verbascum thapsus* L.
378	Purple toadflax	*Linaria purpurea* (L.) Mill.
379	Thyme-leaved speedwell	*Veronica serpyllifolia* L.
380	Green field-speedwell	*Veronica agrestis* L.
381	Ivy-leaved speedwell	*Veronica hederifolia* L.

Rubiaceae

382	Cleavers	*Galium aparine* L.

Caprifoliaceae

383	Elder	*Sambucus nigra* L.
384	Snowberry	*Symphoricarpos albus* (L.) S.F. Blake

Dipsacaceae

385	Wild teasel	*Dipsacus fullonum* L.

Asteraceae	
386 Spear thistle	***Cirsium vulgare*** (Savi) Ten.
387 Creeping thistle	***Cirsium arvense*** (L.) Scop.
388 Goat's-beard	***Tragopogon pratensis*** L.
389 Perennial sow-thistle	***Sonchus arvensis*** L.
390 Smooth sow-thistle	***Sonchus oleraceus*** L.
391 Prickly sow-thistle	***Sonchus asper*** (L.) Hill
392 Dandelion	***Taraxacum*** sect. ***Ruderalia*** Kirschner, H. Øllg. & Stepanek
393 Smooth hawk's-beard	***Crepis capillaris*** (L.) Wallr.
394 Fox-and-cubs	***Pilosella aurantiaca*** (L.) F.W. Schultz & Sch. Bip.
395 Daisy	***Bellis perennis*** L.
396 Yarrow	***Achillea millefolium*** L.
397 Stinking chamomile	***Anthemis cotula*** L.
398 Canadian fleabane	***Conyza canadensis*** (L.) Cronquist
399 Feverfew	***Tanacetum parthenium*** (L.) Sch. Bip.
400 Pineappleweed	***Matricaria discoidea*** DC.
401 Scentless mayweed	***Tripleurospermum inodorum*** (L.) Sch. Bip.
402 Leptinella	***Cotula squalida*** Hook. f.
403 Common ragwort	***Senecio jacobaea*** L.
404 Groundsel	***Senecio vulgaris*** L.
405 Heath groundsel	***Senecio sylvaticus*** L.
406 Colt's-foot	***Tussilago farfara*** L.

Juncaceae	
407 Field wood-rush	***Luzula campestris*** (L.) DC.

Liliaceae	
408 Asparagus	***Asparagus officinalis*** L.

Iridaceae	
409 Stinking iris	***Iris foetidissima*** L.

Thyme-leaved speedwell, *Veronica serpyllifolia* (above), growing in cracks in paving in the Leicester garden alongside spotted dead-nettle, *Lamium maculatum*

Creeping-Jenny, *Lysimachia nummularia* (below left), appeared spontaneously in the author's garden

GRASSES

Nomenclature and order follow Stace (1997)

Poaceae	
410 Chewings fescue	***Festuca rubra*** L. subsp. ***commutata*** Gaudin
411 Hard fescue	***Festuca brevipila*** R. Tracey
412 Perennial rye-grass	***Lolium perenne*** L.
413 Crested dog's-tail	***Cynosurus cristatus*** L.
414 Annual meadow-grass	***Poa annua*** L.
415 Rough meadow-grass	***Poa trivialis*** L.
416 Smooth meadow-grass	***Poa pratensis*** L.
417 Wood meadow-grass	***Poa nemoralis*** L.
418 Cock's-foot	***Dactylis glomerata*** L.
419 Oat	***Avena sativa*** L.
420 Yellow oat-grass	***Trisetum flavescens*** (L.) P. Beauv.
421 Yorkshire-fog	***Holcus lanatus*** L.
422 Creeping soft-grass	***Holcus mollis*** L.
423 Sweet vernal-grass	***Anthoxanthum odoratum*** L.
424 Common bent	***Agrostis capillaris*** L.
425 Black bent	***Agrostis gigantea*** Roth
426 Creeping bent	***Agrostis stolonifera*** L.
427 Velvet bent	***Agrostis canina*** L.
428 Timothy	***Phleum pratense*** L.
429 Upright brome	***Bromopsis erecta*** (Huds.) Fourr.
430 Barren brome	***Anisantha sterilis*** (L.) Nevski
431 Two-rowed barley	***Hordeum distichon*** L.
432 Meadow barley	***Hordeum secalinum*** Schreb.
433 Bread wheat	***Triticum aestivum*** L.

FLOWERLESS PLANTS

ALGAE

434 *Desmococcus olivaceus* (Pers. ex Ach.) Laundon

LICHENS

435 Unidentifed species

LIVERWORTS

436 *Marchantia polymorpha* L.

MOSSES

437 Capillary thread-moss — *Bryum capillare* Hedw.
438 Hair-pointed grimmia — *Grimmia trichophylla* Grev.
439 Cypress-leaved plait-moss — *Hypnum cupressiforme* Hedw. var. *resupinatum*
440 Golden thread-moss — *Leptobryum pyriforme* (Hedw.) Wils.
441 *Lophocolea bidentata* (L.) Dum.
442 Swan's-neck thyme-moss — *Mnium hornum* Hedw.
443 Clustered feather-moss — *Rhynchostegium confertum* (Dicks.) Br. Eur.
444 Springy turf-moss — *Rhytidiadelphus squarrosus* (Hedw.) Warnst.
445 Wall screw-moss — *Tortula muralis* Hedw.

FERNS

446 Male-fern — *Dryopteris filix-mas* (L.) Schott

Male fern, *Dryopteris filix-mas* (above), grows in the rockery in the back garden and in a shady area to the east of the house

Brown roll-rim, *Paxillus involutus*, a large toadstool growing on the lawn (right)

FUNGI

Nomenclature and order follow Phillips (1981)

447 Shaggy parasol — *Lepiota rhacodes* (Vitt.) Quél.
448 *Lyophyllum decastes* (Fr. ex Fr.) Sing.
449 Deceiver — *Laccaria laccata* (Scop. ex Fr.) Cke.
450 Snowy wax cap — *Hygrocybe nivea* (Scop.) Fr.
451 *Mycena flavo-alba* (Fr.) Quél.
452 *Mycena fibula* (Bull. ex Fr.) Kühner
453 *Mycena* species
454 Woolly milk-cap — *Lactarius torminosus* (Schaeff. ex Fr.) S.F. Gray
455 *Lactarius pubescens* (Fr. ex Krombh.) Fr.
456 Wood blewit — *Lepista nuda* (Bull. ex Fr.) Cooke
457 *Pluteus* species
458 *Cortinarius violaceus* (L. ex Fr.) Fr.
459 *Cortinarius evernius* Fr.
460 Brown roll-rim — *Paxillus involutus* (Fr.) Fr.
461 Poison pie — *Hebeloma crustuliniforme* (Bull. ex St. Amans) Quél.
462 *Hebeloma mesophaeum* (Pers.) Quél.
463 *Inocybe tigrina* Heim
464 *Inocybe napipes* Lange
465 *Psathyrella candolleana* (Fr.) Maire
466 Liberty cap — *Psilocybe semilanceata* (Fr. ex Secr.) Kummer
467 Shaggy ink cap — *Coprinus comatus* (Müll. ex Fr.) S.F. Gray
468 Many-zoned polypore — *Coriolus versicolor* (L. ex Fr.) Quél.
469 *Clavulinopsis luteo-alba* (Rea) Corner
470 Earth fan — *Thelephora terrestris* (Ehrh.) Fr.
471 *Peziza vesiculosa* Bull. ex St. Amans
472 *Tarzetta cupularis* (L. ex Fr.) Lamb.

1

2

1 The lawn of the Leicester garden in the mid-1970s, with montbretia (*Crocosmia* x *crocosmiiflora*), sweet tobacco (*Nicotiana alata*) and evening primrose (*Oenothera biennis*).
2 A cold frame used before the greenhouse was erected. To the left is the trunk of a horse chestnut (*Aesculus hippocastanum*), rhubarb (*Rheum* x *hybridum*) at the base of the tree, cabbage and a white *Saxifraga* cultivar

3–6 (opposite) **Shade beneath the apple tree: 3** Oregon grape (*Mahonia aquifolium*) with spotted dead-nettle (*Lamium maculatum*) and crack willow (*Salix fragilis*). **4** Lombardy poplar (*Populus nigra* 'Italica'), blackberry (*Rubus fruticosus*) and spotted dead-nettle. **5** Hollyhock (*Alcea rosea*), lilac (*Syringa vulgaris*) and spotted dead-nettle. **6** The compost heap

3

4

5

6

7 Foxgloves (*Digitalis purpurea*), French marigolds (*Tagetes patula*), peonies (*Paeonia officinalis*) and wallflowers (*Erysimum cheiri*) in a bed on the east side of the lawn

8–10 An area of mixed planting beside the lawn, the so-called 'lawn bed', in different years
8 Sunflowers (*Helianthus annuus*), runner beans (*Phaseolus coccineus*), Mexican aster (*Cosmos bipinnatus*), royal mallow (*Lavatera trimestris*), and snapdragon (*Antirrhinum majus*)

9 Runner beans, youth-and-age (*Zinnia elegans*), French marigolds, African marigolds (*Tagetes erecta*), lettuce (*Lactuca sativa*), sweet tobacco (*Nicotiana alata*) and, to the right, Shasta daisies (*Leucanthemum* × *superbum*)

10 Runner beans, French and African marigolds, sweet tobacco, snapdragons and Shasta daisies

11

12

13

14

11 Royal mallow (*Lavatera trimestris*), pot marigolds (*Calendula officinalis*) and African marigolds, with alpine poppies (*Papaver alpinum*) in the foreground. **12** Daisies (*Bellis perennis*), forget-me-nots (*Myosotis*), biting stonecrop (*Sedum acre*) and garden pinks (*Dianthus* cultivar) growing in cracks.

13 *Begonia* Semperflorens Group, yellow iris (*Iris pseudacorus*), feverfew (*Tanacetum parthenium*) and rosebay willowherb (*Chamerion angustifolium*). **14** Shade on eastern side of house, with flowering currant (*Ribes sanguineum*), holly (*Ilex aquifolium*) and buddleia (*Buddleja davidii*) a year. In addition, there are nine species of moss, at least 26 species of fungi, and many other flowerless plants (see lists, p29). The composition of the flora to some extent changed from year to year – that is the essence of gardening – but in terms of the quantity of vegetation and its rich variety, the flora consistently provided a broad, firm base to support, directly or indirectly, the many animals that are the subject of the following chapters.

MONITORING THE GARDEN'S ANIMALS

The fauna, particularly the insects, of my garden was investigated from 1972 to 2001 in a number of ways. As every ecologist knows, a great deal can be gained from regular observation and surveillance of a study area, and it is in this respect that a private domestic garden is a particularly satisfactory site for investigation, being close and accessible. Information has been gleaned on every walk around the garden, while gardening, and sometimes simply by glancing out of the window. Whenever insect damage to leaves was noticed, or its tell-tale signs, such as frass on a path beneath a plant, was seen, a search was made for the perpetrator. Adults, such as beetles, were collected for identification, and larvae were reared in jars on the appropriate foodplant, until adult, when they, too, could be identified. Predatory hoverfly larvae were also collected, and reared to maturity. Beating vegetation out onto a tray (or upturned umbrella) yielded further larvae for rearing. Sometimes the end result was not the adult expected, say a moth, but instead an ichneumonid, which could also be identified. Rearing of larvae found in the garden produced a wealth of information about feeding relationships, and facilitated recognition of food chains.

Continuous use of a Malaise trap for the period 1972–2001 and of pitfall traps in 1979–1990, regular use of a mercury-vapour light trap and of hand-netting, and occasional use of baited traps resulted in the collection of a great deal of information about invertebrates, particularly insects, in the garden. Abundance, diversity, fluctuations in numbers, seasonality, and appearances and disappearances were monitored. Continual observations and rearing of larvae provided the information that 'fleshes out' the trap data, and gives an idea what insects are doing in the garden.

Large white butterfly, *Pieris brassicae*, indelibly marked on the hind wing using a felt-tip pen

Hand-netting butterflies

Detailed investigation of the fauna, of its abundance, diversity and changes, from season to season or year to year, necessitated use of various trapping methods. The simplest of these was hand-netting of butterflies, which were marked and released during the insect flight seasons of 1972–1979 inclusive. Butterflies were marked on the wing with waterproof ink, using an ordinary felt-tip marker, so that individuals already caught and scored could be recognised. Although prone to some bias, because of variations in time spent catching butterflies, and because netting was unsystematic, this method gave a reasonable estimate of changes, with season and year, in numbers of individuals and species visiting the garden.

Malaise trap (above left) for catching flying insects in the Leicester garden, with lilac, *Syringa vulgaris*, to the right, confused Michaelmas daisy, *Aster novii-belgii*, to the left, and apple tree behind

Beneath the Malaise trap (above right) grew spotted dead-nettle (*Lamium maculatum*), creeping-Jenny (*Lysimachia nummularia*), perennial cornflower (*Centaurea montana*) and various annual flowers

Malaise trap

One of the most useful and profitable trapping techniques was the operation of a Malaise trap (above). This catches flying insects non-selectively. It uses no bait or other attractant, simply intercepting insects in the course of their normal activities. Moreover, it was kept continuously in position, by day and by night, in all weathers. The type of Malaise trap used was that designed and described by Townes (1972). It is an open-sided, tent-like structure of meshed fabric, about 2m long and 1.3m wide, supported on poles held upright by guy ropes.

The pitched roof rises obliquely to a peak almost 2m tall at one end, where there is an opening leading into a collecting jar containing 70% ethanol as killing agent and preservative. An internal baffle of netting hangs down to ground level all along the central axis; it reaches almost up to the opening into the collecting jar. The side openings, about 1.1m tall, run the full length of the trap. Flying insects that wander into the open sides and come up against the central baffle tend to fly or clamber upwards towards the daylight (the roof of the trap is white), and eventually fall into the collecting jar. The collecting jar was emptied and the catch sorted at least once a week.

Not all insects that enter the open sides of the trap end up in the collecting jar. It is generally believed that larger beetles drop to the ground on striking the baffle, and then walk away. An unknown proportion of other groups also find their way out again; the trap is estimated, for instance, to capture only 20% of the Ichneumonidae that enter its 2.9m^3 of air space (Owen, J., Townes & Townes 1981). Trapping is, moreover, generally believed to have an effect on the community sampled, because of the consequences of removing individuals. Removal of each leaves a vacancy which is promptly filled by immigration. Consequently, trapping in itself boosts the size of the trap sample. However, it does provide a large, unselected sample of flying insects in a standardized way, and can be used to compare abundances, diversities and species composition at different times.

The Malaise trap in the garden was operated every year from 1972 to 2001, from 1 April to 31 October in the same site in the back garden. One open side faced a lilac bush and other vegetation close to the eastern fence; the other side was open to the main part of the garden. Below the trap was low vegetation, including spotted dead-nettle, creeping-Jenny, perennial cornflower (*Centaurea montana*) and various annual flowers. Observation suggested that it was situated across a flight path of insects, particularly butterflies. In 1978, a second Malaise trap was operated on the overgrown lawn (meadow) on the western side of the front garden to compare the effectiveness of trapping in the two sites. The comparison was made using hoverflies (Syrphidae). The total catch from the second trap was smaller, the relative frequency of species was different, and a hoverfly new to the garden was caught. This suggests firstly that the fauna differs slightly in different parts of the garden, and secondly, and more importantly, that continued use of the original site was justified.

Pitfall trap filled with 4% formalin. On the right is its cover, which has been removed

Pitfall traps

Another trapping method, used continuously from 1 April to 31 October 1979–1990, inclusive, was operation of pitfall traps. These consisted of plastic drinking beakers, set into permanently positioned lengths of sunken piping, and one-third- to half-filled with 4% formalin solution as killing agent and preservative. Over each trap was positioned a square of hardboard mounted on nails which were driven into the soil to give a clearance of about one centimetre, sufficient to exclude frogs and larger animals and to keep out rainwater and leaves, but high enough to admit carabid beetles, centipedes, spiders and other ground-dwelling invertebrates. In all, 11 pitfalls were operated, two in the back garden rockery, four in back garden herbaceous borders, two below the shrubbery to the east of the house, and three below the Malaise trap. The pots were emptied, cleaned and re-filled with formalin when necessary, usually every two weeks.

Mercury vapour light trap with the light bulb projecting above the main body of the trap, within which are egg cartons (just visible) on which the lured moths will settle

Mercury-vapour light trap

Most of the information about moths in the garden comes from operation of a standard mercury-vapour light trap. This consists of a circular, black box, 60cm in diameter and 25cm tall, topped by a truncated cone of transparent plastic, in the centre of which is the mercury-vapour light bulb mounted in a metal funnel. Moths attracted to the light strike the baffles in the metal funnel as they circle, and drop into the body of the trap, which is packed with papier-mâché egg cartons on which they come to rest.

When the trap is unpacked in the morning, most moths can easily be identified, and only the 'difficult-to-identify' specimens need be collected. The bulk were kept in the covered trap in a cool place, to be released the following evening. The trap was used regularly up until 1992, usually being operated on the back lawn, but occasionally at the north end of the garden near the garage.

Sometimes the mercury vapour light was removed from the trap and suspended against a white sheet for inspection of moths as they were attracted during the night.

A baited trap similar to that described in the text, but in this case designed to be free-standing, with a collecting jar filled with ethanol attached obliquely to the top of the trap to capture insects disturbed from the bait

Baited trap

A baited trap was used occasionally to capture particular sorts of insects. The design used consisted of a cylinder of white netting fabric, about 30cm in diameter and 60cm in height, closed with netting at the top. The netting was supported top and bottom by circles of stiff wire, below which was suspended a board of equal diameter. The trap was hung from a tree branch by a handle attached at the top, and bait was placed on the board. The gap of about 3cm between board and trap was sufficient to admit insects as large as butterflies, but not so large as to be an obvious exit. When insects leave the bait, they tend to fly up into the cylinder of netting, and could be removed by lifting down the trap so that it collapsed. Baited traps were used principally for collecting moths, butterflies and hoverflies, using fermented fruit, and blow-flies (Calliphoridae) and grey flesh flies (Sarcophagidae), using fish heads and skins.

THE WEATHER

The climate for the part of Britain in which the garden is situated tends towards the continental. The county of Leicestershire is the part of Britain farthest from the sea and hence from maritime influences (Pye 1972b). It is, therefore, relatively dry, with well-marked seasonal temperature differences. Weather records for the whole period during which the garden was being studied were supplied by the meteorological station at Newtown Linford, just to the northwest of the city of Leicester, and 11.5km from my garden. The records for the two 15-year monitoring periods are presented separately, for the reasons given at the end of this chapter.

Table 2.1 Monthly minimum and maximum temperatures, rainfall and sunshine for the two 15-year periods 1972–1986/1987–2001 at Newtown Linford, Leicestershire, 11.5km from the garden

	Mean minimum temp. (°c)		Mean maximum temp. (°c)		Mean rainfall (mm)		Mean sun (hours)	
	'72–'86	'87–'01	'72–'86	'87–'01	'72–'86	'87–'01	'72–'86	'87–'01
January	0.2	0.9	5.8	6.7	59.8	62.0	49.3	54.9
February	−0.1	1.2	5.6	8.1	50.8	50.3	50.2	77.1
March	1.6	2.5	8.9	10.5	55.1	49.7	87.2	103.8
April	3.1	3.6	11.7	12.6	42.8	62.9	121.6	132.3
May	5.8	6.2	15.2	16.5	59.1	45.7	159.3	184.5
June	8.7	8.7	18.6	18.8	62.8	60.9	169.1	164.3
July	10.7	11.0	21.0	21.9	42.4	51.4	169.4	192.6
August	10.6	10.8	20.6	21.8	59.2	55.3	163.3	188.3
September	8.6	9.0	17.5	17.9	56.2	68.9	125.7	124.2
October	6.0	6.3	13.4	14.0	51.6	72.8	93.2	101.8
November	3.0	3.1	9.2	9.4	57.4	63.4	63.4	59.4
December	1.4	1.8	7.0	7.2	68.5	69.4	46.4	47.2

Map of the city of Leicester and its suburbs, showing the location of Newtown Linford meteorological station relative to the author's garden

For the period 1972–1986, the mean maximum and minimum July temperatures were 21°c and 10.7°c; for January, the mean maximum was 5.8°c and the mean minimum 0.2°c (Tab. 2.1). The temperature extremes for this period were 32.3°c and –14.6°c, recorded in July 1976 and January 1982, respectively.

The second 15-year period, 1987–2001, was slightly warmer on average, although the overall maximum and minimum temperatures were less extreme (Tab. 2.1.). The mean July temperatures were 21.9°c maximum and 11.0°c minimum; for January, the mean maximum was 6.7°c and the mean minimum 0.9°c. The extremes were 25.7°c, in July 1996, and –2.3°c, in January 1987.

Leicester city temperatures show a marked 'heat island' effect, and my garden is undoubtedly influenced climatically by its proximity to the city, thus the Newtown Linford figures quoted above probably underestimate the actual garden temperatures. Pye (1972b) recognised a growing season for vegetation in the Leicester region of 230 days, from late March to early November, during which mean temperatures exceed 6°c, and evapotranspiration reaches a maximum in July.

The rainfall varies considerably from year to year (Figs. 2.4 and 2.5), although distribution over the year is remarkably uniform (Tab. 2.1). Annual average rainfall for Newtown Linford was 665.5mm for the period 1972–1986, when monthly averages varied from 42.4mm in July to 68.5mm in December. For the period 1987–2001, the annual average was 712.7mm and the monthly averages varied from 45.7mm in May to 72.8mm in October. According to Pye (*op.cit.*), July to January is the wettest period for Leicester. He states that the total duration of rain in a month, and also the number of days with rain, is greater in November, December, January and February than in the rest of the year.

The monthly average rainfall for July (1972–1986) is lower than might be

Figure 2.4 Annual rainfall 1972–1986 and 15-year mean annual rainfall at Newtown Linford

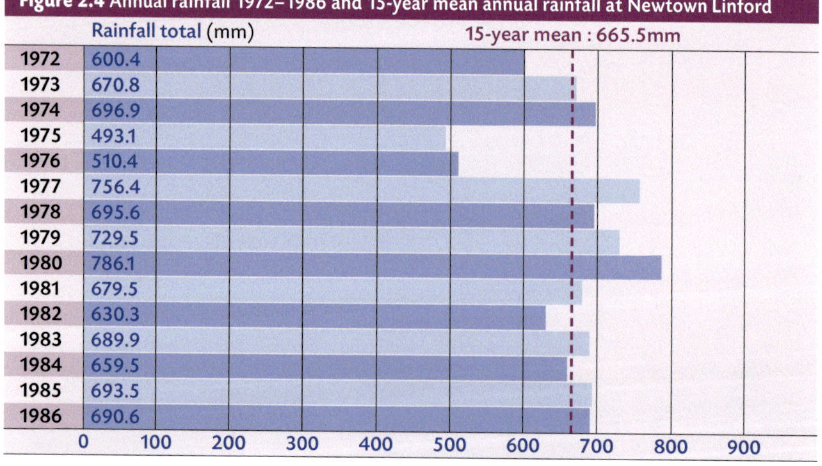

expected, largely because of low rainfall in July in 1975, 1976 and 1977. The summers of 1975 and 1976 were long and hot, and 1976 was widely regarded as a 'drought year', when water use was restricted and supply limited in many parts of the country. Mean maximum monthly temperatures for June, July and August in 1975 and 1976 were markedly higher than the 15-year mean maxima for these months, the mean maximum for August 1975 being 24.7°C. However, 15-year mean maxima were also greatly exceeded in August 1990 (24.2°C), July and August 1995 (24.2°C and 25.4°C, respectively), July 1996 (25.7°C) and August 1997 (24.2°C) (*cf.* Tab. 2.1).

Front page, *Leicester Mercury*, 24 June 1976: the severe drought preoccupied local and national media for much of the summer and autumn of 1976

The years of 1975 and 1976 were also particularly sunny, with total hours of sun of 1,393.2 and 1,454.3, respectively, compared with 15-year mean annual sunshine of 1,298.1 hours. The sunniest year was 1995, with 1,676.1 hours of sun, compared with a 15-year annual mean of 1,430.9 hours.

In the later chapters of this book, when discussing annual fluctuations in numbers of insects and other invertebrates, reference is frequently made to the weather in different years. An attempt has therefore been made to summarize the seasonal weather, year by year, using records from the Newtown Linford meteorological station. Winter is taken as December, January and February; spring as March, April and May; summer as June, July and August; and autumn as September, October and November.

Mean maximum and mean minimum temperatures, total rainfall and total hours of sunshine for each month were compared with 15-year means for each of the two periods, 1972–1986 and 1987–2001, to arrive at an overall description

Figure 2.5 Annual rainfall 1987–2001 and 15-year mean annual rainfall at Newtown Linford

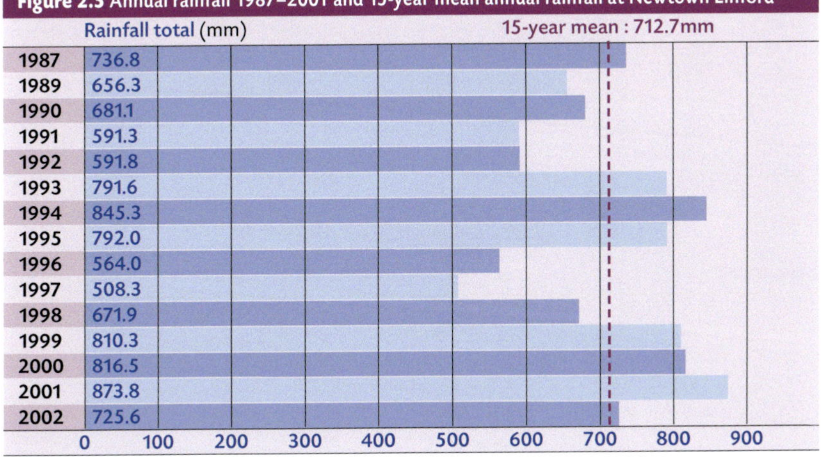

The bird table in winter (above left), an important source of food for birds when the ground is frozen

Hungry woodpigeons damage overwintering garden crops such as broccoli (above right)

Table 2.2 Seasonal weather in the years 1972–2001 at Newtown Linford. Each description refers firstly to temperatures, secondly to rainfall, and thirdly to sunshine (ave = average)

	Winter (Dec, Jan, Feb)	Spring (March, April, May)	Summer (June, July, Aug)	Autumn (Sept, Oct, Nov)
1972	mild, ave, dull	mild, dry, dull	cool, dry, dull	ave, dry, dull
1973	warm, dry, ave	ave, wet, bright	warm, damp, bright	cool, dry, ave
1974	warm, ave, ave	ave, dry, ave	cool, damp, ave	cool, wet, dull
1975	warm, dry, dull	ave, wet, dull	hot, dry, bright	cool, dry, bright
1976	warm, dry, ave	warm, dry, bright	hot, dry, bright	ave, wet, dull
1977	cold, wet, ave	cool, dry, bright	cool, damp, ave	ave, dry, ave
1978	cool, damp, ave	mild, dry, bright	cool, damp, dull	warm, dry, ave
1979	cold, wet, bright	cool, wet, ave	ave, dry, dull	warm, ave, bright
1980	cool, wet, bright	mild, dry, bright	cool, wet, dull	cool, damp, ave
1981	ave, dry, ave	ave, wet, dull	ave, dry, dull	ave, damp, ave
1982	cold, dry, dull	warm, dry, bright	mild, damp, dull	mild, ave, ave
1983	mild, ave, ave	ave, wet, dull	hot, dry, bright	warm, ave, ave
1984	ave, damp, bright	ave, dry, ave	hot, dry, bright	warm, wet, dull
1985	cold, dry, bright	ave, ave, ave	cool, damp, ave	warm, dry, ave
1986	cold, damp, bright	cool, damp, ave	ave, damp, ave	ave, damp, bright
1987	cold, dry, ave	mild, ave, ave	cool, damp, dull	ave, damp, ave
1988	ave, damp, bright	ave, dry, dull	cool, damp, dull	ave, dry, ave
1989	mild, dry, dull	warm, damp, bright	hot, dry, bright	warm, dry, ave
1990	mild, wet, ave	warm, dry, bright	hot, dry, bright	ave, ave, ave
1991	cold, ave, ave	warm, ave, dull	hot, ave, ave	mild, dry, dull
1992	ave, dry, dull	ave, damp, bright	warm, wet, ave	cool, damp, dull
1993	ave, dry, dull	mild, damp, ave	cool, wet, ave	cool, wet, dull
1994	cold, damp, ave	cool, damp, ave	hot, damp, bright	mild, damp, dull
1995	mild, wet, ave	mild, dry, bright	hot, dry, bright	warm, damp, ave
1996	cool, damp, dull	cool, dry, ave	hot, dry, bright	mild, dry, ave
1997	cool, dry, dull	warm, dry, bright	warm, wet, dull	mild, dry, bright
1998	warm, dry, bright	warm, wet, ave	mild, ave, dull	cool, wet, dull
1999	mild, wet, ave	warm, wet, ave	warm, damp, bright	mild, damp, bright
2000	mild, ave, bright	warm, wet, ave	warm, dry, dull	mild, wet, dull
2001	cool, damp, bright	warm, wet, dull	warm, dry, ave	warm, ave, ave

of the seasonal weather for each year. These descriptions are given in Table 2.2. Inevitably, these are generalizations, as each season includes three months, one of which may have been cooler or warmer, wetter or drier, and duller or brighter than the others.

Throughout this chapter the weather information for 1987–2001 has been presented separately from the information already published for 1972–1986 (Owen, J. 1991), because there has been a recognisable change in conditions. Summer mean and maximum temperatures were higher in 1987–2001, and all 15-year mean monthly maximum and minimum temperatures were higher than in 1972–1986.

Although 15-year mean monthly rainfall for any particular month was more, or less, in 1987–2001 than in 1972–1986, the 15-year mean rainfall was higher for the later period. In most months, the 15-year mean total of hours of sunshine was higher in 1987–2001 than in 1972–1986, and the 15-year mean was considerably higher. This means that in the second half of the 30-year period, it was warmer, wetter (particularly in autumn) and sunnier.

Berries of shrubs such as firethorn, *Pyracantha coccinea* (top), and *Cotoneaster* (above) are another important food source for birds in autumn

3 : Butterflies

Any garden, and certainly my garden, is dominated in terms of wildlife by two groups of animals: insects and birds. Among insects, the most familiar, most welcome and best-known group is undoubtedly butterflies. They are conspicuous, readily identified and attractive, and consequently are popularly regarded as indicating the health of a habitat. This is quite a reasonable view, since butterflies are typical insects and, as they have herbivorous larvae, are directly dependent on the vegetation. That most people recognize and remark on butterflies is undoubtedly because they are large, relative to most British insects, and are active by day. Indeed, many people seem to accord butterflies the status of honorary birds, thus admitting them to the 'fur and feathers' category of animals, which is all that some people recognize as wildlife.

A red admiral, *Vanessa atalanta* (opposite), feeding on nectar from a *Hebe*. Most butterflies hold their wings together vertically when at rest, whereas most moths rest with their wings held flat over their backs

Butterflies, together with moths, form the insect order Lepidoptera, characterized by the presence of scales on the wings, body and legs. The mouthparts almost always form a long, suctorial proboscis, coiled when not in use, and the larval form is a caterpillar, most of which are herbivorous. The superfamilies Hesperioidea (skippers) and Papilionoidea are popularly known as butterflies, but, although there is confident consensus about recognition of a butterfly as such, there is no single absolutely trustworthy taxonomic character by which butterflies can be separated from moths.

The difference between butterflies and moths is not clear-cut. Skippers, such as large skipper, *Ochlodes venata* (below), resemble moths in several respects

Most, but not all, moths have one or more bristles, the frenulum, on the hind wing, which locks onto the fore wing, whereas in all butterflies the expanded edge of the hind wing holds the wings together. Butterflies have a club-like swelling at or near the tip of the thread-like antennae; moths have antennae of various forms, but those that have antennae with club-like swellings at the tip also have a frenulum. Most moths rest with the wings folded flat over the back, whereas butterflies tend to rest with the wings held upright, but there are exceptions to both.

Moths are typically drab in colour and nocturnal or crepuscular, whereas butterflies are usually diurnal and tend to be brightly coloured. There are, however, brightly coloured diurnal moths, and dull-coloured butterflies that fly at dusk, in overcast weather and in the shade. Many moth larvae weave a silken cocoon in which they pupate, whereas butterfly pupae are naked and often brightly ornamented. However, moths that pupate underground do not usually form a cocoon, and skippers regularly make a loose cocoon. Skippers are indeed rather moth-like and different from typical butterflies, as is recognized by their classification in a separate superfamily.

The butterfly species most frequently trapped, both in the Malaise trap and by hand-netting (see Table 3.1, below), was small white, *Pieris rapae* (above left)

Another species often caught in the Malaise trap was green-veined white, *Pieris napi*, (above right)

SPECIES DIVERSITY OF BUTTERFLIES

Conservationists and naturalists are usually interested in seeing a variety of species, and put a high value on habitats with a large species list. It has become clear that hand-netting and Malaise trapping tend to catch species in different proportions and, consequently, the species ranking lists for the two trapping methods in the period 1972–1979, when both methods were used, were different (Owen, J. 1991). Over the entire 30-year period, 1972–2001, 17 species of butterfly appeared in Malaise trap samples. The frequencies overall of these are shown in Tab. 3.1, together with the totals of those hand-netted in 1972–1979.

Table 3.1. Ranking by abundance of butterfly species captured in a Malaise trap and hand-netted

Malaise trap (1972–2001)		**Hand-netted** (1972–1979)	
Small white	535	Small white	8,577
Green-veined white	267	Small tortoiseshell	2,860
Large white	115	Large white	2,469
Hedge brown	42	Green-veined white	1,135
Small tortoiseshell	40	Peacock	939
Wall	39	Red admiral	331
Orange tip	24	Meadow brown	78
Meadow brown	20	Wall	64
Peacock	10	Orange tip	54
Holly blue	10	Brimstone	46
Small copper	7	Small copper	16
Red admiral	6	Painted lady	16
Brimstone	4	Small skipper	6
Large skipper	3	Comma	5
Speckled wood	3	Large skipper	2
Common blue	1	Common blue	2
Painted lady	1	Small heath	2
		White-letter hairstreak	1
		Silver-washed fritillary	1
		Marbled white	1
		Hedge brown	1
Total	**1,128**	**Total**	**16,606**

Small tortoiseshells, peacocks, red admirals and painted ladies are more likely to be hand-netted than to enter the Malaise trap, and commas, small heaths, white-letter hairstreaks, silver-washed fritillaries and marbled whites were not taken at all in the trap, although this is likely to be due to their relative rarity and/or the great difference between the total sample sizes. Holly blues and speckled woods became more common in England and first appeared in the garden long after the hand-netting programme had finished.

The difference in representation between species in the two types of samples is a consequence of behavioural differences. The preference for shade, and weak flight, of meadow browns, walls and hedge browns make them more likely to enter the Malaise trap, and then less likely to find a way out, than active, strong fliers such as painted ladies and red admirals. Satyrids (see list, p50) are probably over-represented in the Malaise trap catch; red admirals and painted ladies under-represented.

One advantage in employing two different methods to compile a list of garden butterflies,

evident particularly in the short term, is precisely that different species are likely to be caught. On the other hand, Malaise trapping is unlikely to produce such rare vagrants as white-letter hairstreaks and silver-washed fritillaries, both of which were caught in the drought year of 1976, when nectar sources were scarce, and butterflies wandered far and wide in search of food.

More often trapped by hand were small tortoiseshell, *Aglais urticae* (above left) and large white, *Pieris brassicae* (above right)

ANNUAL FLUCTUATIONS IN BUTTERFLY NUMBERS

The results of 30 years of Malaise trapping, from 1972–2001, are shown in Fig. 3.1 as a histogram of individuals (N). This shows clearly that there was a run of good butterfly years from 1973 to 1976, a good year in 1978, and another run of good years from 1982 to 1984. Thereafter, apart from a relatively good year in 1990, butterfly numbers were low. 1973 was far and away the best year, followed by 1975 (N=172 and N=105, respectively); 1981, 1987, 1991, 1993 and 2000 were particularly poor years with only 4, 6, 12, 9 and 9 individuals, respectively, caught. However, annual fluctuations in numbers of species were not nearly so marked, although most (10) were caught in 1973 and fewest (1) in 1986 and 1987 (Tab. 3.2). This is because the average number of individuals per species (N/S) tended to be highest in years of greatest abundance, and lowest in poor years; N/S was 17.2 in 1973, and only 1.5 in 1993. This means that in good butterfly years, abundance is in large part a consequence of increases in numbers of the species already there, rather than an addition of new species. In every year but ▸*p51*

Fig. 3.1. Annual fluctuations in numbers of butterflies caught in Malaise trap, 1972–2001

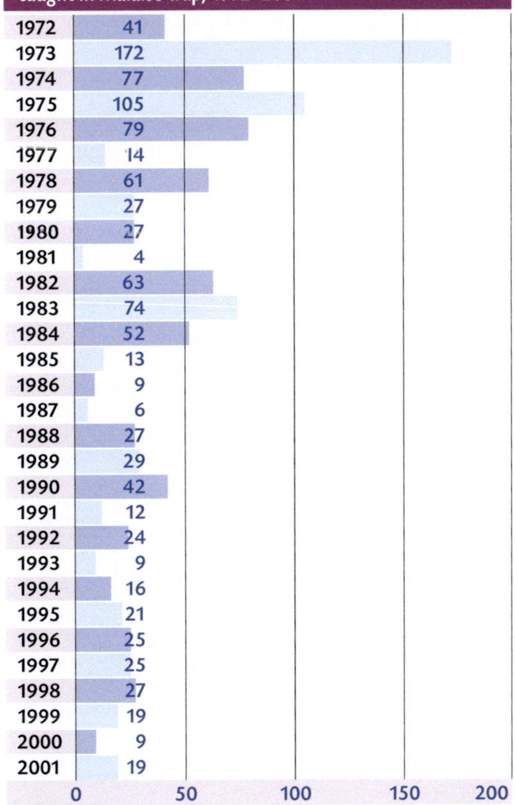

1 A brimstone, its wings marked after capture by the author

2 Brimstone found over-wintering on brambles near the garden, its wings covered in melting frost

3 Comma on kale

4 Red admiral feeding on *Aster novi-belgii*

5 Marbled white, a rare visitor to the garden, was captured only once

6 Painted ladies, though never frequent in the garden, were caught more often by hand than in the Malaise trap

The natural history of the garden's species of butterfly

Butterflies are better known than any other group of insects. There is a large number of books detailing the life histories and habits of British butterflies, so, rather than repeating much easily obtainable information, I shall briefly describe only the species that have been recorded in my Leicester garden. Describing all 23 species recorded there gives reasonable coverage of the range of life styles in the group; the information comes from several sources, particularly Heath, Pollard & Thomas (1984).

Fewer than half of these are regularly and particularly associated with gardens: brimstone, large white, small white, green-veined white, orange tip, red admiral, painted lady, small tortoiseshell and peacock. The others are infrequent, occasional or rare visitors, although holly blue and speckled wood have become more frequent since 1991 and 1996, respectively, in my garden. My late husband, Denis Owen, had for some years predicted that speckled wood would move into the garden as it became more common in England, and the first observation of this species in the garden was made in 1996, the year that he died.

Apart from observation, butterflies were monitored in the garden by two methods: hand-netting, marking and releasing from 1972 to 1979, and Malaise trapping from 1972 to 2001 (see Chapter 2, p36–38).

Hesperiidae

1 **Small skipper**, *Thymelicus sylvestris* (Poda), is a butterfly of rough grasslands, on the wing in July and August, and has one generation a year. The caterpillars feed mainly on Yorkshire fog, *Holcus lanatus*, and also on other grasses, and hibernate when newly hatched. Pupation is in a silken cocoon, covered with grass leaves, at the base of the plant.

2 **Large skipper**, *Ochlodes venata* (Turati), is found on grassland, on grassy rides in woods, along hedgerows and in similar sites. There is one generation a year, adults flying from June to August. Caterpillars feed mainly on cock's-foot, *Dactylis glomerata*, and on other grasses. They hibernate as larvae, complete their growth the following spring, and pupate in cocoons spun in 'tents' of grass blades.

Pieridae

3 **Brimstone**, *Gonepteryx rhamni* (L.), is on the wing from March to September or October; they fly in woods, scrub and wetlands, as well as in built-up areas. They hibernate as adults, which is why they are seen so early and late in the year, but tend to be scarce in June and July when the single generation is at the caterpillar and pupa stages.

In the winter of 1976–77, I found two male brimstones hibernating in a bramble patch in an abandoned railway cutting not far from the garden. On a cold day in January, the butterflies' wings were covered with hoar frost and larger beads of ice, and at a distance they were hard to distinguish from the leaves that still remained on the plant; they were still in the same position in March.

The caterpillars are restricted to alder buckthorn, *Frangula alnus*, and purging buckthorn, *Rhamnus catharticus*, but pupate amongst low-growing vegetation. Brimstones are mobile butterflies, and make local movements from egg-laying sites to feeding and wintering sites.

4 **Large white**, *Pieris brassicae*, (L.) is a widespread and highly mobile butterfly particularly associated with agricultural land, and gardens, where *Brassica* species, and nasturtiums (*Tropaeolum majus*), the larval food-plants, are grown. On the wing from April to October, two or, in some years, three generations are completed. Large whites pass the winter as pupae, formed on tree trunks or walls.

It has long been believed that they are migratory, and that the British population is boosted each year by immigrants from continental Europe, but the butterfly monitoring scheme (Pollard *et al.* 1986) produced no evidence for this.

5 **Small white**, *Pieris rapae* (L.), is also widespread, mobile and particularly associated with gardens and fields where brassicas, a common larval foodplant, are cultivated. The caterpillars also feed on other crucifers, such as garlic mustard (*Alliaria petiolata*) and hedge mustard (*Sisymbrium officinale*). They hibernate as pupae. Adults fly from April to October. There are two or sometimes three generations a year, the earlier ones tending to pupate on the foodplant, the later ones on tree trunks and walls. Migratory flights and directional movements of small whites are often reported, but there is no consensus as to whether or not there is regular directional movement.

6 **Green-veined white**, *Pieris napi* Verity, is another widespread and mobile butterfly. It is on the wing from April to October, passing the winter as a pupa. The caterpillars feed on cuckooflower (*Cardamine pratensis*), on hedge and garlic mustard, and on other crucifers, and pupate on nearby vegetation. There are two or three generations a year, but some of the first-generation pupae persist until the following spring before emergence of the adult.

7 **Orange tip**, *Anthocaris cardamines* Verity, is widespread, particularly in open woods and lanes, and is becoming increasingly associated with gardens. Adults are on the wing in May and June, and there is only one generation a year, hibernation being as a pupa. Caterpillars feed on the flowers and seed pods of cuckooflower, garlic mustard and other hedgerow crucifers, and there are many reports of their using garden arabis (*Arabis alpina* subsp. *caucasica*), and dame's-violet (*Hesperis matronalis*), as they do in my garden. The angular, well-camouflaged pupae are formed on the foodplant or on nearby plant stems.

Lycaenidae

8 **White-letter hairstreak**, *Strymonidia w-album* (Knoch), is usually very local around elms, *Ulmus* species, on which the caterpillars feed. Adults fly in July and August and there is a single generation. They hibernate as eggs, and the larvae feed first on the elm flowers, later on the leaves, and pupate on the elm twigs.

9 **Small copper,** *Lycaena phlaeas* (Fab.), is found from May to October in open situations in woods or along hedgerows and in similar sites. The caterpillars feed on common sorrel (*Rumex acetosa*) and sheep's sorrel (*R. acetosella*), and pupate amongst low vegetation. There are two, or sometimes, three generations a year, hibernation being as larvae.

10 **Common blue,** *Polyommatus icarus* (Rottemburg), flies on open grassland and in large woodland glades from May to September. There are two generations a year, and few adults are seen in July. The caterpillars, the hibernating stage, feed on common bird's-foot-trefoil, *Lotus corniculatus*, and pupate on the ground.

11 **Holly blue,** *Celastrina argiolus* (L.), is found in sunny rides and clearings in woodland, along hedgerows, in gardens and in urban parks. Numbers quite frequently increase then, after several years, decline again. There are two generations a year in most parts of the country, flying in April/May and August. Caterpillars of the early generation feed on the flowers of a variety of shrubs, particularly holly, *Ilex aquifolium*, but summer caterpillars feed on ivy flowers, *Hedera helix*. Pupation takes place on the food plant, and they hibernate as pupae.

Nymphalidae

12 **Red admiral,** *Vanessa atalanta* (L.), is widespread, flying in a range of habitats from April to October. Some individuals hibernate as adults, and these may be seen on the wing in early April, but most red admirals seen early in the summer are migrants from the south. The British butterflies are part of a widespread population extending from northern Europe to the Mediterranean and North Africa, which shifts northwards in spring and early summer and retreats southwards in the summer and autumn. Two generations are passed in Britain, the caterpillars feeding on common nettle, *Urtica dioica*, and spinning the leaves together as a shelter for pupation.

13 **Painted lady,** *Cynthia cardui* (L.), is similarly widespread, and may be seen in a variety of habitats from May to September. Unlike red admirals, however, they seem unable to survive the British winter in any form, and their appearance in early summer is dependent on immigration. They, too, are part of a population extending to North Africa, which ebbs and flows, southwards in autumn and northwards in early summer.

The abundance of painted ladies in Britain varies greatly from year to year, depending on breeding conditions to the south. One or two generations are produced in the British summer, using thistles (*Cirsium* and *Carduus* species) as foodplants, and pupating on the foodplants or nearby.

14 **Small tortoiseshell,** *Aglais urticae* (L.), is widespread in open situations, from March to October, and hibernates as an adult. There are two generations a year, the caterpillars feeding on common nettle and pupating on the foodplant, on tree trunks or on walls.

15 **Peacock,** *Inachis io* (L.), though widespread and mobile, is predominantly a woodland butterfly, flying from March until October, and hibernating as adults. One generation a year is produced, using common nettle as a foodplant and pupating on tree trunks.

16 **Comma,** *Polygonia c-album* (L.), also uses common nettle as a caterpillar foodplant, although it sometimes feeds on elm. They fly from March to October in woodland, amongst shady scrub and along hedges. Pupation is on the foodplant, and they hibernate as adults. Although there are two generations a year, not all adults produced from the first generation of larvae breed immediately, but instead go into early hibernation until the following year.

17 **Silver-washed fritillary,** *Argynnis paphia* (L.), is a woodland butterfly that flies in July and August. Caterpillars feed on violets (*Viola* species) and pupate on any convenient object. There is one generation a year, overwintering as larvae.

Satyridae

18 **Speckled wood,** *Pararge aegeria* (L.), a woodland butterfly that frequents dappled shade, and is increasingly seen in densely planted gardens, flies from April to October. It overwinters either as larvae or pupae, the staggered emergence as adults of these individuals leading to an extended second generation. Caterpillars feed on a variety of grasses, pupating on the foodplant. Males perch in sunny spots and seem to defend their territory from intruding males.

19 **Wall,** *Lasiommata megera* (L.), flies from May to September on unimproved grassland, on roadsides and along woodland edge. There are two, sometimes three, generations a year, and adults are scarce in July. Caterpillars feed on tor-grass (*Brachypodium pinnatum*), false brome (*B. sylvaticum*), cock's-foot (*Dactylis* species) and other grasses, and pupate on grass stems. They hibernate as larvae.

20 **Marbled white,** *Melanargia galathea* Verity, flies on rough grassland from June to August. Caterpillars feed on sheep's-fescue (*Festuca ovina*), tor-grass and other grasses, and pupate on the ground at the base of the foodplant. There is one generation a year, which, as in other satyrids (browns), hibernate as larvae.

21 **Hedge brown,** *Pyronia tithonus* Verity, is on the wing from July to September along hedges, lanes and woodland edge and among scrub. Various fine-leaved grasses are used as larval foodplants, and pupae are formed on grass stems. There is one generation a year, hibernating as larvae.

22 **Meadow brown,** *Maniola jurtina* (L.), flies from June to September on grasslands and other grassy areas. They, too, feed as caterpillars on various fine-leaved grasses and pupate on grass stems. The single generation passes the winter as larvae.

23 **Small heath,** *Coenonympha pamphilus* (L.), is on the wing on unimproved grasslands and other grassy areas from June to September. Two generations are produced a year, using various fine-leaved grasses as food and pupating on grass stems. The second generation overwinter as larvae.

two, up to and including 1986, the commonest butterfly was small white, forming up to 66.9% of the total catch; from 1987 onwards, total numbers were lower, and the predominance of small whites not so marked. In 1988 and 1992, large whites were the commonest; in a number of years, green-veined whites were the commonest; in 1991, wall, and in 1999 and 2001, hedge browns were the commonest.

Effects of weather

The writers of the report of the first 10 years (1976–1985) of the butterfly monitoring scheme organized by the Institute of Terrestrial Ecology (Pollard *et al.* 1986) were confident that high summer temperatures are associated with good butterfly years. While recognizing that annual variations in total Malaise trap catch mask differences in response between species, I have attempted to correlate the size of the butterfly catch with weather conditions in the current and preceding year. Work on the 1972–1986 figures suggested that two of the most important effects of weather on butterfly abundance are that numbers go up in a warm summer, and after a mild winter, especially if that winter was fairly dry (Owen, J. 1991). The 1987–2001 figures to some extent bear this out, as 1990,

Table 3.2. Size and composition of annual catches of butterflies in Malaise trap, 1972–2001

Year	Individuals (N)	Species (S)	N/S	Species taken once only	Commonest	Commonest as % N
1972	41	8	5.1	4	*Pieris rapae*	53.7
1973	172	10	17.2	5	*Pieris rapae*	66.9
1974	77	5	15.4	1	*Pieris rapae*	54.6
1975	105	7	15.0	1	*Pieris rapae*	60.0
1976	79	9	8.8	3	*Pieris rapae*	55.7
1977	14	5	2.8	3	*Pieris napi*	42.9
1978	61	4	15.3	1	*Pieris rapae*	54.1
1979	27	4	6.8	1	*Pieris rapae*	48.2
1980	27	5	5.4	0	*Pieris rapae*	37.0
1981	4	2	2.0	0	*Pieris rapae, P. napi*	50.0=
1982	63	7	9.0	2	*Pieris rapae*	55.6
1983	74	8	9.3	4	*Pieris rapae*	47.3
1984	52	9	5.8	4	*Pieris rapae*	36.5
1985	13	4	3.3	0	*Pieris rapae*	53.9
1986	9	1	9.0	0	*Pieris rapae*	100.0
1987	6	1	6.0	0	*Pieris rapae*	100.0
1988	27	3	9.0	0	*Pieris brassicae*	59.3
1989	29	6	4.8	3	*Pieris napi*	34.5
1990	42	7	6.0	2	*Pieris rapae*	40.5
1991	12	4	3.0	0	*Lasiommata megera*	33.4
1992	24	7	3.4	3	*Pieris brassicae*	37.5
1993	9	6	1.5	3	*Pieris rapae, P. napi, Anthocaris cardamines*	22.2=
1994	16	5	3.2	3	*Pieris napi*	43.8
1995	21	5	4.2	2	*Pieris napi*	42.9
1996	25	8	3.1	4	*Pieris rapae, P. napi, Pyronia tithonus*	24.0=
1997	25	6	4.2	1	*Pieris rapae, P. napi*	24.0=
1998	27	7	3.9	2	*Pieris napi*	40.7
1999	19	6	3.2	2	*Pyronia tithonus*	36.8
2000	9	4	2.3	2	*Pieris napi*	55.6
2001	19	5	3.8	2	*Pyronia tithonus*	57.9
1972–1986	818	15	54.5	3	*Pieris rapae*	55.5
1987–2001	310	16	19.4	5	*Pieris napi*	28.0
1972–2001	1128	17	66.4	3	*Pieris rapae*	47.4

Eggs, caterpillar and pupa of comma. The eggs (top left) are laid on common nettle, *Urtica dioica*, the main foodplant of the distinctively marked caterpillar (top right), which then pupates on the plant within a chrysalis that resembles a dried-up leaf (above)

the only reasonably good year for butterflies, had a hot, dry summer after a mild winter, but that winter was wet. There were other hot and dry summers in the period, but only two followed mild winters, and one of those winters was wet.

General observation reinforces the trapping figures, and of course has been particularly important since the hand-netting programme ended. My garden diary records that butterflies were very common in August 1975, in the hottest weather for several years, and that in 1975 many small coppers were seen (seven were caught). In 1976, what looked like a holly blue was seen flying high over the garden; it was first definitely recorded in 1991, both by observation and in the Malaise trap, and in 1992 became abundant. None was recorded from 1993 to 1996, but subsequently they were seen every year, and in 1997 were common. Speckled woods were first seen in the garden in 1996, and since then have been regular and persistent visitors as they have become increasingly common.

Despite the low Malaise trap catch, the diary records indicate that 1992 and 1996 were reasonable butterfly years, as well as 1990, when the Malaise trap catch was up. Interestingly, in 1993, when few were caught, the diary claims that butterflies were conspicuous by their absence. 1988 was a good year for painted ladies in the garden. Since painted ladies, in particular, and also red admirals, are long-distance migrants, high numbers of these species probably reflect weather and breeding conditions in southern Europe and the Mediterranean rather than in England. Being strong fliers, painted ladies and red admirals are likely to find their way out of the Malaise trap, so numbers caught do not reflect abundance. Satyrids, on the other hand, are behaviourally more likely to enter the Malaise trap, and once in, are too weak to escape.

Holly blue (above left) and speckled wood (above right); both these species appeared in the study garden in the 1990s, long after the programme of hand-netting had ceased

SUMMARY

The changes in numbers of butterflies caught in the Malaise trap over the 30-year period 1972–2001 reinforce the general feeling, based on observation, that butterflies in general are becoming less abundant, although diversity remains good. Wetter winters (and summers) may have something to do with this, but it is more likely to be due to habitat changes and changes in land use. Some species are becoming more abundant, and it is good to be able to document the establishment of holly blues and speckled woods in the garden, because this was predicted (Owen, J. 1991), as they had been recorded elsewhere in Leicester in the 1980s.

4 : Moths

Ghost moth, *Hepialus humuli* (opposite), quite common in the Leicester garden although irregular

To the general public, moths are less familiar than butterflies, simply because the vast majority of them are nocturnal. There are, however, many more British moths than butterflies, and they are taxonomically more variable. Moreover, a great deal is known about British moths, as there have been dedicated and knowledgeable moth collectors for as long as there have been butterfly collectors. Collection of moths (and the now more generally acceptable study of them and their population dynamics without killing them) was made far easier, and indeed revolutionised, with the invention of the mercury-vapour light trap in the late 1940s. Mercury-vapour light does not attract all species of moth equally, but it draws in vast numbers of some species.

To a large extent, lepidopterists still concentrate on the macro-Lepidoptera (macro-moths), the families of larger moths, but there is increasing interest in the families of smaller moths, the micro-Lepidoptera (micro-moths). Most moth books concentrate on the macro-Lepidoptera, considering only the larger and more familiar micro-Lepidoptera, despite the fact that the latter constitute the majority of British moths, which number in total well over 2,000 species. The distinction between 'macros' and 'micros' is neither scientific nor biological but only convenient and traditional. Inevitably, much of this chapter is about macro-moths, partly because they are more familiar and easier to identify, but also because they have been intensively investigated in the garden; micro-Lepidoptera are less well known, but considerable numbers were identified from Malaise trap samples, particularly from 1996 to 2001, and all those that have been recorded are included.

NATURAL HISTORY OF MOTHS

The natural history of moths is varied, depending on habitat, the way in which caterpillars exploit the foodplant, dispersal and mating behaviour, mode of hibernation, and whether or not they are migratory. All, however, have the same basic life history of egg, caterpillar as a feeding and growing stage, immobile pupa in which a massive reorganization of tissues occurs, and mobile, usually winged, adult. Although a moth without wings sounds rather a contradiction in terms, the females of a number of species are wingless.

Eggs are laid singly or in clusters on the foodplant, or in some cases broadcast in flight over suitable vegetation, as in the ghost moth (*Hepialus humuli*). Caterpillars of many garden moths feed by chewing leaves, some feed on roots,

Silver-Y moth, *Autographa gamma* (right), a frequent visitor to the garden, despite being an immigrant species

others on flower buds, and a number bore into the stems of plants, including trees, or, in the case of some micros, mine burrows within the blades of leaves. The pupae tend to be concealed or protected in some way; many of those of macros are formed in the surface soil, often in a cocoon, while those of micros tend to be formed in the leaf mines, stem-tunnels, or wherever else the larva has been feeding.

Newly emerged female moths produce pheromones which males detect using their antennae, which are complex batteries of chemoreceptors, and males of at least some species, such as angle shades (*Phlogophora meticulosa*), have elaborate brush organs on the abdomen which disseminate an aphrodisiac as they approach the female (Birch 1970). A number of species are migratory, the prime example being silver-Y (*Autographa gamma*), which arrives in Britain from southern Europe in late spring and summer, the numbers varying from year to year. Migration of silver-Y is a large-scale population drift north and northwest early in the year, and a southwards retreat in the autumn.

MOTHS IN THE LEICESTER GARDEN AND THEIR RELATIVE ABUNDANCE

The garden list, 1972–2001, includes 375 species of moth, comprising 284 macro-Lepidoptera and 91 micro-Lepidoptera. The majority were attracted to a mercury-vapour light trap, many entered the Malaise trap, and some were attracted to a trap baited with fermenting fruit. The Malaise trap was in operation continuously every year and so the trapping figures can be used to illustrate annual and seasonal fluctuations. Use of the mercury-vapour trap was irregular, and stopped after 1992, so the trapping results are qualitative, not quantitative. The baited trap was only used occasionally, mostly in 1975. Mercury-vapour light trapping started in 1972, and stopped after 1992. Up until 1996, the majority of moths were identified by D.F. Owen; after that, John E. Owen identified the Malaise trap catch, with help from Sean Clancy.

In the list of the species caught, the macro-moths will be described first, and then the micro-moths, using the order of families given in *The Moths and Butterflies of Great Britain and Ireland* (Emmet & Heath, 1991). However, in order to better describe the garden moth fauna, the macro-moths have been divided into four categories, based largely on the records from 1972–1986, when mercury-vapour light trapping was more frequent:

1, common garden moths – those that are common or quite common;

2, uncommon garden moths – those that are uncommon, even if regular, and those that are irregular;

3, scarce garden moths – recorded fewer than ten times in the 30-year period;

4, moths recorded only once in the 30-year period.

On this basis there are 115 common macro-moth species, 35 uncommon, 82 scarce and 50 were recorded once only. The micro-moths have not been divided into categories but are treated in taxonomic order.

Macro-Lepidoptera

The scientific names are as used in Emmet & Heath (1991.) Foodplants and overwintering stages are taken from Skinner (1998), with some additional information from Emmet & Heath (1991) and Stokoe & Stovin (1948). Foodplants used in the garden are given.

MT indicates that the species has been taken in the Malaise trap. BT indicates that it has been taken in a trap baited with fermenting fruit, usually apple.

COMMON GARDEN MOTHS

Hepialidae

1 **Orange swift** *Hepialus sylvina* (L.). Common, July–September. Larvae on roots of wide variety of herbaceous plants. Overwinter as larvae. MT 1972–2001.

2 **Common swift** *Hepialus lupulinus* (L.). Common, especially in June, but struggling on through summer. Larvae on roots of grasses and occasionally herbaceous plants. Overwinter as larvae. MT 1972–2001.

Thyatiridae

3 **Buff arches** *Habrosyne pyritoides* (Hufnagel). Common, June–August. Larvae on *Rubus fruticosus*. Overwinter as pupae. MT 1972–2001.

4 **Figure of eighty** *Tethea ocularis* (L.). Common, June and July. Larvae on *Populus*. Overwinter as pupae. MT 1987–2001.

Geometridae

5 **March moth** *Alsophila aescularia* (D. & S.). Common, March and April. Larvae on wide variety of deciduous trees. Overwinter as pupae. MT 1972–1986.

6 **Riband wave** *Idaea aversata* (L.). Common, June–September. Polymorphic. Larvae on various herbaceous plants. Overwinter as larvae. MT 1972–2001.

7 **Dark-barred twin-spot carpet** *Xanthorhoe ferrugata* (Clerck). Common, July and August. Larvae on various herbaceous plants. Overwinter as pupae. MT 1972–1986.

8 **Silver-ground carpet** *Xanthorhoe montanata* (D. & S.). Common in June. Larvae on various herbaceous plants. Overwinter as larvae. MT 1972–2001.

9 **Garden carpet** *Xanthorhoe fluctuata* (L.). Common, May–September in two broods. Larvae on range of *Brassicaceae*; in garden, on *Arabis alpina* subsp. *caucasica*, *Aurinia saxatilis*, *Brassica oleracea* and (especially) *Iberis sempervirens*. Overwinter as pupae. MT 1972–2001, BT.

10 **Common carpet** *Epirrhoe alternata* (Müller). Common, July and August. Larvae on various species of *Galium*. Overwinter as pupae. MT 1972–2001.

11 **Yellow shell** *Camptogramma bilineata* (L.). Common, July and August, especially in 1980s. Larvae on various herbaceous plants; in garden, on *Arabis alpina* subsp. *caucasica*, *Aubrieta deltoidea*, *Myosotis arvensis*, *Origanum majorana* and *Taraxacum*. Overwinter as larvae. MT 1972–2001.

12 **The spinach** *Eulithis mellinata* (Fab.). Quite common, June and July. Larvae on *Ribes nigrum* and *R. rubrum*; in garden, on *R. uva-crispa*. Overwinter as eggs. MT 1972–2001.

13 **Common marbled carpet** *Chloroclysta truncata* (Hufnagel). Common, June–September in two broods. Almost all melanics. Larvae on various trees, shrubs and herbaceous plants; in garden, on *Alcea rosea*, *Buddleja davidii*, *Fragaria × ananassa*, *Geranium cinereum*, *Hypericum androsaemum*, *Paeonia officinalis*, *Ribes sanguineum*, *Rubus fruticosus* and *Saxifraga*. Overwinter as larvae. MT 1972–2001, BT.

14 **Grey pine carpet** *Thera obeliscata* (Hübner). June and September–October in two broods. Many melanics. Larvae on various conifers. Overwinter as larvae. MT 1972–1986.

15 **July highflyer** *Hydriomena furcata* (Thunberg). Common, July and August. Larvae on various species of *Calluna*, *Salix*, *Vaccinium*, and on *Corylus avellana*. Overwinter as eggs.

16 **November moth** *Epirrita dilutata* (D. & S.). Common, October and November. Larvae on wide variety of trees and shrubs; in garden, on *Prunus cerasifera*. Overwinter as eggs.

17 **Mottled pug** *Eupithecia exiguata* (Hübner). Common, May and June. Larvae on *Acer pseudoplatanus*, *Crataegus*, *Prunus spinosa* and *Sorbus aucuparia*; in garden, on *Buddleja davidii*, *Crataegus monogyna*, *Fraxinus excelsior*, *Iberis sempervirens*, *Paeonia officinalis*, *Philadelphus coronarius*, *Populus nigra*, *Potentilla fruticosa*, *Ribes uva-crispa*, *Rosa* species and cultivars, *Sambucus nigra*, *Spiraea × vanhouttei* and *Weigela florida*. Overwinter as pupae. MT 1972–2001.

18 **Lime-speck pug** *Eupithecia centaureata* (D. & S.). Common, July and August. Larvae on flowers of many herbaceous plants. Overwinter as pupae. MT 1972–1986.

19 **Common pug** *Eupithecia vulgata* (Haworth). Common, May–July. Larvae on leaves and flowers of a variety of trees, especially *Quercus*, shrubs and herbaceous plants; in garden, on *Ribes sanguineum*. Overwinter as pupae. MT 1987–2001.

20 **Tawny speckled pug** *Eupithecia icterata* (de Villers). Common, August and September. Larvae on *Achillea millefolium*, *Taraxacum* and *Clematis*. Overwinter as pupae. MT 1972–2001.

21 **Green pug** *Chloroclystis rectangulata* (L.). Common, June and July. Larvae on flowers of various *Rosaceae*. Overwinter as eggs. MT 1987–2001.

22 **Brimstone** *Opisthograptis luteolata* (L.). Common. May and June, and second brood in August and September. Larvae on various trees, especially *Rosaceae*; in garden, on *Chaenomeles*, *Crataegus monogyna*, *Prunus cerasifera* and *Sorbus aucuparia*. Overwinter as larvae or pupae. MT 1972–2001.

23 **Canary-shouldered thorn** *Ennomos alniaria* (L.). Common, August and September. Larvae on a variety of trees. Overwinter as eggs. MT 1972–1986.

24 **Dusky thorn** *Ennomos fuscantaria* (Haworth). Common, August and September. Larvae on *Fraxinus excelsior*. Overwinter as eggs.

25 **Early thorn** *Selenia dentaria* (Fab.). Common, March–May, and second brood in July and August. Larvae on various trees and shrubs. Overwinter as pupae. MT 1972–2001.

26 **Scalloped hazel** *Odontopera bidentata* (Clerck). Common, May and June. Larvae on various trees; in garden, on *Betula pendula*, *Buddleja davidii*, *Malus domestica*, *Philadelphus coronarius*, *Potentilla fruticosa*, *Prunus cerasifera*, *Spiraea* × *vanhouttei* and *Weigela florida*. Overwinter as pupae. MT 1972–2001.

27 **Scalloped oak** *Crocallis elinguaria* (L.). Common, July and August. Larvae on most deciduous trees and shrubs; in garden, on *Buddleja davidii*. Overwinter as eggs. MT 1972–2001.

28 **Swallow-tailed moth** *Ourapteryx sambucaria* (L.). Common, June and July. Larvae on *Hedera helix* and a variety of trees and shrubs, mainly *Rosaceae*; in garden, on *Ligustrum ovalifolium* and *Ribes sanguineum*. Overwinter as larvae. MT 1972–2001.

29 **Brindled beauty** *Lycia hirtaria* (Clerck). Common, April and May. Larvae on various deciduous trees; in garden, eggs on *Lonicera periclymenum*. Overwinter as pupae.

30 **Peppered moth** *Biston betularia* (L.). Common, May–July. Larvae on wide variety of trees and herbaceous plants; in garden, on *Buddleja davidii*, *Mahonia aquifolium*, *Malus domestica*, *Potentilla fruticosa* and *Salix fragilis*. Overwinter as pupae.

31 **Willow beauty** *Peribatodes rhomboidaria* (D. & S.). Abundant, July–September. Larvae on wide variety of trees and other plants; in garden, on *Cytisus scoparius*. Overwinter as larvae. MT 1972–2001.

32 **Mottled beauty** *Alcis repandata* (L.). Common, July–September. Larvae on a wide variety of trees and other plants; in garden, on *Crataegus monogyna*. Overwinter as larvae. MT 1972–2001.

33 **Light emerald** *Campaea margaritata* (L.). Common, June and July. Larvae on wide variety of deciduous trees. Overwinter as larvae. MT 1972–1986.

Sphingidae

34 **Lime hawk-moth** *Mimas tiliae* (L.). Common May–July in 1970s, but less so subsequently. Larvae on *Alnus*, *Betula*, *Prunus*, *Quercus*, *Tilia* and *Ulmus*. Overwinter as pupae.

35 **Eyed hawk-moth** *Smerinthus ocellata* (L.). Quite common, June and July. Larvae on *Malus sylvestris*, *Populus* and *Salix*; in garden, on *Salix fragilis*. Overwinter as pupae.

36 **Poplar hawk-moth** *Laothoe populi* (L.). Common, May–August. Larvae on various species of *Salicaceae*; in garden, on *Populus nigra* and *Salix fragilis*. Overwinter as pupae. MT 1972–1986.

Notodontidae

37 **Swallow prominent** *Pheosia tremula* (Clerck). Quite common, May and June, and second brood in August and September. Larvae on various species of *Salicaceae*. Overwinter as pupae.

38 **Pale prominent** *Pterostoma palpina* (Clerck). Quite common, May and June. Larvae on various species of *Salicaceae*. Overwinter as pupae.

Lymantriidae

39 **Vapourer** *Orygia antiqua* (L.). Males common, September and October; females wingless. Larvae on most deciduous trees and shrubs; in garden, on *Pelargonium* cultivar. Overwinter as eggs. MT 1972–1986.

40 **Yellow-tail** *Euproctis similis* (Fuessly). Common, July and August in 1970s, less so subsequently. Larvae on various trees; in garden, on flowers of *Rosa* and on *Crataegus monogyna*. Overwinter as larvae. MT 1972–1986.

Arctiidae

41 **Common footman** *Eilema lurideola* (Zincken). Common, June–August. Larvae on lichens. Overwinter as larvae.

42 **Garden tiger** *Arctia caja* (L.). Common, July and August. Larvae on wide variety of herbaceous plants. Overwinter as larvae.

43 **Buff ermine** *Spilosoma luteum* (Hufnagel). Common, June and July. Larvae on various herbaceous plants and occasionally on trees and shrubs. Overwinter as pupae. MT 1972–1986.

Noctuidae

44 **Garden dart** *Euxoa nigricans* (L.). Common, July and August. Larvae on various herbaceous plants. Overwinter as eggs or larvae.

45 **Turnip moth** *Agrotis segetum* (D. & S.). Common, May–October. Larvae on roots and lower stems of variety of herbaceous plants. Overwinter as larvae. MT 1972–2001, BT.

46 **Heart and dart** *Agrotis exclamationis* (L.). One of the commonest moths from May to August. Larvae on wide variety of herbaceous plants. Overwinter as larvae. MT 1972–2001.

47 **Shuttle-shaped dart** *Agrotis puta* (Hübner). Common in June with second brood in August. Larvae on wide variety of herbaceous plants; one larva in garden in 1983, but foodplant not known. Overwinter as larvae or pupae. MT 1972–1986.

48 **The flame** *Axylia putris* (L.). Common, June and July. Larvae on wide variety of herbaceous plants. Overwinter as pupae. MT 1972–2001.

49 **Flame shoulder** *Ochropleura plecta* (L.). Common in June. Larvae on wide variety of herbaceous plants; in garden, on *Malva sylvestris*. Overwinter as pupae. MT 1972–2001.

50 **Large yellow underwing** *Noctua pronuba* (L.). The commonest garden moth, June–October. One of most abundant moths in Malaise trap. Larvae on a wide range of herbaceous plants and grasses; in garden, on *Aubrieta deltoidea*, *Buddleja davidii*,

Datura stramonium, Epilobium ciliatum, Lamium maculatum, Myosotis alpestris, Nicotiana alata, Oxalis debilis, Papaver alpinum, Taraxacum and *Urtica dioica*. Overwinter as larvae.
MT 1972–2001, BT.

51 **Lesser yellow underwing** *Noctua comes* Hübner. Common, August and September. Larvae on wide variety of trees, shrubs and herbaceous plants; in garden, on *Aubrieta deltoidea, Buddleja davidii, Leucanthemum × superbum, Dianthus barbatus, Digitalis purpurea, Hedera helix, Iberis sempervirens, Phlox douglasii, Thymus* and especially *Lamium maculatum*. Overwinter as larvae. MT 1972–2001, BT.

52 **Lesser broad-bordered yellow underwing** *Noctua janthina* (D. & S.). Common in August. Larvae on wide variety of shrubs and herbaceous plants. Overwinter as larvae. MT 1972–2001.

53 **Least yellow underwing** *Noctua interjecta* Hübner. Common, July and August. Larvae on various shrubs and herbaceous plants. Overwinter as larvae. MT 1972–2001.

54 **Stout dart** *Spaelotis ravida* (D. & S.). Quite common, July–September. Larval foodplants in wild unknown; herbaceous plants in captivity. Overwinter as larvae.

55 **Ingrailed clay** *Diarsia mendica* (Fab.). Common, June–August. Larvae on various shrubs and herbaceous plants. Overwinter as larvae.
MT 1972–1986, BT.

56 **Small square-spot** *Diarsia rubi* (Vieweg). Common, May–September. Larvae on wide variety of herbaceous plants. Overwinter as larvae.
MT 1972–1986.

57 **Setaceous Hebrew character** *Xestia c-nigrum* (L.). Common, June–October in two broods. Larvae on many herbaceous plants; in garden, on *Digitalis purpurea*. Overwinter as larvae. MT 1972–2001, BT.

58 **Double square-spot** *Xestia triangulum* (Hufnagel). Common, June and July. Larvae on wide variety of trees, shrubs and herbaceous plants; in garden, on *Buddleja davidii* and *Lamium maculatum*. Overwinter as larvae.

59 **Square-spot rustic** *Xestia xanthographa* (D. & S.). Common, June–September. Larvae on various grasses, herbaceous plants and deciduous trees. Overwinter as larvae. MT 1972–2001, BT.

60 **The gothic** *Naenia typica* (L.). Common, July and August; more often taken in Malaise trap than at light. Larvae on wide variety of herbaceous plants; in garden, on *Buddleja davidii, Leucanthemum × superbum* and *Lamium maculatum*. Overwinter as larvae. MT 1972–2001.

61 **The nutmeg** *Discestra trifolii* (Hufnagel). Common, May–August. Larvae on several species of *Chenopodiaceae*. Overwinter as pupae.
MT 1972–2001, BT.

62 **Cabbage moth** *Mamestra brassicae* (L.). Common, June–October. Larvae on extraordinary range of herbaceous plants and occasionally shrubs; in garden, on *Acer pseudoplatanus, Artemisia dracunculus, Aster novi-belgii, Atropa belladonna, Brassica oleracea, Buddleja davidii, Calendula officinalis, Callistephus chinensis, Calystegia sepium, Crocosmia × crocosmiiflora, Datura stramonium, Delphinium, Echium plantagineum, Epilobium hirsutum, Erigeron, Humulus lupulus, Lamium maculatum, Lathyrus odoratus, Lactuca sativa, Mahonia aquifolium, Mentha spicata, Ocimum basilicum, Origanum majorana, Oxalis corniculata, Papaver somniferum, Petroselinum crispum, Pisum sativum, Prunus cerasifera, Ribes sanguineum, Ribes uva-crispa, Salvia splendens, Spinacia oleracea, Stachys sylvatica, Tagetes patula, Urtica dioica, Vicia faba, Viola, Weigela florida* and *Zinnia elegans*, and three species of pot plants indoors. Usually overwinter as pupae, sometimes as larvae.
MT 1972–2001, BT.

63 **Dot moth** *Melanchra persicariae* (L.). Common, May–August. Larvae on extraordinary range of herbaceous plants and also trees and shrubs; in garden, on *Aquilegia vulgaris, Aruncus dioicus, Aster amellus, Aster novi-belgii, Betula pendula, Buddleja davidii, Calendula officinalis, Centaurea montana, Leucanthemum × superbum, Crocosmia × crocosmiiflora, Dryopteris filix-mas, Forsythia × intermedia, Humulus lupulus, Hypericum androsaemum, Lamium maculatum, Lavatera trimestris, Levisticum officinale, Lonicera periclymenum, Lunaria annua, Lupinus × regalis, Lycopersicon esculentum, Malus domestica, Malva sylvestris, Melissa officinalis, Mentha spicata, Myosotis alpestris, Origanum majorana, Papaver alpinum, Philadelphus coronarius, Potentilla fruticosa, Ribes sanguineum, Ribes uva-crispa, Salix fragilis, Salvia officinalis, Sambucus nigra, Solidago canadensis, Spinacia oleracea, Tagetes erecta, Tanacetum vulgare* and *Urtica dioica*. Overwinter as pupae.
MT 1972–2001.

64 **Pale-shouldered brocade** *Lacanobia thalassina* (Hufnagel). Quite common, June and July. Larvae on various trees and shrubs. Overwinter as pupae.
MT 1972–1986.

65 **Bright-line brown-eye** *Lacanobia oleracea* (L.). Very common, May–July, and on into autumn in small numbers. Larvae on extraordinary range of shrubs and herbaceous plants; in garden, on *Artemisia dracunculus, Aruncus dioicus, Aster amellus, Aster novi-belgii, Betula pendula, Buddleja davidii, Callistephus chinensis, Calystegia sepium, Leucanthemum × superbum, Convolvulus tricolor, Crocosmia × crocosmiiflora, Cucurbita pepo, Dianthus barbatus, Dipsacus fullonum, Geranium cinereum, Hyssopus officinalis, Iberis sempervirens, Lathyrus odoratus, Lactuca sativa, Levisticum officinale, Lycopersicon esculentum, Mahonia aquifolium, Matthiola incana, Melissa officinalis, Mentha spicata, Nicotiana alata, Ocimum basilicum, Origanum majorana, Oxalis debilis, Paeonia officinalis, Papaver alpinum, Petroselinum crispum, Pisum sativum, Potentilla fruticosa, Ribes nigrum, Ribes uva-crispa, Rumex acetosa, Solanum dulcamara, Spinacia oleracea, Symphytum × uplandicum, Tagetes erecta, Tagetes patula* and *Zinnia elegans*. Overwinter as pupae. MT 1972–2001, BT.

66 **Broom moth** *Ceramica pisi* (L.). Common, June and July, occasionally in August. Larvae on wide variety of trees, shrubs and herbaceous plants. Overwinter as pupae. MT 1987–2001.

67 **The campion** *Hadena rivularis* (Fab.). Common, June and July. Larvae on ripening seeds of many *Caryophyllaceae*. Overwinter as pupae. MT 1972–2001.

68 **The lychnis** *Hadena bicruris* (Hufnagel). Quite common, June and August. Larvae on various species of *Caryophyllaceae*. Overwinter as pupae. MT 1987–2001.

69 **Powdered quaker** *Orthosia gracilis* (D. & S.). Fairly common, April and May. Larvae on various trees, shrubs and herbaceous plants. Overwinter as pupae. MT 1972–1986.

70 **Common quaker** *Orthosia cerasi* (Fab.). Common, March–June. Larvae on various trees; in garden, on *Prunus cerasifera* and *Salix fragilis*. Overwinter as pupae. MT 1972–2001.

71 **Clouded drab** *Orthosia incerta* (Hufnagel). Common, March–June. Larvae on most trees and shrubs. Overwinter as pupae. MT 1972–1986.

72 **Hebrew character** *Orthosia gothica* (L.). Common, April and May. Larvae on various trees, shrubs and herbaceous plants; in garden, on *Potentilla fruticosa*. Overwinter as pupae. MT 1972–2001.

73 **Brown-line bright-eye** *Mythimna conigera* (D.& S.). Quite common, July and August. Larvae on many grasses. Overwinter as larvae.

74 **The clay** *Mythimna farrago* (Fab.). Common, July and August. Larvae mainly on grasses, occasionally on herbaceous plants. Overwinter as larvae. MT 1972–2001.

75 **Smoky wainscot** *Mythimna impura* Hübner). Common, July and August. Larvae on many grasses. Overwinter as larvae. MT 1972–2001.

76 **Common wainscot** *Mythimna pallens* (L.). Common, June–August. Larvae on many grasses. Overwinter as larvae. MT 1972–1986, BT.

77 **The shark** *Cucullia umbratica* (L.). Quite common, June and July. Larvae on various species of *Sonchus* and *Lactuca*. Overwinter as pupae. MT 1987–2001.

78 **Early grey** *Xylocampa areola* (Esper). Common, March–June. Larvae on *Lonicera periclymenum*; in garden, on *L. periclymenum*. Overwinter as pupae. MT 1972–2001.

79 **Brown-spot pinion** *Agrochola litura* (L.). Common, August and September. Larvae on various trees, shrubs and herbaceous plants; in garden, on *Salix fragilis*. Overwinter as eggs. MT 1972–1986, BT.

80 **Beaded chestnut** *Agrochola lychidnis* (D. & S.). Common, September and October. Larvae on grasses and various herbaceous plants. Overwinter as eggs. MT 1972–1986, BT.

81 **Lunar underwing** *Omphaloscelis lunosa* (Howarth). Quite common in September. Larvae on grasses. Overwinter as larvae. MT 1987–2001.

82 **The sallow** *Xanthia icteritia* (Hufnagel). Common, August and September. Larvae on *Salix* catkins when young, later on leaves or on herbaceous plants. Overwinter as eggs. MT 1972–1986.

83 **Poplar grey** *Acronicta megacephala* (D. & S.). Common, May–August. Larvae on various *Salicaceae*. Overwinter as pupae.

84 **Grey dagger** *Acronicta psi* (L.). Common, May–July. Larvae on various trees; in garden, on *Betula pendula, Crataegus monogyna* and *Rosa*. Overwinter as pupae. 1972–2001.

85 **Marbled beauty** *Cryphia domestica* (Hufnagel). Common, June–August. Larvae on lichens. Overwinter as larvae. MT 1972–1986.

86 **Mouse moth** *Amphipyra tragopoginis* (Clerck). Common, July–September. Larvae on various shrubs and herbaceous plants. Overwinter as eggs. MT 1972–1986, BT.

87 **Straw underwing** *Thalpophila matura* (Hufnagel). Common, July and August. Larvae on various grasses. Overwinter as larvae. MT 1972–1986.

88 **Small angle shades** *Euplexia lucipara* (L.). Common, June–August. Larvae on various trees, shrubs, herbaceous plants and, especially, ferns; in garden, on *Alcea rosea, Buddleja davidii, Calystegia sepium, Chamerion angustifolium, Epilobium hirsutum, Lupinus* × *regalis, Lycopersicon esculentum, Malva sylvestris, Oenothera biennis, Paeonia officinalis, Solanum dulcamara, Weigela florida* and, especially, male fern, *Dryopteris filix-mas*. Overwinter as pupae. MT 1972–2001.

89 **Angle shades** *Phlogophora meticulosa* (L.). Common, almost all months of year, especially September–November. Larvae on extraordinary range of herbaceous plants, and also trees and shrubs; in garden, on *Achillea millefolium, Alcea rosea, Aurinia saxatilis, Antirrhinum majus, Apium graveolens, Arabis alpina* subsp. *caucasica, Aster novi-belgii, Aubrieta deltoidea, Beta vulgaris, Borago officinalis, Brassica oleracea, Buddleja davidii, Calendula officinalis, Centaurea montana, Erysimum cheiri, Leucanthemum* × *superbum, Convolvulus tricolor, Dianthus* cultivar, *D. barbatus, Dicentra formosa, Digitalis purpurea, Dipsacus fullonum, Dryopteris filix-mas, Foeniculum vulgare, Fragaria vesca, Geranium cinereum, Geum urbanum, Hesperis matronalis, Hyssopus officinalis, Iberis sempervirens, Iris foetidissima, Lamium maculatum, Lactuca sativa, Lunaria annua, Malva sylvestris, Matthiola incana, Mentha spicata, Myosotis alpestris, Origanum majorana, Oxalis debilis, Petroselinum crispum, Rudbeckia hirta, Rumex acetosa, Salvia officinalis, Saxifraga, Sedum acre, Solidago canadensis, Stachys sylvatica, Stellaria media, Symphytum* × *uplandicum, Tagetes patula, Tanacetum parthenium, Taraxacum* and *Veronica hederifolia*, and on pot plant indoors. Overwinter as larvae. MT 1972–2001, BT.

90 **The dun-bar** *Cosmia trapezina* (L.). Quite common, July–September. Larvae on wide variety of trees and shrubs; in garden, on *Buddleja davidii*. Overwinter as eggs.

91 **Dark arches** *Apamea monoglypha* (Hufnagel). Abundant, June–August. Larvae on various grasses. Overwinter as larvae. MT 1972–2001, BT.

92 **Clouded-bordered brindle** *Apamea crenata* (Hufnagel). Common, May and June. Larvae on various grasses. Overwinter as larvae. MT 1972–2001.

93 **Clouded brindle** *Apamea epomidion* (Haworth). Common, June and July. Larvae on various grasses. Overwinter as larvae. MT 1972–2001.

94 **Dusky brocade** *Apamea remissa* (Hübner). Common, June–September. Larvae on various grasses. Overwinter as larvae. MT 1972–2001.

95 **Small clouded brindle** *Apamea unanimis* (Hübner). Fairly common, July and August. Larvae on various grasses. Overwinter as larvae. MT 1972–2001.

96 **Rustic shoulder-knot** *Apamea sordens* (Hufnagel). Common, May–July. Larvae on various grasses. Overwinter as larvae. MT 1972–2001.

97 **Marbled minor** *Oligia strigilis* (L.). Common, June–August. Larvae on various grasses. Overwinter as larvae. MT 1972–1986.

98 **Tawny marbled minor** *Oligia latruncula* (D. & S.). Common, June–August. Larvae on various grasses. Overwinter as larvae. MT 1972–2001.

99 **Middle-barred minor** *Oligia fasciuncula* (Haworth). Common, June. Larvae on various grasses. Overwinter as larvae. MT 1972–1986.

100 **Rosy minor** *Mesoligia literosa* (Haworth). Common, June–August. Larvae on various grasses. Overwinter as larvae. MT 1972–2001.

101 **Common rustic** *Mesapamea secalis* (L.). Common, June–August. Larvae on various grasses. Overwinter as larvae. MT 1972–2001, BT.

102 **Flounced rustic** *Luperina testacea* (D. & S.). Common, August and September. Larvae on various grasses. Overwinter as larvae. MT 1972–2001.

103 **Rosy rustic** *Hydraecia micacea* (Esper). Common, August and September. Larvae on various herbaceous plants. Overwinter as eggs. MT 1972–2001.

104 **The uncertain** *Hoplodrina alsines* (Brahm). Common, June and July. Larvae on various herbaceous plants. Overwinter as larvae. MT 1987–2001.

105 **The rustic** *Hoplodrina blanda* (D. & S.). Common, July and August. Larvae on various herbaceous plants. Overwinter as larvae. MT 1987–2001.

106 **Vine's rustic** *Hoplodrina ambigua* (D. & S.). Possibly common and frequently overlooked, August and September. Larvae on various herbaceous plants. Overwinter as larvae. MT 1987–2001.

107 **Mottled rustic** *Caradrina morpheus* (Hufnagel). Common, June–August. Larvae on various herbaceous plants; in garden, on *Aster amellus, Lamium maculatum* and *Lavatera trimestris*. Overwinter as larvae. MT 1972–2001.

108 **Pale mottled willow** *Caradrina clavipalpis* (Scopoli). Common, July–September. Larvae on various grasses and on various herbaceous plants. Overwinter as larvae. MT 1972–1986.

109 **Burnished brass** *Diachrysia chrysitis* (L.). Common, June–September. Larvae on various herbaceous plants; in garden, on *Aster novi-belgii, Hyssopus officinalis, Mentha spicata, Origanum majorana, Petroselinum crispum* and *Pisum sativum*. Overwinter as larvae. MT 1972–2001.

110 **Silver Y** *Autographa gamma* (L.). Common, May–November, as immigrant. Usually arrive in June, but May in some years. Often commonest moth in Malaise trap. Larvae on various herbaceous plants; in garden, on *Borago officinalis, Hyssopus officinalis, Lactuca sativa, Nepeta × faassenii, Petroselinum crispum* and *Zinnia elegans*. Continuously brooded. MT 1972–2001, BT.

111 **Beautiful golden Y** *Autographa pulchrina* (Haworth). Common, June. Larvae on various herbaceous plants. Overwinter as larvae.

112 **Plain golden Y** *Autographa jota* (L.). Common, June and July. Larvae on various shrubs and herbaceous plants. Overwinter as larvae. MT 1972–1986.

113 **The spectacle** *Abrostola triplasia* (L.). Common, May–August. Larvae on *Urtica dioica*. MT 1972–2001.

114 **The snout** *Hypena proboscidalis* (L.). Common, July and August. Larvae on *Urtica dioica*; in garden, on *U. dioica*. Overwinter as larvae. MT 1972–2001.

115 **The fan-foot** *Herminia tarsipennalis* (Treitschke). Common, July and August. Larvae on withered leaves of various trees and shrubs. Overwinter as larvae. MT 1972–2001.

UNCOMMON GARDEN MOTHS

Hepialidae

116 **Ghost moth** *Hepialus humuli* (L.). Quite common in June and July, but irregular. Larvae on roots of many herbaceous plants and grasses; in garden, on *Leucanthemum × superbum*. Overwinter as larvae.

Cossidae

117 **Leopard moth** *Zeuzera pyrina* (L.). Twelve records, June–September 1973–1983. Larvae in wood of variety of trees and shrubs. Overwinter as larvae.

Lasiocampidae

118 **The lackey** *Malacosoma neustria* (L.). Eleven records 1973–1983, ten in July and one in August. Larvae on various trees and shrubs. Overwinter as eggs.

Drepanidae

119 **Chinese character** *Cilix glaucata* (Scopoli). Uncommon, May–July and second brood in August. Larvae on various trees and shrubs. Overwinter as pupae. MT 1972–1986.

Geometridae

120 **Common emerald** *Hemithea aestivaria* (Hübner). Not seen until 1976, thereafter frequent in July and August. Larvae on various trees and shrubs. Overwinter as larvae. MT 1972–2001.

121 **Blood-vein** *Timandra griseata* (Petersen). Uncommon, May–September. Larvae on various *Chenopodiaceae* and *Polygonaceae*. Overwinter as larvae. MT 1972–1986.

122 **Single-dotted wave** *Idaea dimidiata* (Hufnagel). Infrequent, July and August. Larvae on *Anthriscus sylvestris* and *Pimpinella saxifraga*. Overwinter as larvae. MT 1972–1986.

123 **Barred straw** *Eulithis pyraliata* (D. & S.). Infrequent, July. Larvae on *Galium*. Overwinter as eggs. MT 1987–2001.

124 **Barred yellow** *Cidaria fulvata* (Forster). Infrequent, July. Larvae on *Rosa canina*. Overwinter as eggs. MT 1987–2001.

125 **Juniper carpet** *Thera juniperata* (L.). Infrequent initially, October; increasingly common in 1990s. Larvae on *Juniperus*. Overwinter as eggs. MT 1987–2001

126 **Small rivulet** *Perizoma alchemillata* (L.). Rather infrequent, July and August. Larvae on *Galeopsis*. Overwinter as pupae.

127 **Bordered pug** *Eupithecia succenturiata* (L.). Infrequent, July–September. Larvae on *Artemisia vulgaris*. Overwinter as pupae. MT 1972–2001.

128 **Double-striped pug** *Gymnoscelis rufifasciata* (Haworth). Infrequent, July and August. Larvae on flowers of variety of shrubs; in garden, on *Buddleja davidii*. Overwinter as pupae. MT 1987–2001.

129 **The magpie** *Abraxas grossulariata* (L.). Common in 1970s, June–August, but much scarcer in 1980s and 1990s. Larvae on various shrubs; in garden, on *Ribes sanguineum* and *R. uva-crispa*. Overwinter as larvae. MT 1972–2001.

130 **Feathered thorn** *Colotois pennaria* (L.). Ten in October and November. Larvae on various trees and shrubs; in garden, on *Potentilla fruticosa*. Overwinter as eggs.

Sphingidae

131 **Elephant hawk-moth** *Deilephila elpenor* (L.). Irregular, June and July. Larvae on *Epilobium*, *Fuchsia*, *Galium* and other plants; in garden, on *Chamerion angustifolium*. Overwinter as pupae. MT 1987–2001.

Notodontidae

132 **Coxcomb prominent** *Ptilodon capucina* (L.). Ten records 1970s and 1981, June–August. Larvae on various deciduous trees. Overwinter as pupae.

Arctiidae

133 **Muslin moth** *Diaphora mendica* (Clerck). Common in early 1970s, none from 1977 to 1991; one in 1991, then another in 1992. Larvae on various herbaceous plants. Overwinter as pupae.

134 **Ruby tiger** *Phragmatobia fuliginosa* (L.). Rather uncommon in early 1970s, July and August, and not seen 1976–2001. Larvae on various herbaceous plants. Overwinter as larvae.

Nolidae

135 **Short-cloaked moth** *Nola cucullatella* (L.). Regular but not common in July, occasionally in August, up until 1979, none afterwards. Larvae on various trees in *Rosaceae*. Overwinter as larvae. MT 1972–1986.

Noctuidae

136 **Broad-bordered yellow underwing** *Noctua fimbriata* (Schreber). Uncommon but regular, July and August. Larvae on various shrubs and herbaceous plants. Overwinter as larvae.

137 **Double dart** *Graphiphora augur* (Fab.). Frequent, June–August. Larvae on various trees and shrubs, and occasionally on herbaceous plants. Overwinter as larvae.

138 **Six-striped rustic** *Xestia sexstrigata* (Haworth). Infrequent but regular, August. Larvae on variety of herbaceous plants. Overwinter as larvae.

139 **Shoulder-striped wainscot** *Mythimna comma* (L.). Quite common until 1976, then sporadic, June and July. Larvae on various grasses. Overwinter as larvae.

140 **Green-brindled crescent** *Allophyes oxycanthae* (L.). Common 1978 and 1979, October, otherwise infrequent. Larvae on *Crataegus* and *Prunus spinosa*. Overwinter as eggs. MT 1972–1986.

141 **Red-line quaker** *Agrochola lota* (Clerck). Common 1977–1979, October, otherwise infrequent. Larvae on *Salix*. Overwinter as eggs.

142 **Centre-barred sallow** *Atethmia centrago* (Haworth). Fourteen records in August and September. Larvae on *Fraxinus excelsior*. Overwinter as eggs.

143 **Copper underwing** *Amphipyra pyramidea* (L.). Not seen until1978, thereafter quite common, August and September. Larvae on various trees and shrubs. Overwinter as eggs.

144 **Light arches** *Apamea lithoxylea* (D. & S.). Frequent, but never common, July and August. Larvae on grasses. Overwinter as larvae. MT 1972–2001.

145 **Small dotted buff** *Photedes minima* (Haworth). Twelve records in July and August, none after 1983. Larvae on tufted hair-grass, *Deschampsia caespitosa*. Overwinter as larvae.

146 **Frosted orange** *Gortyna flavago* (D. & S.). Infrequent, mostly September. Larvae on large-stemmed plants, especially *Arctium*, *Carduus* and *Cirsium*, stem. Overwinter as eggs.

147 **Golden plusia** *Polychrysia moneta* (Fab.). Infrequent, July–September. A recent immigrant to Britain. Larvae on *Aconitum* and *Delphinium*. Overwinter as larvae. MT 1987–2001.

148 **Red underwing** *Catocala nupta* (L.). Infrequent, August and September. Larvae on *Salix* and *Populus*. Overwinter as eggs.

149 **The herald** *Scoliopteryx libatrix* (L.). Frequent, August–October, and after hibernation in May and June. Larvae on various *Salicaceae*. Overwinter as adults. MT 1972–2001.

150 **Small fan-foot** *Herminia grisealis* (D. & S.). Uncommon, June and July, none after 1986. Larvae on *Alnus*, *Quercus* and other trees. Overwinter as pupae. MT 1972–1986.

SCARCE GARDEN MOTHS

Sesiidae

151 **Red-belted clearwing** *Synanthedon myopaeformis* (Borkhausen)

Drepanidae

152 **Oak hook-tip** *Drepana binaria* (Hufnagel)

Thyatiridae

153 **Peach blossom** *Thyatira batis* (L.)

Geometridae

154 **Small blood-vein** *Scopula imitaria* (Hübner). MT 1972–2001
155 **Small dusty wave** *Idaea seriata* (Schrank)
156 **The streamer** *Anticlea derivata* (D. & S.)
157 **Dark spinach** *Pelurga comitata* (L.)
158 **Small phoenix** *Ecliptopera silaceata* (D. & S.)
159 **Broken-barred carpet** *Electrophaes corylata* (Thunberg)
160 **Green carpet** *Colostygia pectinataria* (Knoch)
161 **The tissue** *Triphosa dubitata* (L.)
162 **Winter moth** *Operophtera brumata* (L.). Larvae in garden, on *Corylus avellana, Potentilla fruticosa, Ribes sanguineum, Salix fragilis* and *Ulmus glabra*.
163 **Twin-spot carpet** *Perizoma didymata* (L.)
164 **Toadflax pug** *Eupithecia linariata* (D. & S.)
165 **Freyer's pug** *Eupithecia intricata* (Zett.) subsp. *arceuthata* (Freyer). MT 1987–2001
166 **Wormwood pug** *Eupithecia absinthiata* (Clerck)
167 **Currant pug** *Eupithecia assimilata* Doubleday
168 **Grey pug** *Eupithecia subfuscata* (Haworth). MT 1987–2001
169 **Brindled pug** *Eupithecia abbreviata* Steph. MT 1987–2001
170 **Juniper pug** *Eupithecia pusillata* (D. & S.)
171 **The V-pug** *Chloroclystis v-ata* (Haworth)
172 **Treble-bar** *Aplocera plagiata* (L.)
173 **The seraphim** *Lobophora halterata* (Hufnagel)
174 **Clouded border** *Lomaspilis marginata* (L.). MT 1972–2001
175 **Tawny-barred angle** *Semiothisa liturata* (Clerck)
176 **Latticed heath** *Semiothisa clathrata* (L.)
177 **The V-moth** *Semiothisa wavuaria* (L.)
178 **Brown silver-line** *Petrophora chlorosata* (Scopoli)
179 **Scorched wing** *Plagodis dolabraria* (L.)
180 **Lilac beauty** *Apeira syringaria* (L.). MT 1972–1986
181 **September thorn** *Ennomos erosaria* (D. & S.). MT 1987–2001
182 **Lunar thorn** *Selenia lunularia* (Hübner)
183 **Pale brindled beauty** *Apocheima pilosaria* (D. & S.)
184 **Dotted border** *Agriopsis marginaria* (Fab.). Larvae in garden, on *Crataegus monogyna*.
185 **Mottled umber** *Erannis defolaria* (Clerck). Larvae in garden, on *Prunus cerasifera*.
186 **Waved umber** *Menophra abruptaria* (Thunberg)
187 **Bordered white** *Bupalus piniaria* (L.)
188 **Common white wave** *Cabera pusaria* (L.)
189 **Clouded silver** *Lomographa temerata* (D. & S.). Pupa found in garden in May 1975.
190 **Barred red** *Hylaea fasciaria* (L.)

Sphingidae

191 **Humming-bird hawk-moth** *Macroglossum stellatarum* (L.)

Notodontidae

192 **Buff-tip** *Phalera bucephala* (L.)
193 **Puss moth** *Cerura vinula* (L.)
194 **Sallow kitten** *Furcula furcula* (Clerck)
195 **Iron prominent** *Notodonta dromedarius* (L.)
196 **Figure of eight** *Diloba caeruleocephala* (L.)

Lymantriidae

197 **Pale tussock** *Calliteara pudibunda* (L.)
198 **White satin moth** *Leucoma salicis* (L.)

Arctiidae

199 **White ermine** *Spilosoma lubricipeda* (L.)
200 **The cinnabar** *Tyria jacobaeae* (L.)

Noctuidae

201 **Dark sword-grass** *Agrotis ipsilon* (Hufnagel). MT 1972–1986
202 **Dotted rustic** *Rhyacia simulans* (Hufnagel)
203 **Pearly underwing** *Peridroma saucia* (Hübner)
204 **Red chestnut** *Cerastis rubricosa* (D. & S.)
205 **Light brocade** *Lacanobia w-latinum* (Hufnagel)
206 **Broad-barred white** *Hecatera bicolorata* (Hufnagel). MT 1972–1986
207 **Tawny shears** *Hadena perplexa* (D. & S.)
208 **Varied coronet** *Hadena compta* (D. & S.). MT 1972–2001
209 **Antler moth** *Cerapteryx graminis* (L.)
210 **Feathered gothic** *Tholera decimalis* (Poda)
211 **Small quaker** *Othosia cruda* (D. & S.)
212 **Chamomile shark** *Cucullia chamomillae* (D. & S.)
213 **Brindled green** *Dryobotodes eremita* (Fab.)
214 **Grey chi** *Antitype chi* (L.)
215 **Dark chestnut** *Conistra ligula* (Esper)
216 **Dusky-lemon sallow** *Xanthia gilvago* (D. & S.)
217 **The miller** *Acronicta leporina* (L.)
218 **Alder moth** *Acronicta alni* (L.)
219 **Knot grass** *Acronicta rumicis* (L.)
220 **Svensson's copper underwing** *Amphipyra berbera* Rungs
221 **Old lady** *Mormo maura* (L.). MT 1972–1986
222 **Bird's wing** *Dypterygia scabriuscula* (L.). MT 1972–1986
223 **The olive** *Ipimorpha subtusa* (D. & S.)
224 **Dingy shears** *Parastichtis ypsillon* (D. & S.)
225 **Lesser spotted pinion** *Cosmia affinis* (L.)
226 **Large nutmeg** *Apamea anceps* (D. & S.)
227 **Double-lobed** *Apamea ophiogramma* (Esper). MT 1972–1986
228 **Cloaked minor** *Mesoligia furuncula* (D. & S.). MT 1987–2001
229 **Brown-veined wainscot** *Archanara dissoluta* (Treitschke)
230 **Large wainscot** *Rhizedra lutosa* (Hübner)
231 **Scarce silver-lines** *Bena prasinana* (L.)
232 **Dark spectacle** *Abrostola trigemina* (Werneburg)

MOTHS RECORDED ONCE ONLY (1972–2001)

Sesiidae

233 **Currant clearwing** *Synanthedon tipuliformis* (Clerck). 1976 MT
234 **Yellow-legged clearwing** *Synanthedon vespiformis* L. 1976 MT

Lasiocampidae

235 **The drinker** *Euthrix potatoria* (L.) 1975

Drepanidae

236 **Pebble hook-tip** *Drepana falcataria* (L.) 1990

Geometridae

237 **Grass emerald** *Pseudoterpna pruinata* (Hufnagel) 1977
238 **Small fan-footed wave** *Idaea biselata* (Hufnagel) 2000 MT
239 **Flame carpet** *Xanthorhoe designata* (Hufnagel) 1990 MT
240 **Shaded broad-bar** *Scotopteryx chenopodiata* (L.) 1980
241 **Dark umber** *Philereme transversata* (Hufnagel) 1975
242 **Sandy carpet** *Perizoma flavofasciata* (Thunberg) 1992
243 **Foxglove pug** *Eupithecia pulchellata* Steph. 1987 MT
244 **Pinion-spotted pug** *Eupithecia insigniata* (Hübner) 1980
245 **White-spotted pug** *Eupithecia tripunctaria* Herrich-Schäffer 1974
246 **Plain pug** *Eupithecia simpliciata* (Haworth) 1974
247 **Ochreous pug** *Eupithecia indigata* (Hübner) 1976
248 **Ash pug** *Eupithecia fraxinata* Crewe 1981
249 **Larch pug** *Eupithecia lariciata* (Freyer) 1975
250 **Sloe pug** *Chloroclystis chloerata* (Mabille) 2001 MT
251 **Chimney-sweeper** *Odezia atrata* (L.) 1975 MT
252 **Bordered beauty** *Epione repandaria* (Hufnagel) 1981
253 **August thorn** *Ennomos quercinaria* (Hufnagel) 1976
254 **Purple thorn** *Selenia tetralunaria* (Hufnagel) 1991
255 **Small brindled beauty** *Apocheima hispidaria* (D. & S.) 1981
256 **Oak beauty** *Biston strataria* (Hufnagel) 1978
257 **Scarce umber** *Agriopsis aurantiaria* (Hübner) 1977
258 **White-pinion spotted** *Lomographa bimaculata* (Fab.) 1987
259 **Early moth** *Theria primaria* (Haworth) 1980

Notodontidae

260 **Pebble prominent** *Eligmodonta ziczac* (L.) 1978

Noctuidae

261 **True lover's knot** *Lycophotia porphyrea* (D. & S.) 1981
262 **Dotted clay** *Xestia baja* (D. & S.) 1982
263 **Great brocade** *Eurois occulta* (L.) 1976
264 **Grey arches** *Polia nebulosa* (Hufnagel) 1974
265 **Marbled coronet** *Hadena confusa* (Hufnagel) 1973
266 **Twin-spotted quaker** *Orthosia munda* (D. & S.) 1973
267 **Obscure wainscot** *Mythimna obsoleta* (Hübner) 1976
268 **The wormwood** *Cucullia absinthii* (L.) 1974
269 **The mullein** *Cucullia verbasci* L. 1974. Four larvae on *Buddleja davidii* in 1977.
270 **Deep-brown dart** *Aporophyla lutulenta* (D. & S.) 1976
271 **Blair's shoulder-knot** *Lithophane leautieri* (Boisduval) 1990
272 **Merveille du jour** *Dichonia aprilina* (L.) 1979
273 **Dark brocade** *Mniotype adusta* (Esper) 1975
274 **The brick** *Agrochola circellaris* (Hufnagel) 1986
275 **Flounced chestnut** *Agrochola helvola* (L.) 1977 MT
276 **Barred sallow** *Xanthia aurago* (D. & S.) 1979
277 **The sycamore** *Acronicta aceris* (L.) 1979
278 **Brown rustic** *Rusina ferruginea* (Esper) 1979
279 **Dusky sallow** *Eremobia ochroleuca* (D. & S.) 1976
280 **Scarce bordered straw** *Heliothis armigera* (Hübner) 1988
281 **Gold spangle** *Autographa bractea* (D. & S.) 1973
282 **Beautiful hook-tip** *Laspeyria flexula* (D. & S.) 1972

Micro-Lepidoptera

Names and foodplants for 93 species of micro-Lepidoptera from Emmet (1988) and Emmet & Heath (1991), with foodplants recorded in garden. Number of records given. MT indicates that the species has been taken in the Malaise trap. BT indicates that it has been taken in a trap baited with fermenting fruit, usually apple.

Incurvariidae

283 *Incurvaria masculella* (D. & S.) Two. Larvae in leaves of *Crataegus* or *Rosa*.

Tineidae

284 *Tinea trinotella* Thunberg. One. Larvae in birds' nests.

Yponomeutidae

285 **Spindle small ermine** *Yponomeuta cagnagella* (Hübner).Abundant. Larvae on *Euonymus*; in garden, on *Crataegus monogyna* and *Cytisus scoparius*.
286 **Honeysuckle moth** *Ypsolopha dentella* (Fab.).Two. Larvae on *Lonicera*.
287 *Ypsolopha scabrella* (L.) Three. Larvae on *Cotoneaster, Crataegus* and *Malus*.
288 *Orthotaelia sparganella* (Thunberg). One. Larvae in leaves and stems of *Glyceria maxima, Iris* and *Sparganium*.

Oecophoridae

289 **Brown house moth** *Hofmannophila pseudospretella* (Stainton). Two. On various dried organic material.
290 **White-shouldered house moth** *Endrosis sarcitrella* (L.). Larvae on various dried organic material, also indoors. Four, MT
291 *Carcina quercana* (Fab.). Common. Larvae on leaves of variety of trees; in garden, on *Prunus cerasifera*.
292 *Agonopterix arenella* (D. & S.). One. Larvae on leaves of thistles and similar *Asteraceae*; in garden, on *Centaurea montana*.

293 *Agonopterix nervosa* (Haworth). Common. Larvae on *Cytisus, Genista* and *Ulex*; in garden, common on *Cytisus* × *praecox*.

Gelechiidae

294 *Mirificarma mulinella* (Zeller). One. Larvae on flowers of various *Leguminosae*; in garden, on *Cytisus scoparius*.

295 **Juniper webber** *Dichomeris marginella* (Fab.). Two. Larvae in needles of *Juniperus*, including cultivars.

296 *Brachmia rufescens* (Haworth). One. Larvae in leaves of various grasses.

Blastobasidae

297 *Blastobasis decolorella* (Wollaston). 14, MT. Larvae on fresh vegetation, leaf-litter, dead insects etc.

Tortricidae

298 *Agapeta hamana* (L.). Common. Larvae in roots of *Carduus* and *Cirsium*.

299 **Chequered fruit-tree tortrix** *Pandemis corylana* (Fab.). Common, August and September. Larvae on various trees and shrubs.

300 **Barred fruit-tree tortrix** *Pandemis cerasana* (Hübner). Three. Larvae on various trees, shrubs and herbaceous plants; in garden, on *Chamerion angustifolium* and *Rosa*.

301 **Dark fruit-tree tortrix** *Pandemis heparana* (D. & S.). Two. Larvae on various trees and shrubs; in garden, on *Ligustrum ovalifolium, Potentilla fruticosa* and *Prunus cerasifera*.

302 **Large fruit-tree tortrix** *Archips podana* (Scopoli). Common, June and July. Larvae on various trees and shrubs; in garden, on *Buddleja davidii*.

303 **Carnation tortrix** *Cacoecimorpha pronubana* (Hübner). Five, MT. Larvae on various trees and shrubs, especially *Euonymus japonicus*.

304 **Timothy tortrix** *Aphelia paleana* (Hübner). Two. Larvae on various herbaceous plants and grasses

305 **Cyclamen tortrix** *Clepsis spectrana* (Treitschke). Three. Larvae on various trees and herbaceous plants.

306 *Lozotaenioides formosanus* (Geyer). Four. Larvae on *Pinus sylvestris*.

307 *Lozotaenia forsterana* (Fab.). Two. Larvae on various trees, shrubs and herbaceous plants, especially on *Hedera*.

308 **Red-barred tortrix** *Ditula angustiorana* (Haworth). Two, MT. Larvae on various trees and shrubs, occasionally on herbaceous plants.

309 **Grey tortrix** *Cnephasia stephensiana* (Doubleday). One. Larvae on various herbaceous plants; in garden, on *Centaurea montana*.

310 *Cnephasia* sp. 1. One larva on *Buddleja davidii*.

311 *Cnephasia* sp. 2. One moth.

312 **Green oak tortrix** *Tortrix viridana* (L.). Common. Larvae on *Quercus*, occasionally on other trees.

313 *Acleris forsskaleana* (L.). Common, July and August. Larvae on *Acer*.

314 *Acleris sparsana* (D. & S.). Three. Larvae on various trees.

315 **Rhomboid tortrix** *Acleris rhombana* (D. & S.). One. Larvae on various trees and shrubs, especially *Rosaceae*; in garden, on *Prunus cerasifera*.

316 **Garden rose tortrix** *Acleris variegana* (D. & S.). Common, July–September. Larvae on various trees and shrubs in *Rosaceae*, and on herbaceous plants; in garden, on *Rosa*.

317 *Celypha striana* (D. & S.). Two. Larvae on *Taraxacum*.

318 *Olethreutes lacunana* (D. & S.). 19. Larvae on most species of herbaceous plants, occasionally on trees and shrubs.

319 *Apotomis betuletana* (Haworth). One. Larvae on *Betula*.

320 *Epinotia bilunana* (Haworth). One. Larvae on *Betula*.

321 *Epiblema cynosbatella* (L.). Three. Larvae on *Rosa* and *Rubus*.

322 *Epiblema foenella* (L.). Four. Larvae in roots of *Artemisia*.

323 *Eucosma pupillana* (Clerck). Larvae in stem and roots of *Artemisia absinthium*.

324 **Pea moth** *Cydia nigricana* (Fab.). One. Larvae in pods of *Lathyrus, Pisum* and *Vicia*.

325 *Cydia splendana* (Hübner). One. Larvae in nuts of *Castanea, Juglans* and *Quercus*.

326 **Codling moth** *Cydia pomonella* (L.). Common, June–September. Larvae in fruits of various trees; in garden, in *Malus domestica* fruit every year.

327 *Dichrorampha alpinana* (Treitschke) or *D. flavidorsana* Knaggs. Two. Larvae in roots of *Leucanthemum vulgare* (*D. alpinana*) or *Tanacetum vulgare* (*D. flavidorsana*).

Alucitidae

328 **Twenty-plume moth** *Alucita hexadactyla* (L.). Common. Larvae on *Lonicera periclymenum* flowerbuds and leaves; in garden, on *L. periclymenum*.

Pyralidae

329 *Calamatropha paludella* (Hübner). One, MT. Larvae on *Typha*.

330 *Chrysoteuchia culmella* (L.). Abundant, June–August. MT. Larvae in stem bases of grasses.

331 *Crambus pascuella* (L.). One, MT. Larvae on stem bases of grasses.

332 *Crambus lathoniellus* (Zincken). Two. Larvae on stem bases of grasses, especially *Deschampsia caespitosa*.

333 *Crambus perlella* (Scopoli). Six. Larvae on stem bases of grasses, especially *Deschampsia* and *Festuca*.

334 *Agriphila selasella* (Hübner). One. Larvae on various grasses.

335 *Agriphila straminella* (D. & S.). Four. Larvae on various grasses.

336 *Agriphila tristella* (D. & S.). Abundant, July and August. Larvae on various grasses.

337 *Agriphila geniculea* (Haworth). Two, MT. Larvae on various grasses.

338 *Catoptria pinella* (L.). Two. Larvae on various grasses and sedges.

339 *Scoparia subfusca* Haworth. One, MT. Larvae on roots of *Tussilago* and *Picris hieracioides*.

340 *Scoparia pyralella* (D. & S.). One, MT. Larvae on decaying plant material and roots. ▶

341 *Scoparia ambigualis* (Treitschke). One, MT. Larvae thought to be on mosses.
342 *Eudonia truncicolella* (Stainton). One. Larvae on mosses.
343 *Eudonia angustea* (Curtis). One. Larvae on mosses.
344 *Eudonia mercurella* (L.). Four. Larvae on mosses.
345 **Water veneer** *Acentria ephemerella* (D. & S.). One. Larvae on submerged leaves of water plants.
346 **Garden pebble** *Evergestis forficalis* (L.). Common. Larvae on various species of *Brassicaceae*; in garden, on *Brassica oleracea* and cultivars.
347 *Evergestis pallidata* (Hufnagel). Four. Larvae on various species of *Brassicaceae*.
348 *Pyrausta aurata* (Scopoli). Three, MT. Larvae on *Lamiaceae*, especially *Mentha* including cultivars.
349 *Pyrausta purpuralis* (L.). One. Larvae on *Mentha arvensis* and *Thymus*.
350 **Small magpie** *Eurrhypara hortulata* (L.). Common. Larvae on *Urtica dioica* and other herbaceous plants; in garden, on *U. dioica*.
351 *Phlyctaenia coronata* (Hufnagel). Common. Larvae on various shrubs.
352 *Udea prunalis* (D. & S.). One. Larvae on various trees, shrubs and herbaceous plants.
353 *Udea olivalis* (D. & S.). Common, June and July. Larvae on various herbaceous plants.
354 *Udea ferrugalis* (Hübner). Two, MT. Larvae on various herbaceous plants.
355 **Rush veneer** *Nomophila noctuella* (D. & S.). One. Larvae on various herbaceous plants and grasses.
356 **Mother of pearl** *Pleuroptya ruralis* (Scopoli). Common. Larvae on *Urtica dioica*; in garden, on *U. dioica*.
357 *Sceliodes laisalis* (Walker). One in July 1979, the second British record. An African species feeding on *Lycopersicon esculentum* (tomato) fruits.
358 **Gold triangle** *Hypsopygia costalis* (Fab.). Common. Larvae on dry vegetable matter.
359 *Orthopygia glaucinalis* (L.). Frequent, MT. Larvae on dead and decaying vegetable matter.
360 **Meal moth** *Pyralis farinalis* (L.). One. Larvae on stored cereals and cereal refuse.
361 **Large tabby** *Aglossa pinguinalis* (L.). One. Larvae on chaff and refuse of cereals, seeds, hay and sheep dung.
362 **Lesser wax moth** *Achroia grisella* (Fab.). One, MT. Comb, especially old comb in beehive, dried fruit, dead insects.
363 **Bee moth** *Aphomia sociella* (L.). Four. Larvae in nests of social wasps and bumblebees.
364 *Pyla fusca* (Haworth). Three. Larvae on *Erica* and *Vaccinium*.
365 *Acrobasis repandana* (Fab.). One. Larvae on *Quercus*.
366 *Numonia advenella* (Zincken). Two. Larvae on *Crataegus* and *Sorbus aucuparia*.
367 **Thistle ermine** *Myelois cribrella* (Hübner). One. Larvae on various thistles and burdock (*Asteraceae*).
368 *Euzophera pinguis* (Haworth). One. Larvae on inner bark of *Fraxinus*.
369 *Phycitodes binaevella* (Hübner). Four. Larvae on thistles (*Asteraceae*).
370 *Phycitodes maritima* (Tengstrom). One, MT. Larvae on flowers of *Achillea millefolium* and *Senecio jacobaea*.

Pterophoridae

371 *Platyptilia gonodactyla* (D. & S.). One. Larvae on *Tussilago* and possibly *Petasites*.
372 *Platyptilia ochrodactyla* (D. & S.). Two. Larvae in stems of *Tanacetum vulgare*.
373 **White plume moth** *Pterophorus pentadactyla* (L.). Common, July. Larvae on *Calystegia* and *Convolvulus*.
374 *Oidaematophorus lithodactyla* (Treitschke). Larvae on *Inula conyza* and *Pulicaria dysenterica*; one pupa found on *Populus* in garden.
375 *Emmelina monodactyla* (L.). Seven. Larvae on *Calystegia* and *Convolvulus* species; in garden, on *Calystegia sepium*.

ANNUAL FLUCTUATIONS IN NUMBERS OF MOTHS

Mercury-vapour light trapping was not done regularly or consistently, so the results can only be used qualitatively for describing variations from year to year in the abundance of moths. The Malaise trap, however, although not the ideal way of catching moths, was run continuously from 1 April to 31 October for every year 1972–2001. Moreover, since it uses no attractant, it is non-selective in the species that it intercepts in their flight. Tough, fast, active moths, such as hawk-moths, may be less likely to be caught in a Malaise trap, but they constitute only a small proportion of the garden moth fauna. If, however, this argument had much foundation, silver-Ys (*Autographa gamma*) might be expected to evade capture, whereas they were by far the commonest species in the total 30-year sample, outnumbering the next commonest, the large yellow underwing (*Noctua pronuba*), by more than two to one (N=1,469 and 619, respectively). Annual fluctuations are, therefore, represented here by changes from year to

Large yellow underwing, *Noctua pronuba* (above left), the second most frequent species in the Leicester garden

Orange swift moth, *Hepialus sylvina* (above right), was another frequent find

year in the Malaise trap catch. This is for macro-moths only; the figures and identifications are incomplete for micro-moths. The composition, in terms of species, of the annual catches is known completely only for 1974 onwards, largely because of initial problems in identifying moths caught in alcohol.

Annual fluctuations in moth numbers are shown in Fig. 4.1. By far the best years were 1973 and 1975; 1974 and 1976 were also good. The large catches for the period 1973–1976 were not matched later on, although 1985 and 1996 were reasonably good. The period 1973–1976 was also good for butterflies (Fig. 3.1), although butterfly numbers peaked in 1973, moth numbers in 1975. Moth numbers showed only slight improvement in 1978, compared with those for butterflies, and butterflies showed a marked increase in 1982–1984, which was not as marked in moths, although 1985 was a reasonably good moth year. The agreement between Figs. 3.1 and 4.1 is in the exceptionally poor figures for 1981, and this was also a low point in hoverfly annual catches (Fig. 5.1); evidently the dull but dry summer of 1981, following a dull, wet spring, was as unfavourable for moths as for butterflies and hoverflies. Otherwise, there is no clear correlation between moth numbers and weather factors, moth numbers appearing to decline, regardless of weather, from 1996 onwards.

The results of moth trapping by Rothamsted Research showed that the numbers of 338 species of macro-moths in Britain have decreased 32% since 1968, and in urban areas the loss has been 50%. The decline in numbers has been greater in southern than in northern England. In the north, some moth species that had hitherto been more abundant further south, have shown increases, perhaps as a consequence of climate change (Conrad *et al.* 2004, Conrad *et al.* 2006). My garden in the Midlands is situated in what the researchers regard as southern England, and accords with their findings in suffering a decline in moth numbers. In a report on the Rothamsted findings, habitat change in the countryside, particularly in southern England, is blamed for having had the major impact on moth numbers, although climate change is also implicated (Fox *et al.* 2006).

Autographa gamma was the commonest moth in 18 of the 30 annual Malaise trap samples, and joint commonest with *Hepialus sylvina* in one year; *A. gamma* formed 14.0% to 51.9% of the annual catch (Tab. 4.1). Although it is an immigrant every year, *A. gamma* breeds in the garden, thus augmenting the population. High numbers of *A. gamma* are a reflection of favourable conditions in the garden as well as the extent of immigration. Similarly, *Noctua pronuba*, a resident, was

Four common garden moths

1 Bright-line brown-eye, *Lacanobia oleracea*; its larvae eat a very wide variety of different plants in the Leicester garden

2 Common swift, *Hepialus lupulinus*, is similarly common; its larvae feed on roots of grasses and occasionally herbaceous plants

3 Garden carpet, *Xanthorhoe fluctuata*; its larvae feed on golden Alison (*Aurinia saxatilis*), garden arabis (*Arabis alpina* subsp. *caucasica*), cabbage (*Brassica oleracea*) and perennial candytuft (*Iberis sempervirens*) in the author's garden

4 Common rustic, *Mesapamea secalis*; its larvae feed on various grasses

Fig. 4.1. Annual fluctuations in numbers of moths caught in Malaise trap, 1972–2001

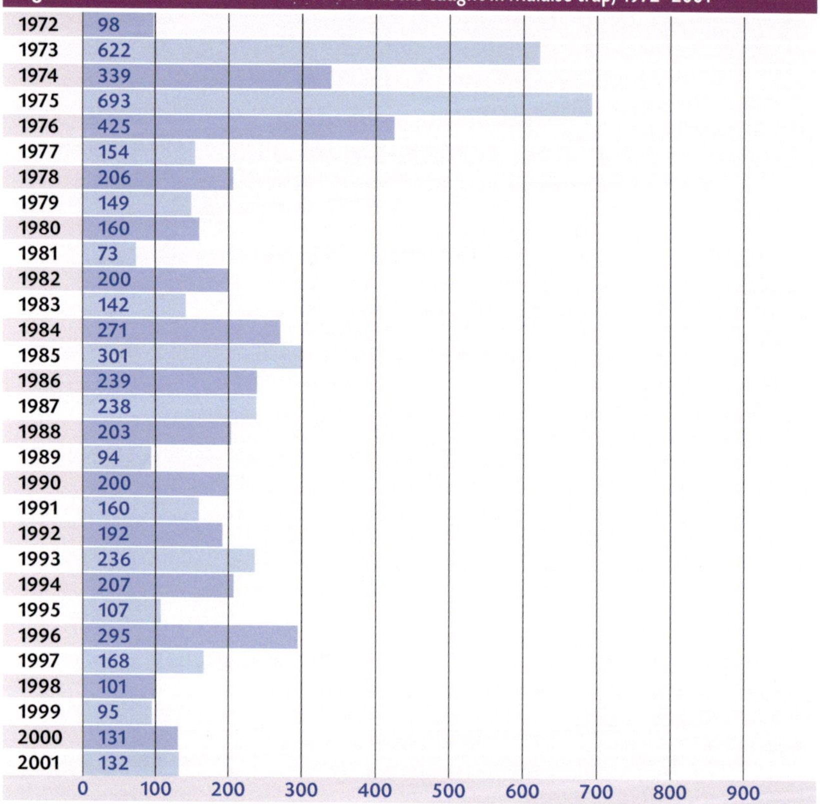

common in 1973 and 1975, 49 and 112, respectively, although it was also common in 1974 (63), 1978 (44), 1987 (41) and 1988 (62), in all of which years it was the commonest moth. *Hepialus lupulinus* and *Lacanobia oleracea* were each the commonest in two years, 1997 and 2001, and 1979 and 1995, respectively. *Mesapamea secalis* was the commonest in 1976, when 154 entered the Malaise trap; *Hepialus sylvina* was joint commonest in 1998, with *Autographa gamma*.

In the second 15-year period, 1987–2001, many fewer moths were caught in the Malaise trap, only 2,560, compared with 4,072 in 1972–1986, although, interestingly, the number of species did not go down. Eleven species other than *Autographa gamma* and *Noctua pronuba* contributed more than 50 to the total of 4,072 in 1972–1986: *Mesapamea secalis* (387), *Lacanobia oleracea* (260), *Xanthorhoe fluctuata* (169), *Noctua comes* (100), *Agrotis exclamationis* (93), *Caradrina morpheus* (85), *Hepialus lupulinus* (72), *Phlogophora meticulosa* (62), *Abrostola triplasia* (58), *Odontopera bidentata* (55) and *Idaea aversata* (51).

In the 15-year period 1987–2001, however, only eight species, other than *Autographa gamma* and *Noctua pronuba*, contributed more than 50; *Lacanobia oleracea* (168), *Hepialus lupulinus* (149), *Xanthorhoe fluctuata* (115),

Table 4.1. Size and composition of annual, 15-year and 30-year catches of macro-moths in Malaise trap, 1972–2001

Year	Individuals (N)	Species (S)	N/S	Species taken once only	Commonest species	As % N
1972	98	>15	n. a.	5	*Autographa gamma*	44.9
1973	622	>16	n. a.	4	*Autographa gamma*	37.5
1974	339	>22	n. a.	6	*Noctua pronuba*	18.6
1975	693	57	12.16	19	*Autographa gamma*	22.5
1976	425	49	8.67	17	*Mesapamea secalis*	36.2
1977	154	30	5.13	11	*Autographa gamma*	24.2
1978	206	39	5.28	16	*Noctua pronuba*	21.5
1979	149	38	3.92	15	*Lacanobia oleracea*	19.5
1980	160	33	4.85	13	*Autographa gamma*	15.0
1981	73	29	2.52	12	*Noctua pronuba*	13.7
1982	200	42	4.76	19	*Autographa gamma*	27.0
1983	142	36	3.94	13	*Autographa gamma*	19.1
1984	271	38	7.13	13	*Autographa gamma*	23.0
1985	301	35	8.60	11	*Autographa gamma*	31.2
1986	239	37	6.46	17	*Autographa gamma*	32.1
1987	238	31	7.68	10	*Noctua pronuba*	18.0
1988	203	28	7.25	11	*Noctua pronuba*	30.7
1989	94	26	3.62	11	*Noctua pronuba*	24.7
1990	200	45	4.44	21	*Autographa gamma*	15.3
1991	160	28	5.71	8	*Autographa gamma*	31.9
1992	192	38	5.05	15	*Autographa gamma*	15.9
1993	236	43	5.49	20	*Autographa gamma*	20.8
1994	207	37	5.59	15	*Autographa gamma*	27.3
1995	107	28	3.82	11	*Lacanobia oleracea*	16.5
1996	295	33	8.94	11	*Autographa gamma*	51.9
1997	168	30	5.60	8	*Hepialus lupulinus*	17.9
1998	101	28	3.61	11	*Autographa gamma* *Hepialus sylvina*	14.0=
1999	95	30	3.17	11	*Autographa gamma*	14.7
2000	131	28	4.68	15	*Autographa gamma*	28.8
2001	132	30	4.40	14	*Hepialus lupulinus*	31.1
1972–1986	4,072	>106	(38.42)	26	*Autographa gamma*	22.7
1987–2001	2,531	107	23.65	38	*Autographa gamma*	21.7
1972–2001	6,603	143	45.17	46	*Autographa gamma*	22.3

Elephant hawk-moth, *Deilephila elpenor*: larvae (showing dramatically different colouring) feeding on rosebay willowherb (right) and adult on lupin (far right)

Camptogramma bilineata (101), *Mesapamea secalis* (90), *Hepialus sylvina* (78), *Noctua comes* (63) and *Chloroclysta truncata* (52), the different composition of the catch suggesting that there were changes in the garden as a habitat.

The greatest annual number of species, 57, was taken in the Malaise trap in 1975, indicating that high catches were caused as much by addition of species as by increase in abundance of species. A total of 138 species was trapped over the 30 years. However, 17 of the species judged by mercury-vapour light trapping to be common in the garden, and listed as such in the previous section, were not caught in the Malaise trap. In the case of some, such as hawk-moths *Mimas tiliae* and *Smerinthus ocellata*, this may be a reflection of behaviour, but in most cases it is no more than chance and a small sample, reflecting the paucity of Malaise trap catches of moths relative to catches at light.

In the previous section, the 50 species recorded once only in the 30-year period, usually by mercury-vapour light trapping, are listed. Thirty-nine of these were caught before 1983, and only seven in the period 1990–2001. The peak year for rarities was 1976, when eight were recorded; this was the drought year, when many nectar-feeding insects wandered far from their usual haunts in search of relatively scarce food, and in the same year three unusual butterfly species were caught in the garden.

One question that is inevitably raised by the lack of good moth years since the mid-1970s is whether there might have been a reduction in the size of moth populations caused by increasing urbanization and disturbance. One of the factors believed by many to affect moths is the increase in outdoor lighting evident in such densely populated and urbanised countries as England. A consideration of the impact of outdoor lighting on moths (Frank 1988) acknowledged that it has many potentially adverse effects, but draws no firm conclusions on its effect on moth numbers.

That there was no marked resurgence in the hot, dry, bright summers of the 1990s, as Fig. 4.1 shows, suggests that there has been a longterm decline in moth numbers. Time alone will show whether this decline is reversible.

Some other common garden hawk-moths

1 Lime hawk-moth, *Mimias tiliae* (far left)

2 Eyed hawk-moth, *Smerinthus ocellata*; its larvae feed on crack willow (*Salix fragilis*) in the Leicester garden (above)

3 Poplar hawk-moth, *Laothoe populi*, a common garden moth, whose larvae feed on Lombardy poplar (*Populus nigra* 'Italica') and crack willow in the Leicester garden (left)

SEASONAL FLUCTUATIONS IN MOTH NUMBERS

For the 150 commoner garden moths, an indication of their seasonality has been given (by species) in the earlier section listing the species. This, however, indicates only their presence by month and not frequency, and for a quantitative view of seasonality it is necessary to look again at the Malaise trap catches. The Malaise trap jar was changed weekly, for convenience on a Sunday, and monthly catches were achieved by summing weeks to give the nearest fit to calendar months; this inevitably means that some months contained four weeks, others five, and the pattern changed from year to year. However, to judge from results obtained when hoverfly catches were summed in the same way, this makes no marked difference to the seasonal pattern that is obtained. The striking feature of this is the lack of consistency and the considerable differences in pattern between years.

During the first fifteen years, 1972–1986, moth numbers reached a peak in July in seven years, in August in six years, in 1984 the July and August figures were equally high, and in 1980 the peak was in June. However, in the second fifteen years, 1987–2001, moth numbers peaked in June in five years, in July in five years, and in August in five years; the June peaks were all in 1994 and later. As a consequence, a summation of the catches for the years 1972–1986 showed an August peak (1,402 out of 4,072), whereas a summation of the catches for the years 1987–2001 showed a June peak (667 out of 2,560). This would certainly suggest that the moth season has become earlier during the thirty years, probably caused by the slight but consistent rise in temperature.

When running a mercury-vapour light trap, it is easy to get the impression that July and August are always good for moths, because the light brings in such a good catch then Looking at the smaller, but perhaps less biased Malaise trap catch, however, it becomes clear not only that there are striking differences between years in the seasonal pattern of moth flight, but also that there have been changes in seasonality.

Larvae of the mullein, *Cucullia verbasci* (above), and of brimstone, *Opisthograptis luteolata* (above right), two moderately specialized feeders. Larvae of *C. verbasci* have been found on *Buddleja davidii*, and those of *O. luteolata* on Japanese quince (*Chaenomeles japonica*), hawthorn (*Crataegus monogyna*), cherry plum (*Prunus cerasifera*) and rowan (*Sorbus aucuparia*) in the Leicester garden

LARVAL FOODPLANTS OF MACRO-MOTHS

In all, 50 species of macro-moth were found breeding in the Leicester garden at one time or another during the 30-year study (1972–2001). The moths and their foodplants are listed and described in the lists on pp57–66. Of these 50, three are specialist feeders, five are moderately specialised, and 42 are more or less polyphagous. The three specialists that breed in the garden are *Eulithis mellinata* on *Ribes uva-crispa*, *Hypena proboscidalis* on *Urtica dioica*, and *Xylocampa areola* on *Lonicera periclymenum*. Moderately specialised are *Xanthorhoe fluctuata* on *Brassicaceae*, *Opisthograptis luteolata* on *Rosaceae*, *Laothoe populi* on *Salicaceae*, *Cucullia verbasci* on *Scrophulariaceae* (and on *Buddlejaceae*) and *Scoliopteryx libatrix* on *Salicaceae*. The group that is conspicuous by its absence from the garden breeding records is the one that uses grasses as foodplants, since lawn grass is usually mown.

Thus 16% of those moths that have bred in my garden are moderately or highly specialised, whereas over 37% of the commoner garden moths are to some extent specialised. This suggests that it is those species with restricted larval food requirements that are unlikely to breed in a garden. Conversely, 84% of the species that have bred in my garden are polyphagous, compared with only just over 60% of the commoner garden moths, implying that polyphages, more or less indiscriminate as to foodplant, are particularly successful in a garden.

Similarly, among the micros that have bred in the garden, eleven are polyphagous, seven are moderately specialised, and only one is specialised (on *Lonicera periclymenum*). *Cucullia verbasci*, however, is particularly interesting, having adopted *Buddleja davidii*, an alien species, as an alternative foodplant in gardens. *Buddleja* belongs to a family, *Buddlejaceae*, unrepresented in the native flora of northwestern Europe, and yet it is apparently acceptable as a foodplant to a specialist species. The reason for this appears to be a chemical resemblance between *Buddlejaceae* and *Scrophulariaceae*, which has led to the moth's acceptance of the alien family as suitable food. Nineteen moths, including two micros, have bred on *Buddleja* in my garden, but all the others are more or less polyphagous.

The four most polyphagous moths in the garden have fed on *Buddleja* at one time or another. Details of their foodplant species are given in the earlier list of common garden moths (pp57–61). *Phlogophora meticulosa* fed on 54 plant

species in the garden (plus one pot plant indoors); *Lacanobia oleracea* on 43, *Melanchra persicariae* on 40, and *Mamestra brassicae* on 38 (plus three pot plants indoors); the plants belong to many different families. These moths are exposed to a whole range of plant chemistry, particularly with regard to secondary plant substances. These are chemicals that play no part in a plant's life processes but have probably evolved as part of its defensive strategy against herbivores; they are its characteristic chemical signature. Species that are able to accommodate such chemically diverse foods seem to be able to do so because of the adaptive nature of the enzyme complement of their digestive systems (see Owen, J. 1991).

Three of the most polyphagous moths in the garden (above, left to right) were dot moth (*Melanchra persicariae*), cabbage moth (*Mamestra brassicae*) and bright-line brown-eye (*Lacanobia oleracea*) (larva above; see p68 for adult)

Why then should not all moth species be polyphagous as larvae? The answer has to be that different species have taken different evolutionary options. There are obvious drawbacks to specialization in that it sets restrictions on where and when a moth can find larval foodplants, but there are almost certainly physiological advantages in the efficiency that comes from being highly adapted to processing one type of food. If, however, competition between herbivores is strong, there is selective advantage in switching to an unrelated foodplant, whose insect fauna is taxonomically different and which has a different complex of predators and parasitoids associated with it.

One of the advantages of extreme polyphagy in the Leicester garden is that foodplants are available all the year round. *Lamium maculatum*, for instance, has green leaves throughout the year, and such moth caterpillars as those of *Phlogophora meticulosa* can be found on it in almost every month of the year. Specialists tend to be restricted to a short breeding season, but polyphages can breed at any time and hence have fewer restraints on population growth.

Foodplants used in the Leicester garden

Buddleja davidii is the most widely used plant species in the garden, despite its alien origin, with moth larvae of 19 species feeding on it. Next most widely used are shrubby cinquefoil (*Potentilla fruticosa*) and cherry plum (*Prunus cerasifera*) with nine each, *Rosa* and spotted dead-nettle (*Lamium maculatum*) with eight, crack willow (*Salix fragilis*), flowering currant (*Ribes sanguineum*) and hawthorn (*Crataegus monogyna*) with seven, Shasta daisy (*Leucanthemum* × *superbum*), gooseberry (*Ribes uva-crispa*) and marjoram (*Origanum majorana*) with six, and Michaelmas daisy (*Aster novi-belgii*), candytuft (*Iberis sempervirens*), common nettle (*Urtica dioica*) and parsley (*Petroselinum crispum*) with five.

1

2

3

4

5

6

7

8

9

10

11

12

Of the most widely used foodplants, on each of which caterpillars of five or more moth species feed, nine are aliens, and only six are native. This is despite the widely held belief that 'Introduced plants, such as garden forms, often have no herbivores at all and therefore no food-chains exploiting them' (Elton 1966). In large measure this belief arises because of the way that books on moths and on foodplants present data on foodplants, and on the way others have used this information. The books were written for naturalists and collectors; they have therefore concentrated on plants, such as blackthorn (*Prunus spinosa*), likely to yield rarities, and have largely ignored alien plants. Furthermore, they tend to single out tree and shrub species, and are much more general about herbaceous plants; the only exceptions are when citing the herbaceous foodplants of specialist-feeders or rarities. Little specific information is given about polyphages, especially those on herbaceous plants.

Unfortunately, it is just such books that have been used as sources by people such as Leather (1986), in compiling lists of herbivorous species utilizing particular plants. For instance, Leather does not record any Lepidoptera using *Potentilla fruticosa* as a foodplant, whereas, in my garden alone, nine moth species feed on it. In large part, this arises because *P. fruticosa*, a native, is rare in the wild and because few lepidopterists consider garden plants. However, it is widely grown in gardens, and, moreover, the cultivars differ little, if at all, from the wild type in foliage characteristics.

Moth foodplants. The plants most widely used by moth larvae in the Leicester garden:

1 *Buddleja davidii*

2 *Potentilla fruticosa*

3 *Prunus cerasifera*

4 *Rosa*

5 *Lamium maculatum*

6 *Salix fragilis*

7 *Ribes sanguineum*

8 *Crataegus monogyna*

9 *Leucanthemum × superbum*

10 *Ribes uva-crispa*

11 *Origanum majorana*

12 *Aster novi-belgii*

13 *Iberis sempervirens*

14 *Urtica dioica*

15 *Petroselinum crispum*

Native and alien foodplants

It has become generally accepted among urban conservation groups that 'native is best' when planting areas for habitat creation. The origins of this belief seem to be investigations by Southwood and others of the insect fauna associated with particular trees. They found that native willows (*Salix*), oaks (*Quercus*) and birches (*Betula*) support many insect species, whereas such common aliens as sycamore (*Acer pseudoplatanus*) and horse chestnut (*Aesculus hippocastanum*), although introduced several centuries ago, support relatively few (Kennedy & Southwood 1984, Southwood 1961).

Tree species abundance, number of years that the species has been present in Britain, and whether it is evergreen or coniferous, were found to account for 74% of the variation in insect species richness between British trees (Kennedy & Southwood 1984). It is assumed that the longer a tree species has been present, the more time herbivores have had to evolve adaptations to overcome its

A *Potentilla fruticosa* cultivar growing in the author's garden. Although *P. fruticosa* is a native species, its restricted distribution in the wild means it is seldom listed by lepidopterists as a foodplant for British native moths. However, nine species of moth larvae were found to be feeding on this single shrub

physical and chemical defences. Aliens are assumed to have defences that are unfamiliar and hence initially impregnable. This may well be true of trees, which are long-lived and therefore might be expected to have evolved and to develop strong defence, but there are at best tenuous grounds for extending this theory to herbaceous plants. Yet this is exactly what has happened, in that advice that 'native is best' when planting for wildlife has been extended to all plants.

Analysis of the use by herbivorous moth larvae of native and alien plants in the Leicester garden has shown that extensive planting of aliens is by no means a disaster for moths (Owen, D.F. 1986). I have shown that 46 species of moth used 40 native plants in the Leicester garden, and 38 species of moth used 75 alien plant species. Moths used more than 27% of the native plant species present in the garden, and 35% of the aliens present, suggesting that aliens are more heavily utilized, although the difference is not statistically significant.

The four most polyphagous of the moths that breed in the garden illustrate well this dependence on aliens. *Phlogophora meticulosa* has fed on 12 native and 42 alien species, *Lacanobia oleracea* on 7 native and 36 alien, *Melanchra persicariae* on 15 native and 25 alien, and *Mamestra brassicae* on 8 native and 30 alien species. These four feed mostly on herbaceous plants, but what of species that utilize trees and shrubs? *Odontopera bidentata* has been found on five shrubs (one native, four alien) and three trees (two native, one alien). *Ourapteryx sambucaria*, which is reported as feeding on a variety of trees and shrubs, mainly *Rosaceae*, occurs in the garden on *Ribes sanguineum* and *Ligustrum ovalifolium*, both aliens, and neither of them in the family *Rosaceae*.

Many of the native plants grown intentionally in the garden are cultivars, but they have usually been changed in either flower or fruit character and not in any way that would affect leaf-feeders. Thus, the cultivars of *Potentilla fruticosa* and *Rosa* that are grown in my garden are probably indistinguishable from wild forms to a herbivore. Since *P. fruticosa* is rare in the wild, it is scarcely surprising that it is rarely mentioned as a foodplant, but, in the Leicester garden, no fewer than nine species of moth use it. *Brassica oleracea* has been changed by selective breeding into what is probably a more palatable form, and at least two of the moth species that feed on it can sometimes become a pest problem in gardens and commercially.

Swallow-tailed moth, *Ourapteryx sambucaria*, a common garden moth whose larvae feed on garden privet (*Ligustrum ovalifolium*) and flowering currant (*Ribes sanguineum*) in the Leicester garden

There are more moth species on native than on alien plants in the garden, despite the fact that more alien than native plants are used by moth larvae. Twenty-five families of garden plants that are widely utilized by moth larvae include 98 native and 151 alien plants. Thirty-three of the native and 64 of the alien species are used as foodplants by a total of 68 species of moth, 46 on native and 38 on alien plants. The most heavily used plant family is *Rosaceae*, of which seven natives and five aliens are used by a total of 27 species of moth.

Second to this, however, is *Buddlejaceae*, an alien family, with just the one species, *Buddleja davidii*, used by 19 species of moth. In contrast, *Asteraceae*, the largest family of garden plants with 18 natives and 40 aliens, is used by 13 moth species, only two on natives and all 13 on aliens. The 24 species of *Lamiaceae* in the garden, 9 native and 15 alien, are used by 12 species of moth, three on native and all twelve on alien species. The families *Caprifoliaceae*, *Grossulariaceae* and *Salicaceae* are heavily used relative to their representation in the garden. Two native and one alien *Caprifoliaceae* (out of a total of four) are used by eight moth species; two native and one alien *Grossulariaceae* (out of a total of four) are used by eight moth species; and the three native *Salicaceae* are used by ten species of moth (Owen, J. 1991).

There is not strong evidence here that 'native is best'. On the contrary, there is every indication that alien plant species are widely acceptable as food for moth caterpillars. Some families of plants, such as *Lamiaceae*, *Grossulariaceae*, *Rosaceae* and *Salicaceae*, are evidently good as foodplants for moth caterpillars, but in general terms, the best way to make a garden attractive as a breeding site for moths is to diversify the vegetation.

5 : Hoverflies

Drone fly, *Eristalis tenax* (opposite), whose rat-tailed maggots feed on decaying organic material in water or in wet situations

Of the many flower-dependent groups of insects that abound in gardens, one of the most abundant and diverse is the family Syrphidae, the hoverflies. Moreover, few discrete, readily recognizable groups are as curious, or as interesting. Hoverfly larvae feed on such diverse foods as plants (including bulbs), aphids and other Homoptera, decaying organic material (including dead wood) and debris (including dead insects) in bee, ant and wasp nests, while the adults are flower-feeders, making them, as a group, excellent indicators of habitat diversity. Here, however, I must admit to some partiality, hoverflies being the garden group in which I am most interested, and on which I have spent most time. Those most often noticed in gardens are the larger species, many of which are brightly coloured, but from April until October a variety of small, slim, dark hoverflies can usually be found feeding at flowers or hovering in sunny spots or dappled shade.

Most hoverflies are glossy-bodied, and many have yellow bands or pairs of yellow spots (lunules) on the black abdomen; some are robust, others elongate in form; a few bear a striking resemblance to solitary or social wasps; some resemble honeybees; and several furry, polymorphic species are mimics of bumblebees. Hoverflies, nevertheless, constitute a family, in contrast to butterflies and moths, which encompass, respectively, several and many families; in other words, hoverflies are taxonomically less variable.

The behavioural feature of the family Syrphidae which leads to their being called hoverflies is the ability to hover seemingly motionless, the transparent wings a shimmering blur, then dart away so fast they seem to vanish, only to re-appear, often in the exact spot they left. They are predominantly nectar- and pollen-feeders, hence their American name of flower flies. Anatomically, members of the family can be distinguished from other Diptera by the vena spuria, a thickening of the wing membrane along the central axis of the wing, looking like a vein but unconnected to any others, and by well-developed marginal cross-veins which are more conspicuous than the flimsy edge of the wing.

NATURAL HISTORY OF HOVERFLIES

So varied are the life histories of hoverflies that it is difficult, and would be misleading, to generalize. The common denominator is the dependence of adults on energy-rich fluids, usually in the form of nectar. However, some smaller hoverflies, such as *Melanostoma* and *Episyrphus* species, seem to satisfy

their energy requirements by eating pollen, and *Xylota* species, whose fidgety behaviour reinforces their resemblance to solitary wasps, apparently collect pollen and other substances from leaf surfaces, and rarely visit flowers. I have often seen *Eupeodes luniger* (syn. *Metasyrphus luniger*) and *Syrphus ribesii* lapping honeydew from leaf surfaces on aphid-infested plants, and suspect that hoverflies, like many other nectar-feeding insects, are indiscriminate as to the source of fluids rich in sugars, salts and nutrients. I have caught ten species in traps baited with rotting fruit, and two of those ten in traps baited with fish offal.

Narcissus fly or large bulb fly, *Merodon equestris* (top), was the commonest hoverfly in the Leicester garden in three years in the late 1990s

The larvae of *Merodon equestris* feed on and in bulbs (above), and so are generally regarded as pests

Different species take nectar and pollen in different proportions when feeding at flowers, and have correspondingly adapted mouthparts. Species such as *Episyrphus balteatus*, which take predominantly pollen, have short probosces with broad, fleshy labella (terminal lobes); species such as *Rhingia campestris*, which more often take nectar, have long probosces with slim, elongate labella; and species such as *Eupeodes corollae*, which take nectar and pollen in more or less equal proportions, have mouthparts intermediate in length and structure (Gilbert 1981). It seems that, in all species, the protein component of pollen is necessary for maturation of eggs and probably also of sperm.

Small hoverflies, when feeding, pollinate plantains once thought to be wind-pollinated (Stellemann 1978), and it is highly likely that the larger, furry species play a major part in pollination, as pollen readily adheres to their hairs. Certainly, hoverflies can be trained to visit artificial flowers of particular colours by the provision of a reward in the form of sugar solution; such behaviour is a necessary prerequisite for a pollinator.

Larval stages

Since adults differing taxonomically and in size have similar food requirements, it is no wonder that gardens, with their abundant cultivated flowers, are good sites for hoverflies. Indeed, males of some species use flowers as courtship territories. There is, however, great divergence in larval habits and hence in habitat typically frequented by adults, particularly egg-laying females. Colyer & Hammond (1968) recognized five categories of larval habits: aphidivorous; frequenters of decaying organic matter; frequenters of sap from tree-wounds or of rotting wood; frequenters of plant stems and roots, or fungi; and dwellers in nests of bees, wasps and ants. I adopted these categories and expressed them in terms of the trophic level at which larvae operate (Owen, J. 1981).

The larval stages of many hoverflies are unknown, however, and such categorization depends on the assumption that taxonomically similar species have similar life histories. Furthermore, there is some overlap between the trophic levels of Colyer & Hammond's five categories, and it may be more realistic to follow Gilbert (1986) in recognizing four groups: larvae that eat plant tissues and plant products (including tree sap); larvae that feed on decaying organic material (including wood in rot-holes in trees); larvae that live in the nests of social insects, mainly as scavengers; and carnivorous larvae, most of which eat aphids.

The details of larval life are best known for the common aphid-feeding species and for those, such as *Eristalis tenax*, which feed on decaying organic material, often in water. It is clear that some species are facultative to some extent in their feeding: all *Platycheirus* (except *P. scutatus*), *Melanostoma mellinum* and *M. scalare* can complete their development on rotting plant material if no aphids are available (Gilbert 1986); and *E. tenax* is found in fully aquatic habitats, in sewage or dung, and in rotting matter of various sorts. Indeed, Samson's riddle in the *Book of Judges* (14:14) 'Out of the eater came forth meat, and out of the strong came forth sweetness' is often assumed to arise from mis-identification of *E. tenax*, breeding in putrid carcasses, as honeybees.

Eupeodes corollae (seen feeding at a Shasta daisy flower, *Leucanthemum* × *superbum*) was the commonest hoverfly overall in the period 1972-86, largely as a consequence of large-scale immigration to the garden in 1975 and 1976

One major habitat absent from suburban gardens is stands of old wood, and as a consequence, those hoverflies that breed in sap runs or rot-holes in trees are under-represented in the garden fauna. The group with herbivorous larvae is represented in gardens mainly by the narcissus fly *Merodon equestris* and the lesser bulb flies *Eumerus strigatus* and *E. tuberculatus*. In more rural localities, the large genus *Cheilosia* would be better represented in this category. *Merodon* and *Eumerus* are the only hoverflies of economic importance, inflicting considerable damage on commercial bulb-growing, but, although they are fairly common, their activities in gardens are rarely evident enough for them to rank as pests. Indeed, *Merodon* is a beautiful fly, and adds considerable interest to the garden fauna as it is highly polymorphic, the different colour forms mimicking different species of *Bombus*.

HOVERFLIES

During the period 1972–2001, 94 species of hoverflies were recorded in the garden. All appeared in Malaise trap samples, although many were never observed alive in the garden.

Chandler (1998) recognized 266 species of British hoverflies, and the following account of garden species follows the taxonomy and classification of this checklist.

The numbers given for each species refer to Malaise trap captures for the 30-year period 1972–2001 in a total sample of 60,736. Hoverflies were identified initially using Coe (1953) and later using Stubbs & Falk (1983).

Baccha. Occasionally seen in shady, overgrown parts of garden. Larvae feed on aphids.

1	***Baccha elongata*** (Fab.)	196

Cheilosia. Rarely seen in garden. Larvae feed on and in roots and stems of plants.

2	***Cheilosia albitarsis*** (Meigen)	9
3	***Cheilosia bergenstammi*** Becker	21
4	***Cheilosia illustrata*** (Harris)	1 (1988)
5	***Cheilosia nigripes*** (Meigen)	1 (1999)
6	***Cheilosia pagana*** (Meigen)	31
7	***Cheilosia proxima*** (Zetterstedt)	2
8	***Cheilosia vernalis*** (Fallén)	175

Chrysotoxum. Closely resemble *Vespula* wasps. Rarely seen in garden. It is assumed that the larvae feed on aphids, possibly root aphids, although there is no proof of this.

9	***Chrysotoxum bicinctum*** (L.)	13
10	***Chrysotoxum festivum*** (L.)	38
11	***Chrysotoxum verralli*** Collin	35

Criorhina. Robust, furry hoverflies resembling bumblebees. Larvae feed in dead wood. An additional individual of *C. berberina* was hand-netted in 1979.

12	***Criorhina berberina*** (Fab.)	1 (1976)
13	***Criorhina floccosa*** (Meigen)	1 (1972)

Dasysyrphus. Never seen alive in garden. Larvae feed on aphids. *Dasysyrphus pinastri* is called *D. lunulatus* in Stubbs & Falk (1983).

14	***Dasysyrphus albostriatus*** (Fallén)	197

15 ***Dasysyrphus pinastri*** (De Geer) 7
16 ***Dasysyrphus tricinctus*** (Fallén) 1 (1974)
17 ***Dasysyrphus venustus*** (Meigen) 5

Didea. Never seen alive in garden. Larvae feed on aphids.
18 ***Didea fasciata*** Macquart 1 (1996)

Epistrophe. *Epistrophe eligans* frequently seen in garden, hovering motionless in sunlight above lawn in early summer. Larvae feed on aphids.
19 ***Epistrophe eligans*** (Harris) 533
20 ***Epistrophe grossulariae*** (Meigen) 1 (1973)
21 ***Epistrophe nitidicollis*** (Meigen) 55

Episyrphus. *Episyrphus balteatus* common in garden and breeds there, the larvae feeding on *Brevicoryne brassicae* and other aphids. Large numbers regularly move into the garden in July and August, immigration being particularly conspicuous in some years.
22 ***Episyrphus balteatus*** (De Geer) 9,414

Eristalinus. Larvae feed on decaying organic material. The aquatic larvae are 'rat-tailed', i. e. the posterior spiracles are extended into a long, telescopic breathing tube.
23 ***Eristalinus sepulchralis*** (L.) 6

Eristalis. Most, particularly the hairier species, bear some resemblance to bees; the drone-fly *E. tenax* is a honeybee mimic, and *E. intricarius*, polymorphic for colour and pattern, resembles various species of bumblebees. They are conspicuous at garden flowers in late summer and autumn. The aquatic, 'rat-tailed' larvae feed on decaying organic material. *Eristalis tenax* and *E. pertinax* are under-represented in Malaise trap samples, perhaps because they are strong fliers. *Eristalis interruptus* is called *E. nemorum* in Stubbs & Falk (1983); name change due to Torp (1984).
24 ***Eristalis abusivus*** Collin 2 (1973)
25 ***Eristalis arbustorum*** (L.) 1,862
26 ***Eristalis horticola*** (De Geer) 47
27 ***Eristalis interruptus*** (Poda) 89
28 ***Eristalis intricarius*** (L.) 55
29 ***Eristalis pertinax*** (Scopoli) 338
30 ***Eristalis tenax*** (L.) 352

Eumerus. Frequently common in Malaise trap samples but rarely seen in garden. Known as lesser bulb flies because the larvae feed on and in bulbs. Although reputedly causing considerable losses, *Eumerus* species are generally regarded as secondary pests with larvae feeding in bulbs already damaged by other pests or diseases.
31 ***Eumerus strigatus*** (Fallén) 894
32 ***Eumerus tuberculatus*** Rondani 1,348

Eupeodes. *Eupeodes* is the *Metasyrphus* of Stubbs & Falk (1983); name change due to Torp (1984). Large-scale immigration of *E. corollae* to the garden during August 1975 and July 1976. Larvae feed on aphids and *E. luniger* breeds in garden, the larvae feeding on potato aphid, *Macrosiphum euphorbiae*, and other aphids.
33 ***Eupeodes corollae*** (Fab.) 7,309
34 ***Eupeodes latifasciatus*** (Macquart) 247
35 ***Eupeodes latilunulatus*** (Collin) 10
36 ***Eupeodes luniger*** (Meigen) 1,388

Ferdinandea. Larvae live in sap runs on trees.
37 ***Ferdinandea cuprea*** (Scopoli) 6
38 ***Ferdinandea ruficornis*** (Fab.) 1 (1973)

Helophilus. *Helophilus pendulus* is frequently seen basking in sunlight beside the garden pond. The aquatic 'rat-tailed' larvae feed on decaying organic material.
39 ***Helophilus hybridus*** Loew 14
40 ***Helophilus pendulus*** (L.) 950
41 ***Helophilus trivittatus*** (Fab.) 2

Heringia. Larvae feed on gall-making aphids.
42 ***Heringia heringii*** (Zetterstedt) 10

Lejogaster. Larvae feed on and in decaying organic material.
43 ***Lejogaster metallina*** (Fab.) 19

Leucozona. Larvae feed on aphids.
44 ***Leucozona lucorum*** (L.) 40

Melangyna. Larvae feed on aphids.
45 ***Melangyna cincta*** (Fallén) 23
46 ***Melangyna lasiophthalma*** (Zetterstedt) 12
47 ***Melangyna umbellatarum*** (Fab.) 10

Melanogaster. Aquatic larvae which feed on decaying organic material, and tap air spaces in underwater plants for oxygen with modified spiracles.
48 ***Melanogaster hirtella*** (Loew) 1 (1982)

Melanostoma. Adults frequent at flowers. Larvae are facultative aphid-feeders, able to complete their development on rotting vegetation; both *M. mellinum* and *M. scalare* breed in garden.
49 ***Melanostoma mellinum*** (L.) 4,915
50 ***Melanostoma scalare*** (Fab.) 1,397

Meligramma. Larvae feed on aphids.
51 ***Meligramma trianguliferum*** (Zetterstedt) 9

Meliscaeva. Larvae feed on aphids; *M. auricollis* breeds in garden, the larvae feeding on *Adelges abietis*, an adelgid.
52 ***Meliscaeva auricollis*** (Meigen) 227
53 ***Meliscaeva cinctella*** (Zetterstedt) 3

Merodon. Hairy species occurring in many colour forms, most of which resemble different species of bumblebee. The inheritance of the colour patterns has been investigated by Conn (1972 a & b). Conspicuous on flowers or hovering above them in June and early July when numbers of bumblebees are at a peak. Known as narcissus flies or large bulb flies and regarded as horticultural pests because the larvae feed on and in bulbs.
54 ***Merodon equestris*** (Fab.) 1,721

Myathropa. The 'rat-tailed' larvae feed on decaying

organic material in rot-holes and other water-filled cavities in trees. Under-represented in Malaise trap samples, perhaps because it is a strong flier.

55 ***Myathropa florea*** (L.) 18

Neoascia. Larvae feed on and in decaying organic material.

56 ***Neoascia podagrica*** (Fab.) 94

Neocnemodon. Never seen alive in garden. Larvae feed on woolly aphids and on gall-making aphids.

57 ***Neocnemodon vitripennis*** (Meigen) 119

Paragus. Adults never seen alive in garden. Larvae feed on aphids. Prior to 1982, Coe (1953) was used for identification, leading to designation of all specimens as *P. tibialis* (Fallén). After that, the key in Speight (1978) was used, leading to designation of all males as *P. haemorrhous*. It is assumed a) that all females captured since 1981 were also *P. haemorrhous*, and b) that all individuals taken prior to 1982 were *P. haemorrhous* – certainly the voucher specimen is.

58 ***Paragus haemorrhous*** Meigen 157

Parasyrphus. Larvae feed on aphids.

59 ***Parasyrphus malinellus*** (Collin) 1 (1980)
60 ***Parasyrphus punctulatus*** (Verrall) 5

Pipiza. Never seen alive in garden. Larvae feed on aphids. The taxonomy of this genus is confused and confusing.

61 ***Pipiza austriaca*** Meigen 12
62 ***Pipiza bimaculata*** Meigen 2
63 ***Pipiza fenestrata*** Meigen 2
64 ***Pipiza luteitarsis*** Zetterstedt 6
65 ***Pipiza noctiluca*** (L.) 177

Pipizella. Larvae feed on root aphids. *Pipizella viduata* is called *P. varipes* in Stubbs & Falk (1983); name change due to Thompson, Vockeroth & Speight (1982).

66 ***Pipizella viduata*** (L.) 23

Platycheirus. Adults of many species are frequent at flowers or basking on leaves. Larvae of all but *P. scutatus* are facultative aphid-feeders, able to complete their development on rotting vegetation; *P. scutatus* is thought to be an obligate aphid-feeder. *P. albimanus* and *P. scutatus* breed in garden; *P. scutatus* found feeding on *Brevicoryne brassicae*, *Myzus cerasi* and other aphids. *Platycheirus albimanus* is called *P. cyaneus* in Owen (1991).

67 ***Platycheirus ambiguus*** (Fallén) 498
68 ***Platycheirus albimanus*** (Fab.) 7,399
69 ***Platycheirus angustatus*** (Zetterstedt) 533
70 ***Platycheirus clypeatus*** (Meigen) 2,472
71 ***Platycheirus immarginatus*** (Zetterstedt) 11
72 ***Platycheirus manicatus*** (Meigen) 545
73 ***Platycheirus peltatus*** (Meigen) 448
74 ***Platycheirus scambus*** (Staeger) 14
75 ***Platycheirus scutatus*** (Meigen) 4,894

Pyrophaena. Adults never seen alive in garden. Larvae feed on aphids.

76 ***Pyrophaena granditarsa*** (Forster) 43

Rhingia. Often seen at flowers, especially *Buddleja*. Larvae feed on and in cow dung.

77 ***Rhingia campestris*** Meigen 516

Riponnensia. Larvae on and in decaying organic material. *Riponnensia splendens* called *Orthonevra splendens* in Stubbs & Falk (1983).

78 ***Riponnensia splendens*** (Meigen) 4

Scaeva. Larvae feed on aphids.

79 ***Scaeva pyrastri*** (L.) 52
80 ***Scaeva selenitica*** (Meigen) 4

Sericomyia. The 'rat-tailed' larvae feed on and in wet decaying organic material.

81 ***Sericomyia silentis*** (Harris) 1 (1987)

Sphaerophoria. Certain identification of most *Sphaerophoria* is possible for males only. Larvae feed on aphids and *S. scripta* breeds in the garden; it is occasionally abundant and conspicuous on flowers, possibly as a result of immigration.

82 ***Sphaerophoria interrupta*** (Fab.) 69
83 ***Sphaerophoria rueppellii*** (Wiedemann) 19
84 ***Sphaerophoria scripta*** (L.) 2,696

Syritta. Common in the garden in summer and early autumn, the males defending territories and courting females over flowers. Larvae feed on decaying organic material.

85 ***Syritta pipiens*** (L.) 2,710

Syrphus. *Syrphus ribesii* is one of the most conspicuous garden hoverflies. Larvae feed on aphids, and females of *S. ribesii* have several times been seen laying eggs among aggregations of aphids on garden plants. Larvae have been found feeding on *Aphis sambuci*, *Betulaphis quadrituberculata*, *Brevicoryne brassicae*, *Myzus cerasi* and other aphids.

86 ***Syrphus ribesii*** (L.) 2,558
87 ***Syrphus torvus*** Osten-Sacken 2
88 ***Syrphus vitripennis*** Meigen 478

Triglyphus. Larvae feed on aphids.

89 ***Triglyphus primus*** Loew 1 (1985)

Tropidia. Larvae feed on decaying organic material.

90 ***Tropidia scita*** (Harris) 1 (1973)

Volucella. *Volucella bombylans* is hairy and occurs in three colour forms mimicking different species of bumblebees. Larvae feed as scavengers, occasionally as predators in nests of social wasps and bumblebees.

91 ***Volucella bombylans*** (L.) 17
92 ***Volucella pellucens*** (L.) 37

Xanthogramma. Larvae feed on aphids.

93 ***Xanthogramma pedissequum*** Harris 27

Xylota. Larvae feed in dead wood and other decaying vegetable matter.

94 ***Xylota segnis*** (L.) 6

Eupeodes luniger lapping honeydew excreted by aphids on rosebay willowherb, *Chamerion angustifolium*

DIVERSITY AND ABUNDANCE OF HOVERFLIES

The abundance and diversity of the garden hoverfly fauna is evident from the total 30-year sample of 60,736 individuals of 94 species taken in the Malaise trap, which collects only a proportion of the flying insects that spontaneously enter its airspace. Differences in flight behaviour may mean that some species are more likely than others to be caught. For instance, large, active species, such as *Eristalis tenax* and *Myathropa florea*, seem to be under-represented in the catch, although this is a subjective view based on the numbers seen in the garden. On balance, however, it is a reasonable assumption that the trap collected an unselected sample.

The Malaise trap catch of hoverflies might be expected to show a sex ratio approximating to one, but overall many more females than males were caught, particularly of species with aphid-feeding larvae, females of which may have been looking for egg-laying sites. The handful of species where males outnumbered females were mainly species that resemble bees or wasps whose males use flowers as courtship territories, and also the well-known migrant *Episyrphus balteatus*.

Table 5.1. Size and composition of annual, 15-year and 30-year catches of hoverflies in Malaise trap, 1972–2001

Year	Individuals (N)	Species (S)	N/S	Species taken once only	Commonest species	As % N
1972	1,339	47	28.5	10	*Platycheirus albimanus*	18.7
1973	3,164	62	51.0	14	*Platycheirus scutatus*	19.5
1974	2,505	53	47.3	11	*Platycheirus albimanus*	31.5
1975	6,363	54	117.8	8	*Eupeodes corollae*	40.6
1976	3,411	55	62.0	16	*Eupeodes corollae*	39.8
1977	2,586	37	69.9	8	*Episyrphus balteatus*	44.2
1978	6,068	50	121.4	8	*Platycheirus albimanus*	20.4
1979	3,488	51	68.4	14	*Melanostoma mellinum*	23.3
1980	1,327	53	25.0	13	*Platycheirus albimanus*	18.8
1981	885	44	20.1	9	*Platycheirus albimanus*	32.1
1982	4,256	52	81.7	10	*Melanostoma mellinum*	27.2
1983	1,283	47	27.3	8	*Eumerus tuberculatus*	18.3
1984	2,064	48	43.0	11	*Sphaerophoria scripta*	19.6
1985	3,553	51	69.7	9	*Episyrphus balteatus*	30.2
1986	1,457	43	33.9	9	*Episyrphus balteatus*	29.9
1987	2,868	41	70.0	4	*Episyrphus balteatus*	29.0
1988	1,017	44	23.1	10	*Platycheirus albimanus*	23.8
1989	965	39	24.7	6	*Platycheirus albimanus*	18.5
1990	1,054	40	26.4	6	*Syrphus ribesii*	22.6
1991	861	36	23.9	8	*Sphaerophoria scripta*	27.6
1992	1,231	39	31.6	3	*Episyrphus balteatus*	27.0
1993	1,513	46	32.9	8	*Platycheirus albimanus*	19.0
1994	815	41	19.9	9	*Episyrphus balteatus*	30.0
1995	422	36	11.7	9	*Merodon equestris*	17.0
1996	476	38	12.5	9	*Merodon equestris*	25.0
1997	3,224	48	67.2	15	*Episyrphus balteatus*	45.5
1998	417	36	11.6	13	*Merodon equestris*	13.4
1999	787	48	16.4	12	*Episyrphus balteatus*	29.2
2000	539	41	13.2	14	*Platycheirus albimanus*	18.2
2001	798	41	19.5	13	*Episyrphus balteatus*	37.2
1972–1986	43,749	91	480.8	11	*Eupeodes corollae*	13.9
1987–2001	16,987	73	232.7	8	*Episyrphus balteatus*	22.6
1972–2001	60,736	94	646.1	13	*Episyrphus balteatus*	15.5

Of the 94 species, 21 were caught only in the 15-year period 1972–86, four were caught only in the 15-year period 1987–2001, and 69 were caught throughout the 30 years. This would tend to suggest that the hoverfly fauna was richer during the first 15 years. Analysis of annual changes in abundance and composition of the catch deals necessarily with common species, but rare species and occasional visitors also become involved in ecological relationships, however transient, and must be taken into account in any description of the fauna.

Myathropa florea, a large hoverfly whose aquatic larvae live in water-filled rot-holes in trees

Fourteen species were each taken in the Malaise trap in one year only, and thirteen of these were represented by only one individual. A hoverfly species represented by a single individual in an unselected sample of 60,736 is an undisputed rarity; 1973 emerges as a year for rarities (see species list) but even in 1999 something different turned up (*Cheilosia nigripes*). It seems that many hoverflies are genuinely rare, at least in a suburban garden.

The rate of addition of new species to the garden list fell dramatically after 1973 (when it reached 61) but the additional 28 years of trapping meant the addition of 33 species, or 35% of the total. Despite the drop in rate of addition of new species after 1973, every year brought some change in the composition of the catch, so that interest and anticipation persisted year after year.

Fig. 5.1. Annual fluctuations in numbers of hoverflies caught in Malaise trap, 1972–2001

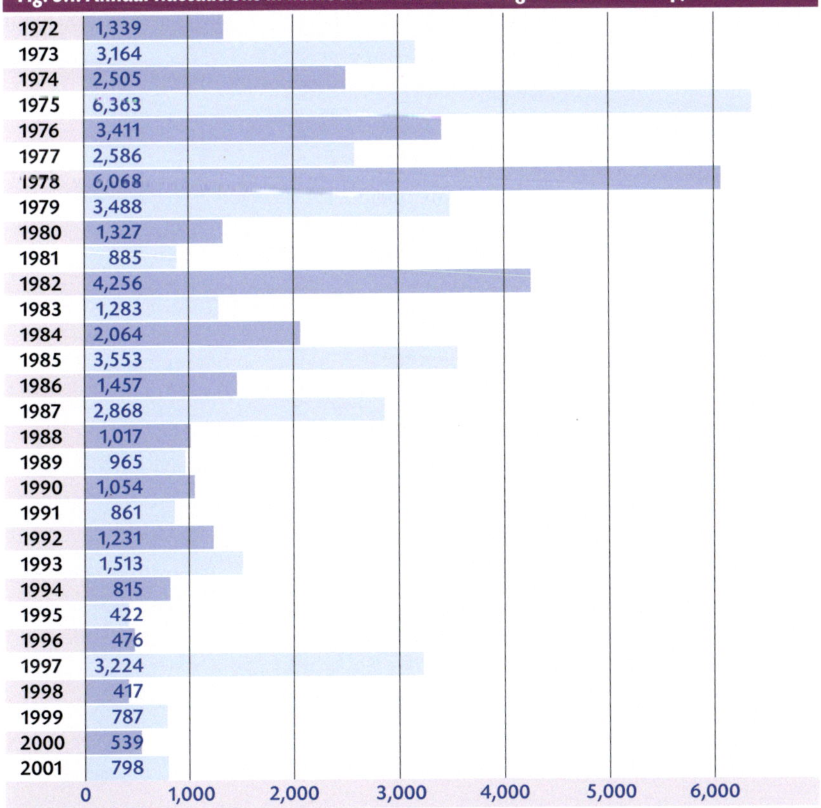

ANNUAL FLUCTUATIONS IN NUMBERS AND IMMIGRATION OF HOVERFLIES

The catch varied considerably from year to year, as shown in Tab. 5.1 and Fig. 5.1. The years 1975 and 1978 stand out as being particularly good for hoverflies, with 6,363 and 6,068 individuals trapped, respectively. In 1975, vast numbers of hoverflies appeared in gardens in early August, when it was unusually hot and sunny, and 'plagues' of them were reported on the east coast of Britain. But there was no conspicuous movement or influx in 1978, which was simply a 'good' year for hoverflies.

The predominant species in August 1975 was *Eupeodes corollae*, which constituted over 40% of the total catch. *Eupeodes corollae* formed almost as large a proportion of the catch in 1976, an unusually dry year. In 1977, however, when the total catch was average and the number of species low, *Episyrphus balteatus*, a well-known migrant, was so abundant that it contributed 44.2% of the catch. The only comparable year was 1997, when *E. balteatus* formed 45.5% of the catch; in 1997, however, N (the number of individuals) was high compared with every year since the mid-1980s and subsequent years, and the number of species (S) was quite high. 1981 (N=885) was a particularly poor year for hoverflies until the late 1980s and subsequently, when numbers declined dramatically, only 417 being trapped in 1998.

There are no clear associations between temperature and rainfall and the annual hoverfly catch. *Episyrphus balteatus* and *Platycheirus albimanus* were each the commonest species in nine years, *Merodon equestris* in three, *Eupeodes corollae, Melanostoma mellinum* and *Sphaerophoria scripta* each in two, and *Platycheirus scutatus, Eumerus tuberculatus* and *Syrphus ribesii* each in one.

Fluctuating numbers

The annual total of species was highest in 1973 (62), a long, hot summer, and has tended to decline since, being only 36 in 1991, 1995 and 1998. The relationship between number of individuals (N) and number of species (S) was not constant, so that N/S, the average number of individuals per species, ranged from 11.6 in 1998 to 121.4 in 1978. However, there is a strong correlation between N/S and N, showing that overall abundance is a result of increase in numbers of species already present, rather than addition of new species. This is supported by the relatively small number of species, eight, taken once only in 1975 and 1978, as opposed to a maximum of 16 in 1976.

1976 was a year of extreme drought conditions: by August, gardens were withered and the countryside parched and brown, and many nectar-feeding insects wandered far from their usual haunts in search of food. However, the number of species taken once only also rose spectacularly to 15 in 1997, which was not a drought year, nor a particularly hot and dry summer.

When the immigrants *Episyrphus balteatus* and *Eupeodes corollae* are excluded from the calculation of N/S in those years when immigration was obvious, the value of N/S, the average number of individuals per species, falls markedly and is probably more meaningful. Thus, in 1975, N/S falls from 117.9 to 58.4, and in 1997 from 67.2 to 33.3. The exclusion of *E. balteatus* and *E. corollae* from the calculations in immigration years meant that *Platycheirus albimanus* was the commonest species in a further four years, making it the commonest garden hoverfly in 13 of the 30 years. It also makes *Platycheirus scutatus* and *Melanostoma mellinum* each the commonest in a further three years, bringing the years in which they were each the commonest hoverfly up to four and five years, respectively, and it makes *Merodon equestris* the commonest in a further one year, bringing to four the number of the 30 years that it was the commonest garden hoverfly.

Episyrphus balteatus feeding at flowers of poppy, *Papaver*. It not only breeds in the study garden but is also involved in mass immigration such that in nine of the study years it was the most frequently observed hoverfly in the garden

Factors affecting the totals

Annual fluctuations in the total number of hoverflies caught may be the result of changes in the abundance of one or many species; several different patterns in annual variation are evident. Immigration was important in some years, particularly in 1975, 1976, 1977, 1985, 1986, 1987, 1992, 1994, 1997, 1999 and 2001, affecting numbers of *Episyrphus balteatus* in general, and of *Eupeodes corollae* in 1975 and 1976 as well as other species with aphid-feeding larvae. Both these species breed commonly on agricultural land, and in fine, warm springs and summers population explosions of their aphid food lead to a build-up in numbers. Depletion of their food supply caused by their predatory activities, or, perhaps in 1975, 1977 and other years by crop-harvesting, and in 1976 by withering of plants and death of aphids, led to mass movements away from the countryside bringing vast influxes to gardens with their abundant food resources. *Episyrphus balteatus* is also common in continental Europe, and some immigration emanates from across the Channel and proceeds northwards.

The garden hoverfly species captured in the Malaise trap more than 1,000 times in the 30-year period 1972–2001 were by no means all common every year (Tab. 5.2). The annual numbers of *Platycheirus albimanus*, for instance, the most abundant garden hoverfly apart from the immigrants, varied from a low of five in 1996, to a high of 1,238 in 1978. The numbers of the ecologically very similar *Platycheirus scutatus* were also at their lowest in 1996 and highest in 1978. *Platycheirus clypeatus* reached a peak of 650 in 1982, but was absent altogether in 1995, 1996 and 1998.

1978 was the best year but one (cf. 6,363 in 1975) for garden hoverflies (N=6,068), and there was no conspicuous immigration. *Melanostoma mellinum*, similar to *Platycheirus*, also reached a high in 1982 which was overall a good hoverfly year, whereas in 1977, only two *M. mellinum* were caught and no *M. scalare*, another usually common garden hoverfly.

Larval habits

Sphaerophoria scripta, *Syrphus ribesii* and *Eupeodes luniger*, all typical garden hoverflies, have, like *Platycheirus* and *Melanostoma*, aphid-feeding larvae. All showed considerable variation in annual catch, *S. scripta* varying from seven in 1995 to 404 in 1984, *S. ribesii* from only one in 2001, to 409 in 1987, and *E. luniger* from six in 1986 and 1994 to 123 in 1978 (Tab. 5.2).

The other species whose annual totals are shown in Tab. 5.2 do not have predatory larva. The larvae of *Eristalis arbustorum* and *Syritta pipiens* feed on decaying organic material. *Eristalis arbustorum* was most abundant in 1973 (362) and least in 2000 and 2001 (1 in each year), whereas *S. pipiens* was most abundant in 1978 (610) and least in 2001 (3). The larvae of *Merodon equestris* and *Eumerus tuberculatus* are herbivores, feeding in bulbs. This is a food source

Table 5.2. Annual variations in numbers of hoverflies caught more than 1,000 times in the 30-year period 1972–2001

Table 5.2 (1). 1972–1981

Species	1972	1973	1974	1975	1976	1977	1978	1979	1980	1981
Platycheirus albimanus	250	465	789	402	129	100	1,238	452	249	284
Platycheirus clypeatus	53	116	136	393	105	3	401	286	16	19
Platycheirus scutatus	205	618	358	407	327	243	1,155	243	169	89
Melanostoma mellinum	200	249	112	65	49	2	266	811	98	35
Melanostoma scalare	72	46	98	20	10	-	89	163	55	66
Sphaerophoria scripta	8	32	28	96	40	52	356	86	27	23
Syrphus ribesii	5	77	16	316	17	77	260	130	51	19
Eupeodes corollae *	17	50	165	2,580	1,359	415	114	272	66	73
Eupeodes luniger	56	60	119	52	38	54	123	102	20	16
Episyrphus balteatus *	82	126	103	746	402	1143	459	299	26	12
Eristalis arbustorum	29	362	10	75	21	53	336	74	69	11
Merodon equestris	10	25	32	34	66	32	48	49	42	13
Syritta pipiens	35	146	106	240	287	146	610	155	56	32
Eumerus tuberculatus	5	118	58	145	68	15	12	8	3	3

Table 5.2 (2). 1982–1991

Species	1982	1983	1984	1985	1986	1987	1988	1989	1990	1991
Platycheirus albimanus	205	182	112	439	95	476	242	178	153	33
Platycheirus clypeatus	650	35	28	16	85	31	11	9	25	1
Platycheirus scutatus	98	108	133	124	77	109	66	28	31	17
Melanostoma mellinum	1,159	46	179	265	305	64	118	10	22	16
Melanostoma scalare	120	16	13	78	37	42	55	13	16	20
Sphaerophoria scripta	65	81	404	270	15	47	25	139	89	238
Syrphus ribesii	20	45	10	364	19	409	5	68	238	46
Eupeodes corollae *	414	58	221	191	70	233	12	83	40	99
Eupeodes luniger	14	11	35	67	6	105	7	41	74	66
Episyrphus balteatus *	473	20	189	1,073	435	831	18	27	33	82
Eristalis arbustorum	164	62	144	117	14	46	62	24	20	52
Merodon equestris	68	34	75	51	100	37	59	79	46	23
Syritta pipiens	273	151	178	39	19	39	25	13	13	6
Eumerus tuberculatus	78	235	119	33	33	6	13	39	40	30

Table 5.2 (3). 1992–2001

Species	1992	1993	1994	1995	1996	1997	1998	1999	2000	2001
Platycheirus albimanus	28	287	122	39	5	171	53	96	98	27
Platycheirus clypeatus	10	13	19	–	–	4	–	5	1	1
Platycheirus scutatus	13	61	46	12	6	54	16	33	34	14
Melanostoma mellinum	157	62	29	46	18	424	16	22	21	49
Melanostoma scalare	22	28	26	5	10	80	38	49	50	60
Sphaerophoria scripta	34	40	9	7	30	312	52	19	56	16
Syrphus ribesii	83	180	18	2	11	32	13	14	12	1
Eupeodes corollae *	238	99	43	41	27	225	29	23	11	41
Eupeodes luniger	76	59	6	18	19	84	14	14	25	7
Episyrphus balteatus *	332	111	243	40	80	1467	4	230	31	297
Eristalis arbustorum	12	49	3	14	17	13	5	2	1	1
Merodon equestris	56	92	63	72	119	110	56	80	56	94
Syritta pipiens	11	28	24	8	11	36	4	5	11	3
Eumerus tuberculatus	18	28	43	42	24	41	7	31	22	31

* immigrant species

that probably does not vary much from year to year, unlike aphids and decaying organic material, supply of which depend in different ways on weather conditions. Annual numbers of *M. equestris* vary less than those of other common garden hoverflies, and the range is only from 10 (1972) to 100 (1986); it was one of very few species to remain abundant in 2001. However, numbers of *E. tuberculatus* did fluctuate widely, from 3 in 1980 and 1981 to 235 in 1983. Since the 14 species shown in Tab. 5.2 are the most abundant garden hoverflies, making up more than 85% of the total 30-year catch, it is axiomatic that some or most of them were common in years when the total catch (N) was high, and that at least some were very scarce in years when N was low. However, annual variation was by no means the same for all species.

Less-abundant species

There were marked changes in the abundance of some of the less-common garden species. The overall picture is of a decline in numbers in some species, such as *Chrysotoxum* species and *Eumerus strigatus* since the 1970s, and in others, such as *Eristalis tenax*, *Rhingia campestris* and *Platycheirus ambiguus*, since the 1980s. However, *Epistrophe eligans* was abundant every year in early summer, and *Epistrophe nitidicollis* became more regular in the 1990s. One of the most interesting occurrences was that of *Xanthogramma pedissequum*, a beautifully marked black and yellow hoverfly, which was taken once in 1972, then not again until 1992, after which it was regularly captured in the Malaise trap.

Xanthogramma pedissequum was caught once in 1972, then not again until 1992, after which it was caught regularly

The influx of *Eupeodes corollae* in 1975 and 1976 and of *Episyrphus balteatus* in 1977 was associated with a dramatic decline in numbers of *Melanostoma* species and, to a lesser extent, of *Platycheirus* species. In 1978, when the numbers of *E. corollae* and *E. balteatus* fell, those of *M. mellinum* and, particularly, of *P. albimanus* and *P. scutatus* reached unprecedented levels. However, despite this appearance of reciprocal fluctuations between different species whose larvae are predators of aphids, a detailed computer analysis of population dynamics and ecomorphological relationships has shown that hoverfly species respond independently to fluctuations in essential resources (Gilbert & Owen, J. 1990). There is little if any evidence of interactions between species and, indeed, the concept of a garden 'community' of hoverflies seems at best notional, the so-called community being merely a coincidence of species in space and time. Assessing the 15-year period 1972–86, morphologically similar species (that are, presumably, in competition) were not inversely correlated in population size (Gilbert & Owen, J., *op. cit.*).

Relative stability

The variation from year to year of the total catch and its composition means that it is hard to generalize about the garden hoverfly fauna. Certainly it would be misleading to base assumptions on two or three years' investigation, which suggests that the results of student projects, spanning three years or less, have to be treated with some caution. Detailed computer analysis of the first 15 years' garden hoverfly catch, 1972–86, led to the somewhat unexpected conclusion that hoverfly populations are more stable than those of other invertebrate groups. There was no difference in the abundance and stability in general of those whose larvae are predators, phytophages and saprophages. Specialization of larval habitat leads to lower abundance, but has no effect on stability.

Among predators, however, extreme specialists, such as *Platycheirus ambiguus*, are fewest, and generalists, such as *Syrphus ribesii*, are most abundant; both have less stable populations than moderate specialists such as *Platycheirus scutatus*. This may be because the numbers of specialist predators are tied to the variability of their prey, whereas generalists tend to exploit irregular outbreaks of potential prey (Owen, J. & Gilbert 1989).

Eristalis arbustorum, a common garden hoverfly, the rat-tailed maggots of which feed on decaying organic material in aquatic situations

SEASONALITY OF HOVERFLIES

The hoverfly season in central England runs effectively from the beginning of April to the end of October, slightly earlier in some years, although most species fly for only part of the season. Those which hibernate as adults, such as *Eristalis tenax*, may occasionally be seen basking in the sun on a mild winter's day, but this is exceptional.

August is usually the month of greatest activity, numbers trapped gradually building up from April onwards and declining rapidly in the autumn (Tab. 5.3). This seasonal pattern was evident in both 15-year periods, 1972–1986 and 1987–2001, although the overall catch was smaller in the later period. However, there were differences from year to year. While the total catch peaked in August in 20 of the 30 years, in 10 years the peak was in July, the typical pattern of steadily increasing abundance to a summer peak being present even in years of overall small catch, such as 1995 and 1998.

Up until 1979, I regarded the drought year, 1976, as an exception in that hoverfly numbers reached a maximum in July and few were caught in September and October; in essence, the usual seven-month season was compressed into the period April–August. However, as indicated, N peaked in July in many years subsequently, although September and October catches remained reasonably high. The exception was in 1995, when, despite numbers peaking in August, September and October catches were each only two, the annual N being low. 1995 was a hot, dry, bright summer, after a dry, bright spring, and autumn remained warm, although it was damp and had only average amounts of

Table 5.3. Mean monthly captures (N) of hoverflies in 1972–1986 and 1987–2001

	1972–1986	1987–2001
April	7	13
May	168	80
June	265	106
July	602	410
Aug	1,529	518
Sept	291	97
Oct	54	15
Mean Annual N	**2,916**	**1,132**

Table 5.4. Mean number of species (S) of hoverflies in 1972–1986 and 1987–2001

	1972–1986	1987–2001
April	3	5
May	19	19
June	25	17
July	29	23
Aug	34	25
Sept	23	15
Oct	11	6
Mean Annual S	**50**	**41**

sunshine. In both 1975 and 1978, the total catch (N) exceeded 6,000, but the seasonal pattern differed fundamentally: more than two thirds of the 1975 total was caught in August, when hoverflies entered the garden in large numbers, whereas in 1978 monthly catches were consistently high from May to October.

The number of species trapped reached a peak in August in 21 of the 30 years, so that the mean number of species in each 15-year period, 1972–86 and 1987–2001, peaked in August (Tab. 5.4). However, in eight years, the number of species trapped (S) peaked in July, and in one year, 1986, in June. In the majority of years, S peaked when N did, whether in July or August, which resulted in the average number of individuals per species (N/S) also peaking in the same month. In three years when N and N/S peaked in July, the highest number of species was caught in August indicating greater diversity. In 1986, when most species were caught in June, the August catch was so much greater that both N and N/S peaked in the later month. Interestingly, after years of sorting and identifying hoverflies from Malaise trap catches, June was the month when I anticipated most variety and unusual species for the garden.

Mean monthly totals of individuals and species as presented in Tabs. 5.3 and 5.4 obscure differences in seasonality between species. Some species are on the wing for several months, others for only restricted periods. Over the 30 years, most of the common species were recorded in every month from May to October, and many from April to October. However, abundance varied greatly from month to month and year to year, and the 30-year records show seasonal patterns of occurrence that indicate the number of generations a year. This is especially true for the 15 years 1972–1986, when catches were larger, and analysis of seasonality was feasible for 37 species in this period (Owen, J. 1991).

The species fall into six groups: one generation a year and over-wintering as larvae, e.g. *Epistrophe eligans, Merodon equestris* and *Platycheirus ambiguus*; one generation a year and overwintering as adults, e.g. *Episyrphus balteatus, Eristalis tenax* and *Helophilus pendulus*; two generations a year and overwintering as larvae, e.g. *Eristalis arbustorum, Eumerus tuberculatus* and *Melanostoma mellinum*; several overlapping generations a year and overwintering as larvae, e.g. *Platycheirus albimanus, P. scutatus* and *Syritta pipiens*; two generations a year and overwintering as pupae, e.g. *Rhingia campestris*; two generations a year and overwintering stage variable and facultative, e.g. *Sphaerophoria scripta* and *Syrphus ribesii*.

The value of long-term research

Many research projects, particularly those funded by grants, have a duration of three years. In only 10 of the 37 species used in the analysis of seasonal life histories would I have drawn the same conclusions after three years' trapping as I did after 15 years'. Seven species would have been too scarce to yield any results; four would have given results difficult to interpret; six, including *Melanostoma mellinum* and *Platycheirus ambiguus*, would have given a misleading idea of seasonality or abundance; and ten, including *Episyrphus balteatus, Helophilus pendulus, Platycheirus albimanus* and *Syrphus ribesii*, would have given results leading to misinterpretations of life histories. This emphasizes the value of long-term studies, and confirms that conclusions drawn from short-term projects should be treated with caution.

LARVAL HABITS OF HOVERFLIES

Adult hoverflies of different species are broadly similar in their feeding behaviour, but the food, feeding places, and hence life styles, of larvae are diverse. There are four categories: group A, consumers of living roots, bulbs, leaf-miners, stems, tree sap or fungi; group B, predators of aphids and other Homoptera; group C, dwellers in and feeders on soft or liquid decaying organic material, including rotting wood; and group D, scavengers in the nests of ants, bees or wasps. Larvae in each of the four categories are thus operating at different trophic levels (Wiegert & Owen, D. F. 1971). Those that eat live plant material are primary consumers; predators of Homoptera are secondary consumers; those that eat decaying organic material are primary decomposers if they eat plant material, secondary or higher order decomposers if they eat dead material that is of animal origin; and scavengers in nests of social Hymenoptera are secondary or higher order decomposers. Members of most groups are probably opportunist feeders: aphid-feeders occasionally attack other insects and each other; larvae in decaying organic material feed on faeces, on rotting plants, and on decomposing animal remains; and scavengers in Hymenoptera nests occasionally eat live larvae and pupae. As a family, hoverflies are unusual in exploiting the environment in such diverse ways. All four groups are represented by adults caught in the Malaise trap, and the numbers of individuals and species, and the trophic levels of the groups are shown in Tab. 5.5. The larvae of a number of garden species are undescribed, but they are assumed to have habits similar to those of closely related species. Compared with the British list of hoverflies, group B (the species with aphid-feeding larvae) is over-represented in the garden, and group A is under-represented; groups C and D form about the same proportions of the British and garden lists.

Hoverfly groups in the Leicester garden

Group A. Four genera trapped in the garden, *Cheilosia, Ferdinandea, Merodon* and *Eumerus*, fall into group A. The foodplants of two (and possibly three) of the seven species of *Cheilosia* (see Stubbs & Falk 1983) grow in the garden: *C. bergenstammi* was last caught in 1988, by which time its foodplant, common ragwort, had been eliminated from the garden, and *C. vernalis* was first caught in 1978, after its foodplant, yarrow, spread from the clipped confines of the lawn to a flowerbed. Both species of *Ferdinandea* are reported from sap runs associated with damage to trees caused by goat moth (*Cossus cossus*), and only seven individuals have been trapped, all but one of them being *F. cuprea. Merodon equestris* and the two species of *Eumerus* feed as larvae by burrowing in living bulbs, and almost certainly breed in the garden. Group A is represented by 4,210 individuals, almost 7% of the total catch.

Group B. Most garden hoverflies, and the majority of the common species, have aphid-feeding larvae, which I often found among aggregations of aphids on such plants as cabbages, beans and elder. I was able to raise nine species of Syrphinae from larvae collected in the garden: *Episyrphus balteatus, Eupeodes luniger, Melanostoma mellinum, Melanostoma scalare, Meliscaeva auricollis, Platycheirus albimanus, Platycheirus scutatus, Sphaerophoria scripta* and *Syrphus* *ribesii*. As far as is known, the larvae of all species of Syrphinae and Pipizini feed on aphids or other herbivorous insects. Fifty-seven species, and over 81% of all individuals trapped, fall into group B.

Group C. The species in group C are taxonomically diverse but the larvae of all are decomposers. Since, however, the nature of their food varies, they operate on several trophic levels. Larvae of *Eristalis* and *Helophilus* are usually found in water, feeding on minute particles of organic material, but *Eristalis tenax* also occurs in dung, sewage and wet carrion. *Sericomyia silentis* is associated with boggy and peaty areas, and only one was captured in the 30 years. *Myathropa florea* is particularly associated with wet rot-holes in trees. The larvae of *Syritta pipiens* occur in wet garden compost, but they have also been recorded from dung of cows, horses and humans; other Xylotini feed as larvae on and in rotting wood, and adults are rare in the garden. Of the Chrysogastrini, only *Neoascia podagrica* is regularly caught. Many of the species in this group undoubtedly breed in the garden or its immediate vicinity, either in compost, on dung or carrion, or in ponds or tree-holes, and the proximity of farmland accounted for the regular capture of *Rhingia campestris*, which breeds in cow-pats. Group C, represented by 23 species and 7,088 individuals, nearly 12% of the total catch, ranks second in importance to the aphid-feeders in group B.

Group D. *Volucella bombylans* and *V. pellucens* are the only representatives of group D in the garden. Their larvae scavenge in bee and wasp nests, not only eating dead Hymenoptera and detritus, but also stimulating the host larvae to produce excrement, on which they feed. Occasionally, however, especially towards the end of the season, they eat live larvae, pupae, or weakened adults. Group D constitutes less than 0.1% of the total sample.

Table 5.5. Feeding sites and trophic levels of larvae of hoverflies trapped in the Leicester garden, 1972–2001

Trophic group	Feeding site	Trophic level	Representatives	Species (S)	Individuals (N)	% total N
A	Stems, roots, bulbs, tree sap	Primary consumers	*Cheilosia* (7) *Ferdinandea* (2) *Eumerus* (2) *Merodon* (1)	12	4,210	6.9
B	Plants infested with aphids	Secondary consumers	*Syrphinae* (48) *Pipizini* (9)	57	49,384	81.3
C	Decaying organic material, rotting wood	Primary to n^{th} order decomposers	*Rhingia* (1) *Eristalini* (12) *Sericomyia* (1) *Chrysogastrini* (4) *Xylotini* (5)	23	7,088	11.7
D	Nests of wasps & bumblebees	Usually 2nd to n^{th} order decomposers	*Volucella* (2)	2	54	0.1

Note: Separation into trophic levels is based on the system suggested by Wiegert & Owen, D. F. (1971). This differs from the table of feeding sites in Owen, J. (1981) in that species are grouped in four not five categories: to minimize confusion, the trophic groups have therefore been designated by letters not numbers, as in Owen, J. (1981).
Subfamilies and tribes as in Stubbs & Falk (1983).

Syritta pipiens, a common garden hoverfly. Males defend territories and court females above flowers

Trophic frequencies

The trophic composition of the Malaise trap catch varied to some extent from year to year. Although hoverflies with aphid-feeding larvae (group B) constituted the majority of the sample in every year, their relative frequency was considerably lower in 1973, 1983, 1984, 1988, 1989, 1995, 1996 and 1998. In 1973, this was associated with an increase in relative frequency of species with larvae that feed on decaying organic material (group C); the late summer in 1973 was warm and sunny, and species such as *Eristalis arbustorum*, which continue flying late in the season, were particularly abundant. In 1983, the relative frequency of group A was far higher than at any time until 1995, largely because of the numbers of *Eumerus tuberculatus*, which was the commonest species. In the same year, large catches of *Eristalis arbustorum* and *Syritta pipiens* relative to species with aphid-feeding larvae resulted in a high relative frequency of group C. Again in 1984, the relative frequency of groups A and C was greater than usual, largely because of the numbers of *E. tuberculatus* and *Merodon equestris*, and of *E. arbustorum* and *S. pipiens*. 1988 was a cool, damp, dull summer, when the hoverflies with aphid-feeding larvae evidently fared poorly, but the relative frequency of group C, such as *E. arbustorum* and *Helophilus pendulus*, reached its highest value. In the next year, 1989, which was a hot, dry, bright summer, the relative frequencies

Table 5.6 (1). Hoverflies with predatory larvae as percentages of monthly Malaise trap catches, by year

	April	May	June	July	Aug	Sept	Oct
1972	100.0	99.4	82.4	80.1	88.7	91.5	61.8
1973	100.0	98.3	86.1	86.9	66.3	38.8	51.4
1974	83.3	90.7	62.2	87.8	92.1	90.8	64.6
1975	100.0	90.7	51.0	68.5	91.0	66.2	62.7
1976	100.0	93.1	45.2	85.6	76.9	71.4	66.7
1977	-	97.1	63.3	50.9	91.3	90.6	72.7
1978	100.0	96.9	78.8	55.1	85.5	85.7	62.4
1979	100.0	96.1	73.0	64.0	93.8	88.7	65.4
1980	100.0	91.5	76.5	87.7	89.5	56.9	75.0
1981	88.9	93.1	84.4	89.1	70.2	66.3	60.0
1982	95.2	85.7	48.7	84.6	87.9	54.7	47.8
1983	100.0	100.0	67.1	60.3	47.9	51.2	42.9
1984	-	81.5	53.0	58.3	71.6	86.6	66.7
1985	100.0	89.1	72.7	64.6	94.4	66.8	8.7
1986	-	97.0	50.5	57.1	97.2	95.0	62.5
1987	100.0	100.0	96.3	78.5	93.8	72.4	43.6
1988	100.0	64.4	46.9	85.1	69.5	37.0	27.3
1989	100.0	78.5	43.6	82.2	72.5	68.3	62.5
1990	100.0	84.8	65.2	92.7	75.5	73.3	66.7
1991	100.0	95.4	70.0	82.7	84.1	69.9	20.0
1992	100.0	80.7	41.8	91.9	94.3	88.6	100.0
1993	100.0	91.8	55.2	96.5	73.1	56.1	60.6
1994	100.0	92.9	53.7	81.2	86.7	65.6	40.0
1995	66.7	63.6	22.5	67.3	76.7	50.0	100.0
1996	100.0	100.0	29.8	21.1	77.6	77.9	53.3
1997	100.0	76.5	28.7	88.9	96.7	77.0	57.6
1998	100.0	84.5	46.4	80.0	70.5	45.4	33.3
1999	87.9	71.6	24.0	82.3	87.4	53.2	100.0
2000	100.0	88.5	41.8	87.8	73.0	77.4	87.5
2001	100.0	91.3	28.8	85.0	79.0	53.3	66.7
Mean 1972–2001	**97.1**	**90.1**	**60.7**	**79.7**	**86.5**	**80.5**	**59.8**

of both groups A and C were higher than usual. In 1995 and 1996, both hot, dry and bright summers, the relative frequency of group A reached unprecedented levels, largely because of the abundance of *M. equestris*, which in both years was the commonest garden species (see Tab. 5.1). In 1998, both groups A and C were relatively more abundant than usual, in a mild, averagely damp and dull summer. The crashes in some years in relative frequency of hoverflies with aphid-feeding larvae do not seem to correlate with weather.

Large numbers of hoverflies with aphid-feeding larvae, particularly *Eupeodes corollae* and *Episyrphus balteatus*, moved into the garden during the first two weeks of August 1975 and 1977, and during the last two weeks of July in 1976; *Eupeodes corollae* formed about 40% of the total catch in 1975 and in 1976, and *Episyrphus balteatus* over 44% in 1977. This is not reflected in the relative frequency of group B hoverflies in these years, largely because the numbers of resident *Melanostoma* and *Platycheirus* in the catch were far lower than usual; *M. mellinum*, for instance, which breeds in the garden, was represented by only two individuals in 1977, and *M. scalare*, which also breeds, was absent altogether. In 1978, when the numbers of *Eupeodes corollae* and *Episyrphus balteatus* fell, those of *M. mellinum*, and particularly of *Platycheirus albimanus* and *P. scutatus* reached unprecedented levels. Thus the trophic composition of the garden hoverfly fauna was apparently unaffected by immigration of *Eupeodes corollae* and *Episyrphus balteatus* to the garden in 1975, 1976 and 1977. The only comparable year afterwards was 1997, when very large numbers of *Episyrphus balteatus* entered the garden (Tab. 5.6 (2)), but this was not accompanied by a reduction in numbers of other hoverflies with aphid-feeding larvae.

Table 5.6 (2). Relative frequencies of *Eu. corollae*, *Ep. balteatus* and all hoverflies with predatory larvae, by year

	Eu. corollae	*Ep. balteatus*	Total predators
1972	1.3	6.1	87.2
1973	1.6	4.0	70.1
1974	6.6	4.1	87.3
1975	40.5	11.7	84.4
1976	39.8	11.8	79.7
1977	16.0	44.2	87.2
1978	1.9	7.6	80.0
1979	7.8	8.6	89.4
1980	5.0	2.0	81.3
1981	8.2	1.4	81.9
1982	9.7	11.1	81.4
1983	4.5	1.6	54.9
1984	10.7	9.2	69.4
1985	5.4	30.2	86.3
1986	4.8	29.9	85.9
1987	8.1	29.0	89.9
1988	1.2	1.8	64.5
1989	8.6	2.8	73.7
1990	3.8	3.1	85.2
1991	11.5	9.5	79.7
1992	19.3	27.0	88.8
1993	6.5	7.3	73.8
1994	5.3	29.8	72.7
1995	9.7	9.5	62.8
1996	5.7	16.8	57.4
1997	7.0	45.5	91.2
1998	7.0	1.0	69.0
1999	2.9	29.2	75.2
2000	2.0	5.8	74.9
2001	5.1	37.2	75.1
Mean 1972–2001	**12.0**	**15.5**	**81.3**

Seasonal patterns of abundance

Seasonal changes in the relative abundance of species with different larval habits is expressed in Table 5.6 (1) as the percentage of each month's catch composed of aphid-feeders (group B). They were first to appear and increase in numbers in spring, and their frequency relative to those with other larval habits was high in April and May. It fell in June, when the catch increased in variety, but had usually risen again in August. By October, species whose larvae feed on decaying organic material, e.g. *Eristalis* and *Helophilus*, were common, and the relative frequency of aphid-feeders was lower, particularly in 1985.

The greatest departures from this pattern were in 1973, 1977, 1983, 1988, 1991, 1992, 1995, 1996 and 1998. In 1973, *Eristalis arbustorum* and other group C species became particularly abundant in August, such that, by September, the relative frequency of aphid-feeders had fallen to 38.8%. In

Helophilus hybridus (right), an infrequent visitor to the study garden

Table 5.7. Status of hoverflies recorded in the Leicester garden, 1972–2001

Known or assumed to breed in garden (25 species)	Assumed to breed in surrounding area (25 species)	Casual visitors (30 species)	Recorded in one year only (14 species)
Baccha elongata	*Cheilosia bergenstammi*	*Cheilosia albitarsis*	*Cheilosia illustrata*
Cheilosia vernalis	*Cheilosia paganus*	*Cheilosia proxima*	*Cheilosia nigripes*
Dasysyrphus albostriatus	*Chrysotoxum festivum*	*Chrysotoxum bicinctum*	*Criorhina berberina*
Epistrophe eligans	*Chrysotoxum verralli*	*Dasysyrphus pinastri*	*Criorhina floccosa*
Episyrphus balteatus	*Eristalis arbustorum*	*Dasysyrphus venustus*	*Dasysyrphus tricinctus*
Eumerus strigatus	*Eristalis interruptus*	*Epistrophe nitidicollis*	*Didea fasciata*
Eumerus tuberculatus	*Eristalis intricarius*	*Eristalinus sepulchralis*	*Epistrophe grossulariae*
Eupeodes corollae	*Eristalis pertinax*	*Eristalis horticola*	*Eristalis abusivus*
Eupeodes luniger	*Eristalis tenax*	*Eupeodes latilunulatus*	*Ferdinandea ruficornis*
Melanostoma mellinum	*Eupeodes latifasciatus*	*Ferdinandea cuprea*	*Melanogaster hirtella*
Melanostoma scalare	*Helophilus pendulus*	*Helophilus hybridus*	*Parasyrphus malinellus*
Meliscaeva auricollis	*Leucozona lucorum*	*Helophilus trivittatus*	*Sericomyia silentis*
Merodon equestris	*Melangyna cincta*	*Heringia heringii*	*Triglyphus primus*
Pipiza noctiluca	*Myathropa florea*	*Lejogaster metallina*	*Tropidia scita*
Platycheirus albimanus	*Neoascia podagrica*	*Melangyna lasiophthalma*	
Platycheirus ambiguus	*Neocnemodon vitripennis*	*Melangyna umbellatarum*	
Platycheirus angustatus	*Paragus haemorrhous*	*Meligramma trianguliferum*	
Platycheirus clypeatus	*Platycheirus immarginatus*	*Meliscaeva cinctella*	
Platycheirus manicatus	*Platycheirus scambus*	*Parasyrphus punctulatus*	
Platycheirus peltatus	*Pyrophaena granditarsa*	*Pipiza austriaca*	
Platycheirus scutatus	*Rhingia campestris*	*Pipiza bimaculata*	
Sphaerophoria scripta	*Scaeva pyrastri*	*Pipiza fenestrata*	
Syritta pipiens	*Sphaerophoria menthastri*	*Pipiza luteitarsis*	
Syrphus ribesii	*Volucella bombylans*	*Pipizella viduata*	
Syrphus vitripennis	*Volucella pellucens*	*Riponnensia splendens*	
		Scaeva selenitica	
		Sphaerophoria rueppellii	
		Syrphus torvus	
		Xanthogramma pedissequum	
		Xylota segnis	

A handsome hoverfly, *Volucella pellucens* (above left), the larvae of which live as scavengers and occasional predators in wasp nests

Helophilus pendulus (above right), a common hoverfly of gardens whose rat-tailed aquatic maggots feed on decaying organic material and have a posterior breathing tube

1977, the total catch was smaller than in previous years and those with aphid-feeding larvae relatively scarce, such that in July they fell to 50.9% of the catch. In 1983, high numbers of particularly *Eumerus tuberculatus*, which has bulb-feeding larvae, and to a lesser extent of *Eristalis arbustorum* and *Syritta pipiens*, depressed the relative frequency of aphid-feeders from July onwards. In 1988, the relative frequency of hoverflies with aphid-feeding larvae fell to 37.0% in September as a consequence of a good year for *Eristalis*, *Helophilus* and *Rhingia campestris*. In 1991, very few group B hoverflies were caught in October, depressing their relative abundance to 20.0%, whereas in 1992 and 1995, they were the only trophic group caught in October. Numbers of hoverflies with aphid-feeding larvae were generally low in 1996, so that they formed only 57.4% of the annual catch, and only 21.1% of the July catch, when *Merodon equestris*, the year's commonest species, was particularly abundant. In 1998, group B formed only 33.3% of a generally poor October catch.

The usual situation in any month was for hoverflies with aphid-feeding larvae to outnumber those with other larval habits in the Malaise trap sample, although in late summer, those with larvae feeding on decaying organic material became relatively common. Aphid-feeders were not only more abundant than the others, but also showed a more pronounced seasonal peak in numbers. Of the 49,384 hoverflies with predatory larvae, 26,576 (53.8%) were caught in August, whereas of the 11,352 with larvae of other habits, 4,127 (36.4%) were caught in August.

The habitat disturbance, of which gardening is a part, has almost certainly increased the abundance of aphids in the study garden. This is because gardening introduces and encourages tender plant growth on which aphids feed. Consequently, there is an increase of those hoverflies with predatory larvae that feed on aphids. This had led to the anomalous situation where mimetic hoverflies, albeit poor mimics, like most hoverflies with aphid-feeding larvae, are more abundant than their wasp models. Possibly, habitat disturbance has occurred much more rapidly than natural selection can perfect the match between model and mimic, and mimicry patterns seen in disturbed habitats, such as gardens, may be in the process of being lost (Azmeh *et al.* 1998). Many more long-term investigations will be needed to show whether or not this is so.

Goldenrod (*Solidago canadensis*) and Shasta daisy (*Leucanthemum* x *superbum*) are both attractive to hoverflies (above left)

Hoverflies can feed easily at the flowers of pot marigold, *Calendula officinalis* (above right), because of their open structure

HOVERFLIES AS INDICATORS OF VARIETY OF HABITAT

The size and diversity of the hoverfly sample from the Malaise trap indicates that gardens offer not only an abundance of resources but also a varied habitat, and lead to speculation as to what so many different hoverflies are doing in the garden. The dependence of adults on nectar and pollen for food accounts for large catches of resident species, and no doubt explains the presence of so many casual visitors. All the common medium-to-large species are regularly seen at flowers such as shrubby cinquefoil (*Potentilla fruticosa*), *Geranium cinereum*, *Buddleja davidii* and composites including Shasta daisies (*Leucanthemum* × *superbum*), pot marigolds (*Calendula officinalis*), goldenrod (*Solidago canadensis*) and various species and cultivars of *Aster*. Some of the smaller hoverflies, such as *Melanostoma* species, *Platycheirus* species, *Sphaerophoria scripta* and *Syritta pipiens*, are also abundant at flowers, although others, such as *Paragus haemorrhous* and *Pipiza* species, are only known from trap samples.

There is abundant food for adults in gardens, which are especially important in late summer and autumn, when hedgerows and meadows offer few flowers to compete with the colourful displays of herbaceous borders. Females of species whose larvae feed on aphids find a multitude of egg-laying sites in gardens, which can also accommodate, although to a lesser extent, those whose larvae inhabit decaying organic material, bulbs or Hymenoptera nests.

Some garden hoverflies are resident, others visitors, sometimes casual, but probably seeking food or egg-laying sites. Of the 94 species, 33 (such as *Baccha*, *Ferdinandea*, *Criorhina* and *Xylota* species) are usually associated with woodland or woodland edge, 11 (such as *Rhingia* and *Merodon* species) with more open places, and 50 range over both types of habitat (compiled from Pollard 1971, Steele & Welch 1973, Stubbs & Chandler 1978, and Stubbs & Falk 1983).

Gardens are a patchwork of open and shaded places, and so might be expected to accommodate species with no marked habitat preferences. On the other hand, species characteristic of open ground may be attracted to lawns and paths, and woodland species to sheltered and shady corners. On the whole, however, gardens are open habitats, and most of the woodland species are rare visitors.

SUMMARY

On the basis of captures and breeding records, together with the statements about habitat given above, the garden hoverflies can be divided into four categories: 25 species are regular, common and are either known to breed or assumed to do so; a further 25 species are regularly present and are assumed to breed within surrounding gardens and suburban areas; 30 species have been recorded in more than one year, but only infrequently, and are probably casual visitors; and 14 species have been recorded in one year only, 13 of them represented by single individuals, and rank as chance visitors (Tab. 5.7).

To summarize, 50 of the 94 species are regarded as part of the resident fauna of the garden and surrounding area, while 44, including most of the woodland species, are casual or chance visitors. The diversity of larval lifestyles among hoverflies resident in the garden or surrounding area demonstrates the complexity and variety of the garden habitat.

The 30-year investigation has shown that hoverfly abundance in the garden fluctuates markedly from year to year, caused in part, although not entirely, by immigration. Overall, however, it is evident that after 1984 there was a marked decline in the abundance of hoverflies (Fig. 5.1 and Tab. 5.1), a situation in line with that of other groups of insect in the garden.

6 : Bees and wasps

THINK OF A RICH AND PRODUCTIVE GARDEN and you think of one alive with the diligent activity of bees at flowers, the air vibrant with their busy humming. In my garden, the mauve, hooded flowers of spotted dead-nettle (*Lamium maculatum*) dip and sway with the attentions of visiting bumblebees, from early in spring to late in the autumn, the tall violet-blue stems of sage flowers (*Salvia officinalis*) shimmer with their comings and goings, and the trumpet-like hollyhock flowers (*Alcea rosea*) seem to amplify the buzz of the fat, furry visitors. From high summer on into autumn, alfresco meals in the garden are enlivened by the uninvited visits of wasps lapping sweet drinks, licking up jam and the juices of soft fruits, and chewing morsels from cold meats.

A bumblebee, *Bombus terrestris* (opposite), collects pollen from a hollyhock flower (*Alcea rosea*) to take back to its nest as food for the larvae

Common wasp, *Vespula vulgaris*, in flight (above). Common in gardens, they have a two-year (and possibly seven-year) cycle of abundance

The bumblebees so conspicuously active at flowers and the yellow wasps that disrupt picnics are social species. Colonies are founded in spring by over-wintered queens, most of whose offspring are non-breeding female workers that remain associated with the nest, building new cells, tending the queen's eggs and larvae, defending the colony and foraging for food. These social species are the commonest species of bees and wasps in the garden, but Malaise trapping has revealed an unexpected diversity of less-conspicuous solitary species.

Females of solitary bees and wasps make small nests in burrows in the soil, or in hollow plant stems, or in old beetle borings in wood. They make a few cells, usually end to end, each of which is stocked in turn with larval food on which an egg is laid. The majority have no further contact with their offspring. Although several nests of a particular species may be close together, there is no contact or collaboration on the part of the adults.

Solitary bees may be quite conspicuous at flowers, particularly in early summer, when mining bees are also actively digging burrows in the light soil of vegetable beds or worn patches of lawn. Solitary wasps, however, are less common and are rarely noticed. Were it not for trapping, their diversity in the garden would not have been appreciated. In total, 59 species of bees and 62 species of wasp were recorded in the Leicester garden during the period 1975–2001. However, these bees and wasps belong to 13 different families and so their diversity is not easily comparable with that of, say, hoverflies, which belong to just one family.

NATURAL HISTORY OF BEES AND WASPS

Bees and wasps comprise the subdivision Aculeata of the suborder Apocrita in the insect order Hymenoptera. The Aculeata are best known as insects in which the female's ovipositor has evolved into a venomous sting, the eggs instead reaching the outside through an aperture near its base. There are, however, many parasitic species in which the ovipositor retains its egg-laying function, although on other anatomical grounds they are regarded as aculeates. To conform with usually accepted classification and recognise these parasitic forms as aculeates gives a grand total of 549 British species of bees and wasps (Fitton *et al.* 1978). Of these, 25 species belonging to the family Apidae are social bees, including bumblebees and the honeybee, and seven, belonging to the family Vespidae, are social wasps. Of the rest, 227 are solitary bees of seven different families, and 290 are solitary wasps, of ten different families.

Diet

The crucial difference between bees and wasps is in what they eat, and this has led to differences in their structure and appearance. Bees are entirely vegetarian, or first order consumers, in that they eat pollen and nectar, and feed them to their larvae. Adult wasps, however, although themselves feeding on nectar, honeydew and other sweet substances, catch animal food for their larvae. They are predators of other invertebrates, particularly insects, and hence operate as secondary or higher order consumers; they sometimes also eat carrion (when they function as decomposers). Solitary wasp larvae are supplied with live but paralysed (*i.e.* usually stung, but occasionally bitten), prey, whereas social wasp larvae are fed masticated pellets of 'meat'.

Common wasps, *Vespula vulgaris*, feeding on an apple. Although adults feed on sweet substances, they collect animal food for their larvae

As adaptations to collecting and carrying pollen, bees in general are covered with branched hairs giving them a furry appearance, and the majority have particular areas of the body modified as pollen baskets for transporting pollen. Some species have become cuckoo bees, in that they appropriate the nests and food stores of other species for their own larvae, and such bees are much less hairy and lack pollen baskets. Wasps, on the other hand, tend to be smooth-surfaced, even shiny in appearance. In general, bees, as nectar-feeders, have a much longer tongue, formed from the labium and associated maxillae, than do wasps.

Warning colouration

Since they can sting, bees and wasps are distasteful to predators, and have consequently tended to develop conspicuous warning coloration. Usually, this involves some form of yellow and black banding. An inexperienced predator may try to eat a brightly patterned bee or wasp, but learns from the unpleasant experience, and avoids them in future. Not only do many bees and wasps share similar warning coloration (Müllerian mimicry), but also many palatable insects, such as hoverflies, mimic their appearance and thus gain protection (Batesian mimicry). A wasp pattern signifies danger whatever animal is wearing it, and is so effective as a bluff (in Batesian mimicry) because it is so widespread as a genuine warning (Müllerian mimicry).

Behavioural studies

Bees and wasps attract particular interest because of the various stages they display in the development of social behaviour. Researchers use them to study the evolution of social behaviour. For instance, one of the British solitary wasps, *Ammophila sabulosa*, practices progressive provisioning of its larvae. Instead of supplying all the food a developing larva will need, and sealing the cell after laying an egg (mass provisioning), it returns to the nest, inspects the growing larva and its food supply, and provides more food as it is needed, an essential step towards social behaviour. On the other hand, wandering wasps, such as *Tiphia* species, use no nest, merely paralysing and laying eggs on beetle larvae where they find them in the soil; behaviourally, this is a link with the Parasitica, such as Ichneumonidae (see Chapter 9).

Bombus lapidarius, an occasional visitor to the Leicester garden

Some solitary wasps, such as *Chrysis* species, are parasitic on solitary bees and wasps, invading the host's nest and laying an egg which hatches into a larva that consumes the rightful occupant. Others, such as *Sapyga* species, are cleptoparasites in the nests of solitary bees, in that the invasive larva destroys the rightful occupant and then devours the food store; this blurs the distinction between wasps and bees, since the parasitic larva consumes stored pollen and nectar. It is, however, possible to make general statements about the many typical bees and wasps.

Bumblebees

There are 18 species of British bumblebees, *Bombus* (although some authorities also include a Continental species, captured in Kent in the 19th century, and a form of *B. lucorum* that is regarded as a separate species on the Continent). These, together with the honeybee (*Apis mellifera*) and six species of cuckoo bumblebees (*Psithyrus*), make up the family Apidae, characterized by the development of social life. Queen bumblebees that have hibernated in various sites, such as in burrows in well-drained banks (*B. lapidarius*) or in holes in logs (*B. lucorum*), emerge in the spring and start building nests. *Bombus pratorum* is the earliest, often being on the wing in March, whereas queens of *B. lapidarius* rarely emerge before late May or early June. Species differ in their nest site preferences; *Bombus pascuorum*, for instance, nests beneath clumps of grass or moss on the soil surface, whereas *B. terrestris* takes over the abandoned nest of a small mammal at the end of a burrow a metre or more in length.

The queen hollows out a chamber, collects a mass of pollen, and around it constructs a cell from wax extruded between her abdominal segments. She lays several eggs on the pollen mass and also makes a wax honey-pot near the nest

Bombus lucorum, a common garden bumblebee, on a flower of scabious, *Scabiosa*

entrance as a bad-weather food store. As the larvae feed and grow, she supplies them with more pollen and honey and also incubates them by straddling the brood chamber with her warm, furry body. When fully grown, the larvae spin individual silken cocoons for pupation, and the queen recycles their wax brood chamber into new brood cells on the upper, outer edge of the cocoon cluster. The first new bees to emerge are all females and all workers, *i.e.* non-breeding and usually sterile. After a day or two, once the workers are ready to start foraging for nectar and pollen, the queen remains in the nest all the time, constructing new cells, laying eggs, and feeding and incubating larvae.

Female and worker bumblebees are highly adapted for provision of honey and pollen to the nest. Nectar is collected from flowers using the long tongue, which differs in length between species, such that the nectaries of different flowers are accessible to different species. Collected nectar is transported in a specialized length of the digestive tract, the honey stomach, where complex sugars are converted to simpler ones by enzymes. So capacious is the honey stomach that a bumblebee can transport a weight of honey almost equal to her own. Pollen adheres to the hairy body of a flower-visiting bumblebee; she uses her fore legs to brush this towards her mouth, moistens it with nectar, and crams it into pollen baskets, or corbiculae, on the tibiae of her hind legs. The tibia of each hind leg is concave and highly polished, and is surrounded by a fringe of long, inward-curving hairs; male *Bombus* lack these pollen baskets.

The colony steadily increases in size; those of *B. terrestris*, for instance, producing 300–400 workers from a comb 20–25 cm in diameter. As the colony grows, the ratio of workers to larvae, and hence the larval food supply, increases and it seems to be this that leads to development of new queens. Males are produced at the same time, sex determination in Hymenoptera being genetic and depending on fertilization: fertilized (and therefore diploid) eggs produce females; unfertilised, haploid eggs produce males. Males leave the nest, wait around the entrances of other nests of the same species, and mate with virgin queens. No more workers are produced, young queens take no part in nest activities, and the colony gradually disintegrates. Males, workers and the original, founding queen die, and only young, impregnated queens survive the winter.

Cuckoo bumblebees

Psithyrus, cuckoo bumblebees, are different in that they develop no workers but instead exploit the social system of *Bombus* species. Female *Psithyrus* emerge from hibernation when nests of their hosts already contain workers. They infiltrate *Bombus* nests, often (although not always) killing the queen, and start laying eggs. They depend on the *Bombus* workers to tend and feed their young. Since they collect no pollen, *Psithyrus* females have no pollen baskets, and are also generally less hairy than *Bombus*. Each species resembles the *Bombus* species it habitually parasitizes, but it is still uncertain how they manage to gain entry to a host nest.

Solitary bees

Solitary bees, like their social relatives, feed their young on nectar and pollen, although a few, such as *Hylaeus* species, have no pollen baskets and transport pollen in the crop, an enlarged area of the fore gut. Others, such as *Colletes* species, have short, blunt tongues, more like those of wasps. Many do not look particularly like bees, and are small, although others, such as *Anthophora* species, are hairy and as large as the smaller bumblebees, for which they may be mistaken. Female solitary bees typically construct one or more cells in a burrow in soil or in a hollow plant stem; some just compact the soil, *Hylaeus* species use a secretion from the mouth, while *Megachile* species use discs cut from plant leaves. Each cell is mass-provisioned with pollen, moistened with nectar, on which an egg is laid. Pollen may be carried on special brushes of hairs on the thorax, abdomen or hind legs, although the latter are never as highly modified as in social bees. The resulting larva eats the stored food, pupates, and emerges in the following year, long after its mother, with which it has had no contact, has died.

The excavated soil (above) around the entrance to an *Andrena* bee's burrow

An aerial nest (below) of a tree wasp *Dolichovespula media*, a social species

A few solitary bees, however, show the rudiments of social life. In some species of Halictidae, the female remains near the nest and guards it. The bees that emerge from the first cells are, in effect, worker females which build and provision more cells in which the original female, perhaps better called the queen, lays eggs. These produce males and normal females, which, after mating, hibernate until the following year. In employing the activities of non-breeding workers, females of these species are able to raise more offspring than typical solitary bees, such as *Andrena* species.

There are also cuckoo bees or cleptoparasites, such as *Nomada* species, among the solitary bees. Less hairy, since they never collect pollen, they lay their eggs in the nests of other solitary bees, such as *Andrena* species, and thus appropriate for their larvae the stored food, usually destroying the host's egg.

Social wasps

The life cycle of social wasps is not unlike that of bumblebees; a nest is founded by an overwintered queen, produces several batches of non-breeding workers, and then dies out after production of males and new queens. Their nests differ from those of bumblebees in several respects, however. The nest is constructed of paper carton, formed by masticating fragments chewed from sound or slightly rotten wood, depending on species; it may become very large, consisting of several separate horizontal combs of cells, and is enclosed in several overlapping protective envelopes of carton.

Vespula species nest underground or in enclosed cavities, such as roof spaces or hollow trees, whereas *Dolichovespula* species build aerial nests in exposed or semi-exposed positions on trees or the outside of buildings. Nests of *Vespula* species may become particularly large, those of *V. vulgaris* sometimes consisting of as many as 12 combs composed of more than 15,000 cells. One species, *Vespula austriaca*, is a cuckoo wasp, in that over-wintered queens usurp *V. rufa* queens, appropriate their nests and workers, and lay eggs that produce only males and new queens.

Solitary wasps

Most species of wasp, however, are solitary. Each female constructs a small group of individual cells in the soil, in hollow plant stems, or in holes in wood or masonry, depending on species. She stocks each with prey, paralysed usually by stinging or occasionally by biting, and lays an egg on the last item stored before sealing the nest. The prey is characteristic of the type of wasp. *Ectemnius* species, for instance, provision their nests with flies, including hoverflies, *Trypoxylon* species with spiders, *Pemphredon* species with aphids, and *Ancistrocerus* species with caterpillars.

The majority of solitary wasps never return to the nest after completion, and have no contact with their offspring. As already indicated, however, some *Ammophila* species practice progressive provisioning, and in *Ancistrocerus* species there is a tenuous link between mother and offspring, as the egg is laid before the cell is stocked. Non-British species in some genera, such as *Ectemnius* and *Ammophila*, also show the rudiments of social behaviour, in that two or more females share and provision a common nest (Yeo & Corbet 1983).

A few members of the Chrysididae are cleptoparasites of solitary wasps and, as already indicated, species of Sapygidae are cleptoparasites of solitary bees. Other chrysidids are ectoparasites of other sorts of insects, particularly solitary bees and wasps; they lay their eggs individually in occupied cells, the resulting larvae eating the occupants and occasionally the stored food as well.

DIVERSITY AND ABUNDANCE OF BEES AND WASPS

Social and solitary bees and wasps, representing many of the types described, occur in the study garden. Information on their numbers and diversity and on annual fluctuations and seasonality come from Malaise trapping. Social wasps were sorted from the trap sample during 1972–2001, social bees during 1972–1977 and also 1982–2001, and solitary bees and wasps during 1975–2001.

The garden species are given here in four categories: social bees, social wasps, solitary bees and solitary wasps. Under each heading, some idea is given of what the insects are doing in the garden, and the frequency given for each species is the total Malaise trap catch.

Social wasps and solitary bees and wasps were identified by M. E. Archer. Sequence within each category, classification and taxonomy follow Fitton *et al.* (1978).

SOCIAL BEES

1	***Bombus lucorum*** (L.)	607
2	***Bombus terrestris*** (L.)	761
3	***Bombus lapidarius*** (L.)	305
4	***Bombus pratorum*** (L.)	1,319
5	***Bombus hortorum*** (L.)	1,116
6	***Bombus ruderatus*** (Fab.)	86
7	***Bombus pascuorum*** (Scopoli)	5,052
8	***Bombus ruderarius*** (Müller)	114
9	***Psithyrus barbutellus*** (Kirby)	3
10	***Psithyrus campestris*** (Panzer)	1
11	***Psithyrus rupestris*** (Fab.)	1
12	***Psithyrus sylvestris*** Lepeletier	4
13	***Psithyrus vestalis*** (Geoffroy in Fourcroy)	2
14	***Apis mellifera*** L.	not counted

Records of social bees

Bombus pascuorum, the commonest bumblebee in the Leicester garden, nested there in 1993 and 2000

The social bees belong to the family Apidae. Eight species of bumblebees, *Bombus*, were recorded, five of cuckoo bumblebees, *Psithyrus*, and the honeybee, *Apis mellifera*. Bumblebees were abundant in the garden, feeding and collecting pollen from flowers, and occasionally a cuckoo bumblebee was seen. *Bombus lapidarius* nested in the garden in 1994, and *B. pascuorum* in 1993 and 2000, all nests in cavities in walls. Most *Bombus* species were recorded in every year that they were sorted from Malaise trap samples, 1972–1977 and 1982–2001, a total of 26 years, but *B. lapidarius* was absent in 1989 and 1990, *B. ruderarius* was absent in 1983 and after 1988, was only caught once, in 1997, and *B. ruderatus*, although recorded every year 1972–1977, was absent in 1982–2001. Captures of *Psithyrus* were sporadic in the 1970s and early 1980s, but none was caught after 1983. *Apis mellifera* was frequent, and was regularly captured, but no attempt was made to monitor numbers, as it is not a truly wild species. Bumblebees were identified using Alford (1973) and Free & Butler (1959).

Records of social wasps

All eight British social wasp species are Vespidae; seven of them were caught in the Malaise trap. *Vespula vulgaris* was abundant in most years in the garden, and *V. germanica* in many years, both particularly in August and September, and overwintered queens were conspicuous in late spring and early summer in some years; both species were caught every year, 1972–2001, in the Malaise trap. In 1973, *V. vulgaris* nested beneath a mat of saxifrage spreading over a path from a rockery, in 1996, behind slate cladding on the front of the house, and in a retaining wall in 1992. *Vespula rufa* was only captured twice, in 1973 and 1985, and its cuckoo, *V. austriaca* only once, in 1978. *Dolichovespula sylvestris* was taken in 13 years, but *D. norvegica* only once, in 1980. *Dolichovespula media* is a recent immigrant to the country from continental Europe, and is spreading north; it was first captured in the garden in 1995.

SOCIAL WASPS

15	***Dolichovespula norvegica*** (Fab.)	1
16	***Dolichovespula sylvestris*** (Scopoli)	43
17	***Dolichovespula media*** (Retzius)	7
18	***Vespula austriaca*** (Panzer)	1
19	***Vespula rufa*** (L.)	2
20	***Vespula germanica*** (Fab.)	2,124
21	***Vespula vulgaris*** (L.)	5,743

Records of solitary bees

45 species of solitary bees of five families were captured in the Malaise trap, 1975–2001, some of them quite commonly, there being a total of 6,686 individuals. However, it is important to appreciate that trap captures reflect flight activity rather than abundance *per se*. The commonest and most abundant species were *Colletes daviesanus, Hylaeus communis, Andrena fulva* and *Anthophora plumipes*, all caught more than 750 times during the 27-year period. Seven species were taken once only in the 27-year period.

Females of *Andrena fulva* were regularly seen foraging from *Ribes uva-crispa*, females and males of *Colletes daviesanus* from *Tanacetum parthenium*, and megachilids took pollen from a variety of flowers. Females of *Hylaeus communis*

SOLITARY BEES

C common – captured in 22–27 years
F frequent – captured in 15–21 years
O occasional – captured in 6–14 years
R rare – captured in 1–5 years.

Colletidae			
22	***Colletes daviesanus*** Smith, F	1,286	C
23	***Hylaeus communis*** Nylander	819	C
24	***Hylaeus hyalinatus*** Smith, F	23	O
Andrenidae			
25	***Andrena fucata*** Smith, F	5	R
26	***Andrena fulva*** (Müller in Allioni)	1,338	C
27	***Andrena praecox*** (Scopoli)	1	R
28	***Andrena jacobi*** Perkins, R. C. L	385	C
29	***Andrena bicolor*** Fab.	130	F
30	***Andrena nigroaenea*** (Kirby)	151	C
31	***Andrena haemorrhoa*** (Fab.)	295	C
32	***Andrena chrysosceles*** (Kirby)	1	R
33	***Andrena minutula*** (Kirby)	7	R
34	***Andrena subopaca*** Nylander	1	R
Halictidae			
35	***Halictus rubicundus*** (Christ)	6	R
36	***Halictus tumulorum*** (L.)	1	R
37	***Lasioglossum albipes*** (Fab.)	3	R
38	***Lasioglossum fulvicorne*** (Kirby)	1	R
39	***Lasioglossum villosulum*** (Kirby)	33	O
40	***Lasioglossum cupromicans*** (Pérez)	4	R
41	***Lasioglossum leucopum*** (Kirby)	37	O
42	***Lasioglossum morio*** (Fab.)	45	F
43	***Lasioglossum smeathmanellum*** (Kirby)	298	C
44	***Sphecodes epihippus*** (L.)	3	R
45	***Sphecodes fasciatus*** von Hagens	16	O
46	***Sphecodes gibbus*** (L.)	1	R
47	***Sphecodes puncticeps*** Thomson	2	R
Megachilidae			
48	***Anthidium manicatum*** (L.)	18	O
49	***Chelostoma campanularum*** (Kirby)	2	R
50	***Osmia rufa*** (L.)	333	C
51	***Osmia coerulescens*** (L.)	136	C
52	***Osmia leaiana*** (Kirby)	21	O
53	***Megachile centuncularis*** (L.)	109	C
54	***Megachile ligniseca*** (Kirby)	21	O
55	***Megachile versicolor*** Smith, F	4	R
56	***Megachile willughbiella*** (Kirby)	114	C
57	***Megachile circumcincta*** (Kirby)	1	R
58	***Coelioxys inermis*** (Kirby)	2	R
59	***Coelioxys rufescens*** Lepeletier & Serville	3	R
Anthophoridae			
60	***Nomada flava*** Panzer	44	O
61	***Nomada goodeniana*** Kirby)	3	R
62	***Nomada marshamella*** (Kirby)	27	O
63	***Nomada panzeri*** Lepeletier	76	F
64	***Nomada ruficornis*** (L.)	3	R
65	***Anthophora plumipes*** (Pallas)	779	C
66	***Anthophora furcata*** (Panzer)	98	C

Andrena fulva, the commonest solitary bee in the Leicester garden, which nests in undisturbed soil

fed from *Rubus fruticosus* flowers and *Anthophora plumipes* from *Lamium maculatum*, and in both, male-male aggressive interactions occurred over flower patches (Archer 1990a). Females of *Andrena fulva* nested in undisturbed soil, and a dozen neat leaf cells of *Megachile willughbiella* were found in 1985 beneath the compost in a seed-tray in the greenhouse.

Bees consistently foraged in the garden although few species were known to nest. Eleven species of solitary cuckoo bees have been recorded; *Nomada* species eat the stored food of *Andrena* species, *Sphecodes* species that of *Halictus* species, and *Coelioxys* species that of *Megachile* species, each cuckoo species often being associated with one species of host.

Records of solitary wasps

55 species of solitary wasps of six families were captured in the Malaise trap in 1975–2001. As might be expected, as secondary or higher order consumers they were far less abundant than were solitary bees; a 27-year total of 3,610 wasps was trapped compared with 6,686 bees. The commonest and most abundant species, trapped more than 200 times, were *Omalus auratus, Crossocerus annulipes, Pemphredon lugubris, P. lethifer, Passaloecus gracilis* and *P. singularis*. Nine species were taken once only in the 27-year period. Solitary wasps were rarely evident in the garden, although *Ancistrocerus gazella* was seen hunting for caterpillars beneath clover and lettuce leaves (Archer 1990a).

Wasps of the family Chrysididae are parasitic on other insects or cleptoparasites, consuming the stored food of solitary wasps. *Omalus auratus*, for instance, feeds on aphids stored by *Pemphredon*

SOLITARY WASPS

C common – captured in 22–27 years
F frequent – captured in 15–21 years
O occasional – captured in 6–14 years
R rare – captured in 1–5 years.

Chrysididae			
67	***Omalus auratus*** (L.)	229	C
68	***Omalus violaceus*** (Scopoli)	72	C
69	***Chrysis angustula*** Schenk	39	O
70	***Chrysis ignita*** (L.)	17	O
71	***Chrysis impressa*** Schenk	8	O
72	***Trichrysis cyanea*** (L.)	3	R
73	***Cleptes semiauratus*** (L.)	9	O
Tiphiidae			
74	***Tiphia minuta*** Vander Linden	5	R
Sapygidae			
75	***Sapyga clavicornis*** (L.)	2	R
76	***Sapyga quinquepunctata*** (Fab.)	14	O
Pompilidae			
77	***Dipogon variegatus*** (L.)	6	R
Eumenidae			
78	***Ancistrocerus gazella*** (Panzer)	89	C
79	***Ancistrocerus parietinus*** (L.)	2	R
80	***Ancistrocerus trifasciatus*** (Müller)	9	O
81	***Symmorphus connex*** (Curtis)	4	R
82	***Symmorphus gracilis*** (Brullé)	1	R
Sphecidae			
83	***Trypoxylon attenuatum*** Smith, F	68	F
84	***Trypoxylon clavicerum*** Lepeletier	75	F
85	***Trypoxylon figulus*** (L.)	2	R
86	***Crossocerus elongatulus*** (Vander Linden)	123	C
87	***Crossocerus pusillus*** Lepeletier & Brullé	7	R
88	***Crossocerus annulipes*** (Lepeletier & Brullé)	214	C
89	***Crossocerus capitosus*** (Shuckard)	22	F
90	***Crossocerus megacephalus*** (Rossius)	76	C
91	***Crossocerus nigritus*** (Lepeletier & Brullé)	12	O
92	***Crossocerus walkeri*** Shuckard	1	R
93	***Crossocerus podagricus*** (Vander Linden)	12	O
94	***Crossocerus quadrimaculatus*** (F.)	1	R
95	***Crossocerus binotatus*** Lepeletier & Brullé	1	R
96	***Crossocerus dimidiatus*** (Fab.)	9	O
97	***Crossocerus distinguendus*** (Morawitz)	32	R
98	***Ectemnius cavifrons*** (Thomson)	126	C
99	***Ectemnius ruficornis*** (Zetterstedt)	1	R
100	***Ectemnius sexcinctus*** (Fab.)	3	R
101	***Ectemnius continuus*** (Fab.)	1	R
102	***Ectemnius cephalotes*** (Olivier)	54	O
103	***Rhopalum clavipes*** (L.)	64	F
104	***Rhopalum coarctatum*** (Scopoli)	2	R
105	***Oxybelus uniglumis*** (L.)	1	R
106	***Psenulus pallipes*** (Panzer)	101	C
107	***Psenulus concolor*** (Dahlbom)	100	C
108	***Spilomena beata*** Bluthgen	3	R
109	***Spilomena troglodytes*** (Vander Linden)	16	R
110	***Stigmus solskyi*** Morawitz, A	30	O
111	***Pemphredon lugubris*** (Fab.)	385	C
112	***Pemphredon inornatus*** Say	33	F
113	***Pemphredon lethifer*** (Shuckard)	212	C
114	***Pemphredon morio*** Vander Linden	6	R
115	***Diodontus tristis*** (Vander Linden)	26	O
116	***Passaloecus corniger*** Shuckard	54	F
117	***Passaloecus gracilis*** (Curtis)	276	C
118	***Passaloecus insignis*** (Vander Linden)	32	O
119	***Passaloecus singularis*** Dahlbom	917	C
120	***Mellinus arvensis*** (L.)	2	R
121	***Philanthus triangulum*** (Fab.)	1	R

lethifer, P. inornatus and *Passaloecus gracilis* for their larvae, and occasionally on spiders or flies stored by *Trypoxylon* species and *Corynopus coarctatus* respectively (Danks 1971). *Cleptes semiauratus* is parasitic on sawflies, and *Chrysis* species parasitize solitary bees and wasps, eating host eggs and larvae and then, often, devouring the stored food. *Sapyga quinquepunctata* parasitizes *Osmia* and *Hylaeus* bees, the parasitic larva destroying the host larva and then consuming the stored food.

A solitary wasp, *Ectemnius cavifrons*, common in the Leicester garden, which stocks its burrows with flies, including hoverflies, for its larvae to eat

All of the social bees recorded in the garden are widely distributed in the British Isles, the least common being *Bombus ruderatus*, which has a southerly distribution. Cuckoo bumblebees are never anything like as common as *Bombus* species, but all species of *Psithyrus* recorded in the garden have wide distributions. Bumblebee ranges contracted post-1960 compared with pre-1960 records, the Midlands being particularly affected, and Leicester is in the 'central impoverished region' where only 'mainland ubiquitous

Anthophora plumipes, a common solitary bee often seen flying early in the summer, when it may be mistaken for a bumblebee

species' persist (Williams 1982), although the records that Williams analysed do not include *Bombus ruderatus*.

In the late 1990s and early years of the twenty-first century, bumblebee numbers in general were giving cause for concern, the decline being attributed to modern agricultural methods. No cuckoo bumblebee was caught after 1983, and *B. lapidarius* and *B. ruderarius* became scarce from the late 1980s onwards. Pre-1960 county records include *Bombus jonellus, B. humilis, B. sylvarum, B. distinguendus* and *B. subterraneus*, and *P. bohemicus*. Including the honey bee, the garden list comprises 56% of the national list of Apidae; possibly this is the maximum that can be expected, as the only other species recorded in the county post-1960 is *B. soroeensis* (Archer, M. E. pers. comm.).

Of the social wasps, *Vespula germanica, V. vulgaris, V. rufa, Dolichovespula sylvestris* and *D. norvegica* have wide distributions, the last named being the least common, particularly in the English Midlands. *Vespula austriaca*, a cuckoo wasp, however, is far less widespread, being mainly a northern and western species; the garden record was the first not only for the county of Leicestershire, but also for the Midlands. *Dolichovespula media*, first caught in 1995 (five individuals) and again in 1999 (two), is a recent immigrant from continental Europe and is spreading northwards. The only vespid not taken in the garden is the hornet, *Vespa crabro*, which is confined to the southern half of England. They do occasionally occur in the county, and I saw one in 1979 elsewhere in the city of Leicester.

Fig. 6.1. Annual fluctuations in numbers of *Bombus* species caught in Malaise trap, 1972–2001

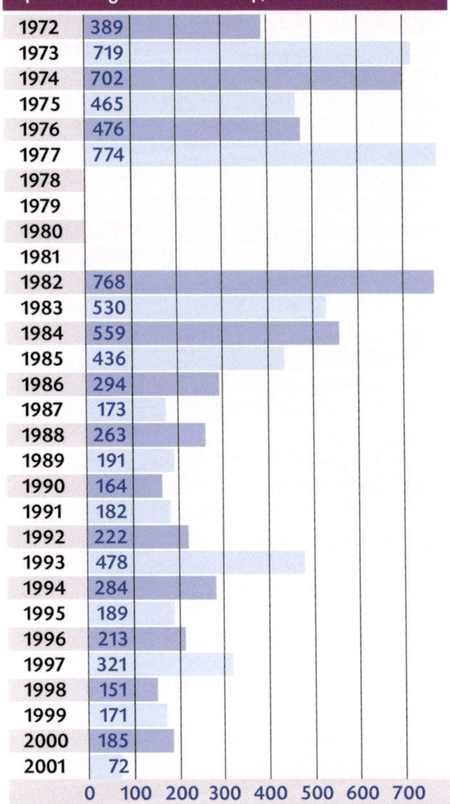

Abundance of solitary bees and solitary wasps

There are 517 species of solitary aculeates in the British Isles, and 185 species (92 bees and 93 wasps) have been recorded in the county of Leicestershire (Archer 1990b); 45.7% of the county's solitary bees and 52.7% of the solitary wasps were recorded in the garden, 1975–2001. Of the bees, Megachilidae may be over-represented and Andrenidae and Anthophoridae under-represented in the garden; of the wasps, Chrysididae and Sphecidae may be over-represented (Archer 1990a).

Megachile versicolor, Omalus violaceus, Ectemnius ruficornis and *E. sexcinctus* are new records for the county, although not all records of other species have been published. In addition, three species of solitary bee (*Andrena minutula, Sphecodes puncticeps* and *Chelostoma campanulorum*) and eight of solitary wasp

(*Chrysis angustula, Tiphia minuta, Symmorphus connex, Symmorphus gracilis, Crossocerus walkeri, Crossocerus distinguendus, Rhopalum coarctatum* and *Philanthus triangulum*) were first caught in the garden after 1988, so do not appear in Archer (1990b), and are new to the county.

Only about 68% of the solitary species had been caught by the twelfth year of trapping (1986), and new species continued to be added until 2000, apparently as they dispersed seeking new habitats; some, such as *Philanthus triangulum* (one in 1996) evidently found none, whereas *Crossocerus distinguendus* did (32 from 1998 onwards).

The percentage of garden solitary wasp species, up until 1986, that were cleptoparasites, using the host's food store, was similar to the percentage of cleptoparasitic bees but, whereas the percentage of cleptoparasitic wasps was nearly the same in the garden and the county, the percentage of cleptoparasitic bees in the garden (18.9%) was considerably lower than that in the county (28.1%).

More than 92% of the solitary wasps (other than cleptoparasites) in the garden, and more than 36% of the solitary bees, are aerial nesters, nesting in stems, walls or wood, as opposed to subterranean nesters, which nest in soil. The percentage of aerial nesters in the county is much lower than this for both bees and wasps. The garden evidently provided a more satisfactory habitat for aerial than for subterranean nesters, almost certainly because the soil is not sandy and is frequently disturbed by cultivation. This would also explain why the percentage

Fig. 6.2. Annual fluctuations in numbers of *Vespula germanica* caught in Malaise trap, 1972–2001

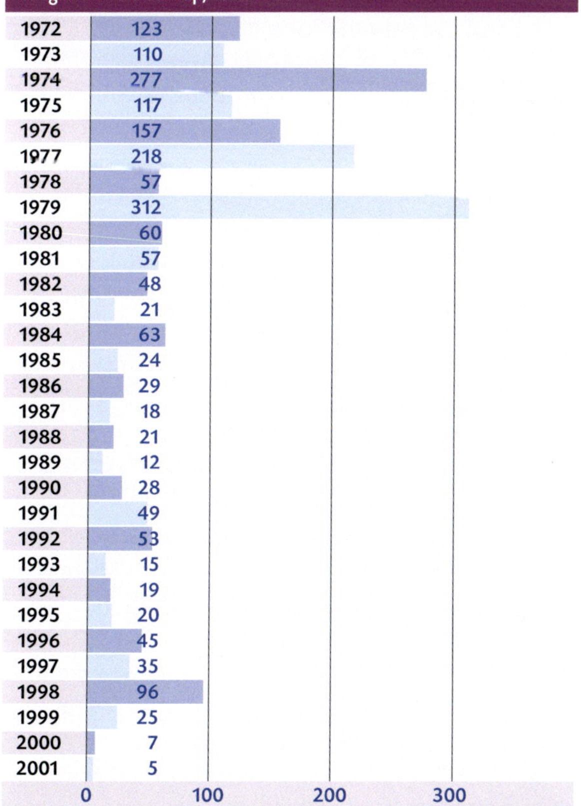

Fig. 6.3. Annual fluctuations in numbers of *Vespula vulgaris* caught in Malaise trap, 1972–2001

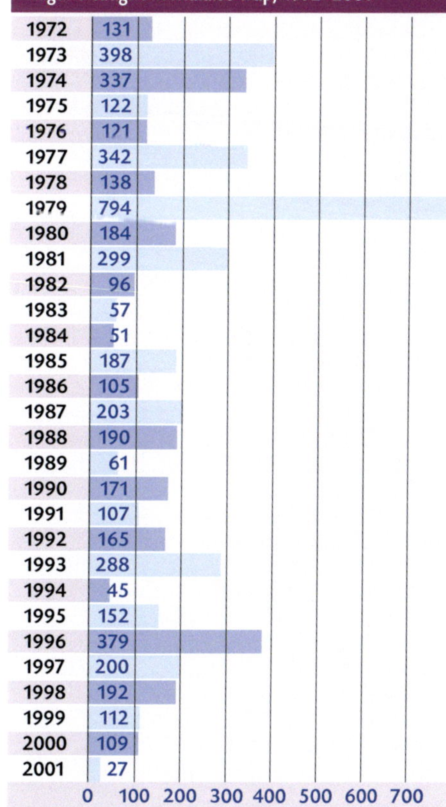

of cleptoparasitic bees was lower than in the county, because many are associated with subterranean nesters (Archer 1990a).

Archer (1990a) showed that the rates of appearance of species in the Leicester garden and disappearance from it do not fit with the immigration and extinction curves that would be expected if the garden was regarded as an island. On the contrary, its solitary aculeate fauna is part of that of adjoining gardens, so that species move in and out by chance from the pool of species in the neighbourhood. Solitary bees and wasps in the garden are both diverse, and reflect the heterogeneity of habitat of the neighbourhood of which the garden is a part. Nevertheless, the garden catch of solitary aculeates represents only 54.6% of the species recorded in the county as a whole.

The situation for social bees and wasps is markedly different; not only are some of the garden species much more numerous than any solitary species, but there are few additional species that could be caught.

ANNUAL FLUCTUATIONS IN NUMBERS OF BEES AND WASPS

Although it is possible, over a time span of 30 years, to generalize about absolute and relative numbers of species of bees and wasps, there are considerable differences in abundance from year to year in the Malaise trap sample. Often, although not always, these fluctuations can be attributed to weather differences from one year to the next, which affect not only the abundance but also the flight activity, and hence ease of trapping, of insects.

Fig. 6.4. Annual fluctuations in numbers of solitary bees caught in Malaise trap, 1975–2001

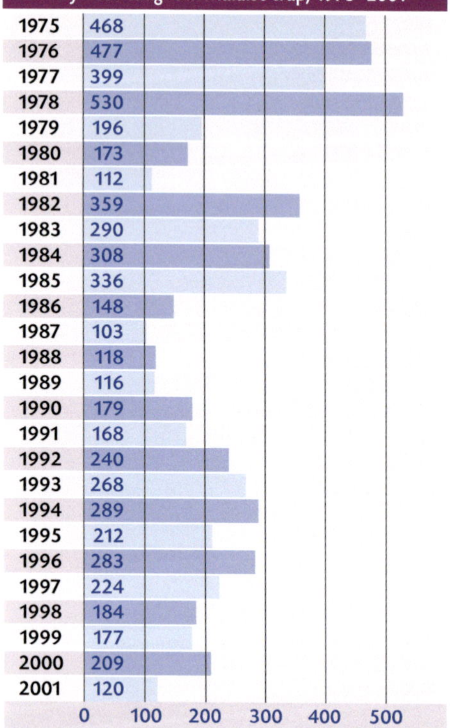

Bumblebees were sorted from Malaise trap samples in 1972–1977 and in 1982–2001. Annual numbers of *Bombus* species trapped are shown in Fig. 6.1. *Bombus pascuorum* was by far the commonest species, and the overall annual pattern largely reflects changes in the abundance of that species, except in 1977, when numbers of *B. hortorum*, *B. lucorum* and *B. pratorum* were relatively high. *Bombus* numbers were high in 1973, 1974, 1977 and 1982, and showed a sudden increase in 1993, being generally low from 1986 onwards, and particularly in 2001. When identification of bumblebees was resumed in 1982, after a four-year gap, *B. ruderatus* had disappeared, and none was trapped 1982–2001. This suggests that the species had become locally extinct after 1977, lending support to the contention of Williams (1982) that it is a southern local species and hence only marginal in the English Midlands. Numbers of *B. lapidarius* were relatively high in 1982, but it was absent in 1989 and 1990, and became scarce in the last four years of trapping, 1998–2001. *Bombus ruderarius* was absent in 1983, and scarce in the period 1984–1988, after which only one was caught, in 1997.

As Fig. 6.1 shows, whatever bumblebee numbers were responding to, it was not quite the same factor or factors that affected the numbers of butterflies, moths or hoverflies. 1973, one of the best bumblebee years, was also good for butterflies (Fig. 3.1) and moths (Fig. 4.1), and 1982 was also good for hoverflies (Fig. 5.1), but 1974 and 1977 were poor years for butterflies, moths and hoverflies. 1993 was a poor year for butterflies, although moths and hoverflies showed something of a resurgence. Of the increasing number of poor bumblebee years, 1972 was also poor for butterflies, moth and hoverflies, and 1986 was also poor for butterflies and hoverflies. Since then, numbers of butterflies, moths and hoverflies, like bumblebees, seem to have been in general decline. As reference to the weather data in Tab. 2.2 shows, bumblebees tended to do well in rather damp summers with average amounts of sunshine, and to be unaffected by temperature.

Relative abundance of *Vespula germanica* and *V. vulgaris*

Numbers of the two common social wasps, *Vespula germanica* and *V. vulgaris*, caught in the Malaise trap fluctuated considerably from year to year (Fig. 6.2 and 6.3). Both species were abundant in 1974, 1977, 1979 and 1998, and *Vespula vulgaris* also in 1973, 1981, 1993 and 1996; 1983, 1985 and most subsequent years were poor for *V. germanica*, and 1983, 1984, 1989 and 2001 for *V. vulgaris*.

Table 6.1. Size and composition of annual, 12-year, 15-year and 27-year Malaise trap catches of solitary bees in the Leicester garden, 1975–2001

Year	Individuals (N)	Species (S)	N/S	Species taken once only	Commonest species	As % N
1975	468	24	19.5	10	*Colletes daviesanus*	47.6
1976	477	27	17.7	2	*Colletes daviesanus*	24.7
1977	399	23	17.6	4	*Colletes daviesanus*	45.1
1978	530	22	25.0	7	*Andrena fulva*	26.4
1979	196	18	10.9	3	*Andrena fulva*	22.5
1980	173	18	9.6	5	*Andrena fulva*	24.9
1981	112	12	9.3	3	*Colletes daviesanus*	27.7
1982	359	19	18.9	5	*Andrena fulva*	38.2
1983	290	17	171	4	*Colletes daviesanus*	39.0
1984	308	20	15.4	5	*Hylaeus communis*	25.6
1985	336	20	16.8	8	*Colletes daviesanus*	51.8
1986	148	16	9.3	4	*Anthophora plumipes*	30.4
1987	103	11	9.4	5	*Andrena fulva* & *Anthophora plumipes*	31.1=
1988	118	15	7.9	3	*Andrena fulva*	40.7
1989	116	16	7.3	7	*Andrena fulva*	26.7
1990	179	17	10.5	6	*Hylaeus communis*	34.6
1991	168	16	10.5	4	*Andrena fulva*	22.0
1992	240	20	12.0	2	*Lasioglossum smeathmanellum*	14.6
1993	268	18	14.9	5	*Colletes daviesanus*	17.2
1994	289	20	14.5	4	*Anthophora plumipes*	28.4
1995	212	27	7.9	8	*Andrena fulva*	25.5
1996	283	21	13.5	2	*Andrena fulva*	31.0
1997	224	23	9.7	4	*Anthophora plumipes*	19.6
1998	184	18	10.2	9	*Andrena fulva*	27.7
1999	177	19	9.3	7	*Andrena fulva*	17.5
2000	209	21	10.0	4	*Andrena haemorrhoa*	23.4
2001	120	20	6.0	5	*Anthophora plumipes*	24.2
1975–1986	3,796	37	102.6	7	*Colletes daviesanus*	30.5
1987–2001	2,890	38	76.1	3	*Andrena fulva*	20.6
1975–2001	**6,686**	**45**	**148.9**	**7**	***Andrena fulva***	**20.0**

German wasp, *Vespula germanica*, on an apple. Once common, German wasps abruptly declined in numbers nationwide in the late 1970s and early 1980s

1979 was a particularly good wasp year while 1983, 1989, 1994 and 2001 were particularly poor years. In 1979, spring was wet and cool, and the summer dry with average temperatures, and it was neither particularly good nor particularly bad for other insects. It is, however, noteworthy that 1974 and 1977, when wasps were abundant, were good years for bumblebees, although 1996 was not; 1979 cannot be compared, as there are no bumblebee figures for that year.

Archer (1985) analysed the population dynamics of *Vespula vulgaris* and *V. germanica*, using, in part, data from the Leicester garden. He identified several characteristics of yearly abundance: large variations in yearly abundance; a two- and possibly a seven-year cycle of yearly abundance; exceptional years of abundance and scarcity tend to occur in pairs; during exceptional years of summer abundance, spring queens are scarce and vice versa for exceptional years of summer scarcity.

In the Leicester garden, a scarce year for *Vespula vulgaris*, as measured by the number of workers captured, (1978), preceded the most abundant year (1979), while for *V. germanica*, the most abundant year (1979) preceded a scarce year (1980). Furthermore, the year when spring queens were most abundant (1978 for *V. vulgaris* and 1980 for *V. germanica*) had few workers, and the year when fewest spring queens were caught (1979 for both species) had the most workers.

Archer (1985) concluded that when spring queens are abundant, competition for nest sites and the risk of usurpation have an adverse effect, leading to production of fewer workers in the population, but when there are fewer queens, more nests succeed, leading to more workers. He postulated that this endogenous factor generates the characteristics of population dynamics listed above, and is probably reinforced by differences in queen quality resulting from different rearing conditions, in that a queen reared in a nest full of workers tends to be better fed and vice versa. This endogenous mechanism is over-compensating because of the potentially large queen production of each colony, and tends to damp to equilibrium, taking about seven years to do so. It is then, however, as the endogenous factor weakens, that the effects of an exogenous factor, probably a function of summer and autumn weather, becomes more evident. Hence the two- and seven-year cycles of abundance. To a great extent, the figures plotted in Fig. 6.2 and 6.3 disguise these cycles, because queens, males and workers of both species are included, although it clearly shows the exceptional abundance of 1979, the key to the suggested two-year cycle.

In the 1985 publication, Archer was dealing with slightly less than half of the 30-year sample. In a subsequent publication (Archer 2001), he showed that the abundance of *V. germanica* declined abruptly during the late 1970s and early 1980s, whereas that of *V. vulgaris* did not. This was almost certainly caused by increased use of pesticides on agricultural land which acted directly by killing wasps and indirectly by affecting their insect food supply. *Vespula vulgaris* was not so susceptible as *V. germanica* because of its better foraging ability (Archer 2001). Archer also showed that during the early 1980s, the two-year cycle of abundance of *V. vulgaris* changed to a nearly perfectly damped pattern of annual abundance.

Abundance of solitary bees

The annual Malaise trap catch of solitary bees varied considerably, between a high of 530 in 1978 and a low of 103 in 1987, although numbers remained good in general (Fig. 6.4 and Tab. 6.1). The periods 1975–1978, 1982–1985 and, to a lesser extent, 1992–1997 were evidently good for bees, whereas 1979–1981, 1986–1989 and 2001 were poor. Forty-five species were caught in total, seven of them represented by only one individual; one of these was caught in 1975, one in 1976, one in 1977, and two each in 1980 and 1981. Numbers of species were high in 1975–1978, 27 species being caught in 1976, the drought year, and 27 species were also caught in 1995, which was also a hot, dry, bright summer. The smallest number

A jewel wasp, *Omalus auratus* (top). In 1976 this was the commonest solitary wasp in the Leicester garden. It is a cleptoparasite, feeding on aphids stored by *Pemphredon* species

Digger wasp, *Pemphredon lugubris* (above), emerging from a burrow in a tree. In three of the study years this was the commonest solitary wasp in the Leicester garden

Table 6.2. Size and composition of annual, 12-year, 15-year and 27-year Malaise trap catches of solitary wasps in the Leicester garden, 1975–2001

Year	Individuals (N)	Species (S)	N/S	Species taken once only	Commonest species	As % N
1975	99	20	5.0	7	*Pemphredon lugubris*	18.2
1976	192	23	8.4	5	*Omalus auratus*	17.2
1977	109	17	6.4	4	*Passaloecus singularis*	19.3
1978	75	20	3.8	10	*Pemphredon lethifer*	13.3
1979	53	17	3.1	7	*Pemphredon lugubris*	30.2
1980	80	18	4.4	8	*Pemphredon lugubris*	25.0
1981	73	14	5.2	4	*Passaloecus singularis*	26.0
1982	166	20	8.3	4	*Passaloecus singularis*	25.9
1983	182	23	7.9	6	*Pemphredon lethifer*	17.1
1984	185	24	7.7	6	*Passaloecus singularis*	28.0
1985	109	18	6.1	3	*Passaloecus singularis*	30.3
1986	78	21	3.7	6	*Passaloecus singularis*	20.5
1987	56	17	3.3	7	*Passaloecus gracilis*	25.0
1988	89	18	4.9	5	*Passaloecus singularis*	32.6
1989	115	22	5.2	8	*Passaloecus singularis*	24.4
1990	193	25	7.7	7	*Passaloecus singularis*	15.5
1991	162	21	7.7	5	*Passaloecus singularis*	30.9
1992	206	29	7.1	12	*Passaloecus singularis*	22.8
1993	226	31	7.3	8	*Passaloecus singularis*	24.3
1994	110	21	5.2	6	*Passaloecus singularis*	35.5
1995	101	23	4.4	9	*Passaloecus singularis*	47.5
1996	191	30	6.4	11	*Passaloecus singularis*	35.1
1997	230	26	8.8	7	*Passaloecus singularis*	44.8
1998	137	25	5.9	5	*Passaloecus singularis*	24.1
1999	176	30	5.9	6	*Passaloecus singularis*	33.5
2000	115	25	4.6	10	*Passaloecus singularis*	32.2
2001	103	27	3.8	13	*Passaloecus singularis*	21.4
1975–1986	1,401	34	41.1	4	*Passaloecus singularis*	18.7
1987–2001	2,209	54	40.9	8	*Passaloecus singularis*	29.7
1975–2001	**3,610**	**56**	**64.5**	**9**	***Passaloecus singularis***	**25.4**

of species was caught in 1981 and in 1987, but the average number of individuals per species (N/S) was at its lowest in 1988, 1989, 1995 and 2001, poor years for bees, showing that the species that were present were less common.

Colletes daviesanus was the commonest species in seven years, and commonest overall in 1975–1986, but suffered a dramatic decline in numbers from 1986 onwards. It nests colonially in masonry and sandstone walls and has never nested in the garden; presumably in the late 1980s, its local nesting site was destroyed. *Andrena fulva* was the commonest species in eleven years and joint commonest with *Anthophora plumipes* in one; *A. plumipes* was commonest in a further four years. *Hylaeus communis* was the commonest species in two years, and *Lasioglossum smeathmanellum* and *Andrena haemorrhoa* each in one. *Andrena fulva* was the commonest species for the period 1987–2001, and overall 1975–2001.

Fig. 6.4 and Tab. 6.1 do not differentiate between spring and summer species, yet these may be fluctuating in different ways. Numbers of spring bees peaked in 1978 and 1982, because of large numbers of *Andrena fulva*, which was the commonest species in both years, whereas summer species peaked in 1976 and 1985, in both of which years *Colletes daviesanus* was the commonest species. The first twelve years, 1975–1986, gave the impression that weather affects numbers of solitary bees, there tending to be more when it was warm and sunny, and, particularly, fewer when spring and summer were cool and dull, as in 1986, although there were exceptions, implying that other factors are important.

In the subsequent 15 years, 1987–2001, however, there seemed to be no clear link between spring and summer weather conditions and number of bees caught. Admittedly, weather conditions were poor in 1987, the worst bee year, but in 1989, which was nearly as bad for bees, spring was warm, damp and bright and summer hot, dry and bright. However, in 1994 and 1996, when there was some resurgence of bee numbers despite the general decline, the summers were hot and bright. Hot, bright weather may increase flight activity and hence increase numbers caught in the Malaise trap, whether or not population sizes are small.

Fig. 6.5. Annual fluctuations in numbers of solitary wasps caught in Malaise trap, 1975–2001

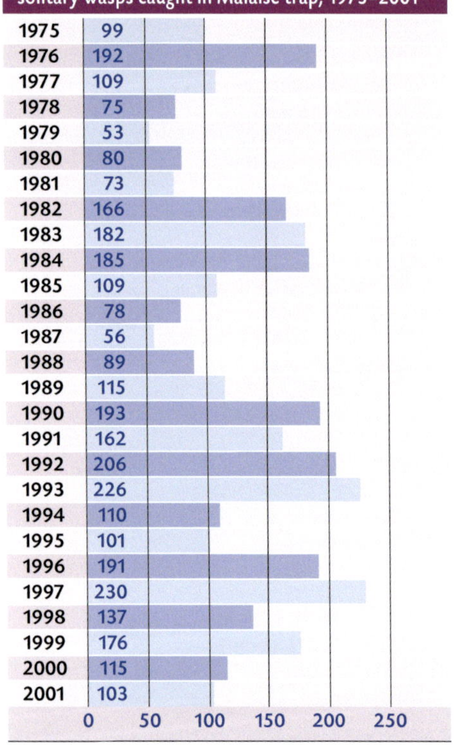

Abundance of solitary wasps

The total 27-year catch of solitary wasps, 3,610, was markedly smaller than that of bees, 6,686. Annual catches varied from 53 in 1979 to 229 in 1997 (Tab. 6.2 and Fig. 6.5). The years 1976, 1982–1984, 1992–1993 and 1997 were good wasp years, and 1979 and 1987 particularly poor, although the pattern of variation was not as marked as in solitary bees, and did not match it particularly well. The good wasp years were also years when the number of species was high; 55 were taken altogether, with a maximum of 31 in 1993, and a minimum of 14 in 1981. Eight species were represented by only one individual in the 27 years, one each in 1980,

1982, 1997, 1998, 1999 and 2000, and two in 1996. Solitary wasps are represented in the garden by a high number of species, many of them rare, and all less abundant than herbivorous species. This pattern is typical of predatory insect groups. As predators, they have to be wide-ranging as individuals, hence the high number of species taken once only.

In the early years of Malaise trapping, I assumed that good wasp years were associated with hot, dry, bright summers, but this certainly did not hold good for subsequent years, and there seems to be no clear link between summer weather and wasp numbers, although fine, warm springs often led to good wasp years. Solitary wasps did not exhibit the overall decline in numbers that solitary bees and many other groups of insect experienced. It is noteworthy that they had their best year as late as 1997; furthermore, the annual average catch in the first twelve years was 116.75, whereas that in the final 15 years was 147.27.

The average number of individuals per species (N/S) was highest in the good wasp years (1976, 1982–1984, 1992–1993, 1997) indicating that abundance was a consequence of species already there being commoner as well as addition of species. It is noteworthy that the average number of individuals per species was considerably lower than for solitary bees, in line with wasps being at a higher trophic level. *Passaloecus singularis* was the commonest species in 20 years and overall in the 27-year catch; *Pemphredon lugubris* was the commonest in three years, *Pemphredon lethifer* in two years, and *Passaloecus gracilis* and *Omalus auratus* each in one. *Passaloecus singularis* became more common during the 27 years 1975–2001.

SUMMARY

Bees and wasps were abundant and conspicuous in the garden for most of the 30-year period, although continuous trapping confirmed the subjective impression that overall numbers were declining. Representation of social species that occur in the English Midlands was good, and few additional species could have been expected. In addition, more than half of the species of solitary aculeates recorded in the county of Leicestershire were caught; relatively few species were known to breed in the garden, many visiting, albeit regularly, to feed and to collect food for their larvae.

The most striking finding of the 30-year investigation was that solitary wasps, unlike most other insect groups, far from declining in numbers, were even more abundant in the later part of the period. The most likely explanation of this is that weather in the second 15 years, 1987–2001, was on average warmer and sunnier, conditions that suited solitary wasps. That it was also wetter probably had no effect because the increased rainfall was in autumn, after the season when solitary wasps are active.

7 : Sawflies, psocids, bugs, lacewings and their allies

Sawflies (order Hymenoptera: suborder Symphyta) are distinguished from other Hymenoptera by the lack of a 'waist' at the base of the relatively broad abdomen. Almost all have two roughened protuberances called cenchri, on the dorsal metathorax, which engage with rough areas of the forewings when the wings are closed. In most, the ovipositor is toothed like a saw, the pattern of the teeth being characteristic for a particular species, and giving rise to the common name of the group. The saw-like ovipositor is used to cut slits in stems and leaves, inside which the eggs are laid. The majority of adult sawflies are small to medium-sized insects with biting mouthparts and a relatively long tongue-like labium, used for lapping nectar. They feed at flowers, usually on pollen and nectar, although some species are at least partly carnivorous.

Gregarious larvae of *Arge pagana* (opposite, top left), a sawfly caught only once in the Leicester garden, feeding on rose leaflets

A psocid, *Stenopsocus stigmaticus* (opposite, top right), rare in the Leicester garden

A parent bug, *Elasmucha grisea*, with young (opposite, centre left). This is one of the common Heteroptera known as shield bugs

Eupteryx aurata (opposite, bottom left), a common garden Homopteran

Green lacewing, *Chrysoperla carnea* (opposite, bottom right), a common lacewing often attracted to light

Natural history of sawflies

Most sawfly larvae feed on leaves and resemble caterpillars in appearance, although they have more pairs of abdominal pro-legs (at least six). Wood-eating sawfly larvae, those that tunnel in leaves and stems, and the few gall-making

Fig. 7.1. Annual fluctuations in numbers of sawflies caught in Malaise trap, 1981–2001

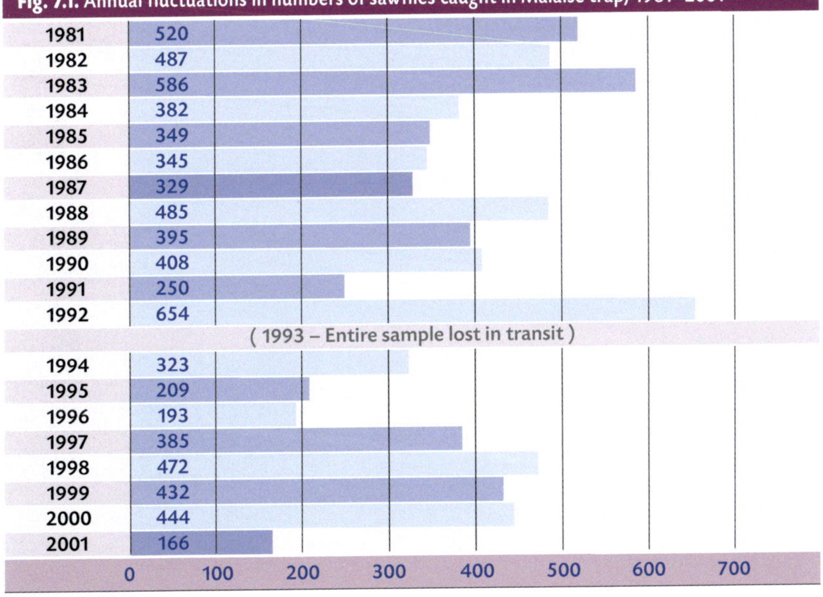

A sawfly, *Selandria serva*, (above), one of the scarcer Tenthredinidae in the Leicester garden

species are more like beetle larvae, with no abdominal legs. The larvae of *Orussus abietinus* (not a UK species) are unusual in being parasites of timber beetles. The largest British family, with about 390 species, is the Tenthredinidae, most of whose leaf-eating larvae are ecologically very like lepidopteran caterpillars, and some species, such as the gooseberry sawfly (*Nematus ribesii*) are commercial pests. The kidney-bean-like galls of *Pontania* species were abundant on willow leaves in the garden in some years; they are induced by the egg-laying puncture, and the larva feeds inside its gall. Most species of sawfly complete their development in a single season and over-winter as a pre-pupa, although a few species over-winter as eggs and wood wasps may remain as larvae for three or four years (Betts 1986, Chinery 1976).

Diversity and abundance of sawflies

Sawflies were sorted from Malaise trap catches in the period 1981–2001; until 1989, they were identified by L. R. Cole, and thereafter by D. A. Sheppard. A total of 7,814 individuals of 87 species was caught, mostly Tenthredinidae, together with a few individuals of four species of Argidae. Over the 21-year period (the entire sample for 1993 was lost in transit), 24 species were represented by only a single individual. Annual catches varied markedly, with the largest (654) in 1992, and the smallest (166) in 2001, although there was no continuous decline,

Table 7.1. Size and composition of annual and 20-year Malaise trap catches of sawflies in the Leicester garden, 1981–2001

Year	Individuals (N)	Species (S)	N/S	Species taken once only	Commonest species	As % N
1981	520	22	23.6	7	*Ametastegia carpini*	73.7
1982	487	34	14.3	11	*Cladius pectinicornis*	27.3
1983	586	25	23.4	7	*Cladius pectinicornis*	45.7
1984	382	27	14.2	11	*Ametastegia carpini*	59.2
1985	349	22	15.9	7	*Ametastegia carpini*	81.7
1986	345	19	18.2	4	*Ametastegia carpini*	67.5
1987	329	14	23.5	5	*Ametastegia carpini*	77.8
1988	485	25	19.4	12	*Ametastegia carpini*	74.2
1989	395	27	14.6	13	*Ametastegia carpini*	58.2
1990	408	36	11.3	15	*Ametastegia carpini*	47.1
1991	250	24	10.4	8	*Ametastegia carpini*	37.2
1992	654	33	19.8	8	*Ametastegia carpini*	62.5
		(1993 – entire sample lost in transit)				
1994	323	27	12.0	9	*Ametastegia carpini*	51.1
1995	209	23	9.1	9	*Ametastegia carpini*	64.6
1996	193	22	8.8	6	*Pristiphora rufipes*	41.5
1997	385	32	12.0	19	*Pristiphora rufipes*	57.7
1998	472	29	16.3	8	*Ametastegia carpini*	38.8
1999	432	24	18.0	11	*Ametastegia carpini*	62.3
2000	444	18	24.7	5	*Pristiphora rufipes*	62.6
2001	166	20	8.3	6	*Ametastegia carpini*	36.1
1981–2001	7,814	87	89.8	24	*Ametastegia carpini*	49.6

1998–2000 being good years (Fig. 7.1 and Tab. 7.1). Numbers of species caught fluctuated, with a peak of 36 in 1990, and the smallest number (14) in 1987. The average number of individuals per species (N/S) was highest in 2000, a good year for sawflies, and lowest in 2001, when the total catch was at its lowest, but diversity remained high.

SAWFLIES

Nomenclature and sequence follow David Sheppard's working checklist manuscript (c.2001).

Numbers show total Malaise trap captures, 1981–2001. A few additional species, collected, or reared from larvae found in the garden, are shown in green.

Annual and overall totals (Tab 7.1) exceed species listings because of specimens that could only be identified to genus.

Argidae

1	*Sterictiphora geminata* (Gmelin in L.)	5	
2	*Arge cyanocrocea* (Forster)	1	(1984)
3	*Arge pagana* (Panzer)	1	(1998)
4	*Arge ustulata* (L.)	1	(1988)

Cimbicidae

5	*Trichiosoma lucorum* (L.) (Larvae on *Crataegus monogyna*, 1977 and 1986)		

Tenthredinidae

6	*Selandria serva* (Fab.)	2	
7	*Strongylogaster macula* (Klug)	8	
8	*Aneugmenus padi* (L.)	48	
9	*Dolerus aericeps* Thomson	1	(1995)
10	*Dolerus planatus* Hartig	1	(2001)
11	*Athalia circularis* (Klug)	25	
12	*Athalia cordata* Lepeletier	169	
13	*Athalia cornubiae* Benson	110	
14	*Athalia liberta* (Klug)	268	
15	*Athalia lugens* (Klug)	4	
16	*Athalia rosae* (L.)	1	(1997)
17	*Empria excisa* (Thomson)	1	(1992)
18	*Empria tridens* (Konow)	10	
19	*Ametastegia equiseti* (Fallén)	4	
20	*Ametastegia glabrata* (Fallén)	8	
21	*Ametastegia carpini* (Hartig)	3,872	
22	*Ametastegia pallipes* (Spinola)	5	
23	*Ametastegia tener* (Fallén)	12	
24	*Allantus calceatus* (Klug)	1	(1989)
25	*Allantus cinctus* (L.)	94	
26	*Allantus cingulatus* (Scopoli)	3	
27	*Allantus rufocinctus* (Retzius in Degeer)	2	
28	*Caliroa annulipes* (Klug)	1	(1994)
29	*Endelomyia aethiops* (Fab.)	12	
30	*Heterarthrus aceris* (Kaltenbach)	1	(1988)
31	*Phymatocera aterrima* (Klug)	5	
32	*Monophadnus pallescens* (Gmelin in L.)	8	
33	*Ardis brunniventris* (Hartig)	4	
34	*Blennocampa phyllocolpa* Viitassari & Vikberg	1	(1994)
35	*Halidamia affinis* (Fallén)	89	
36	*Metallus albipes* (Cameron)	3	
37	*Metallus lanceolatus* (Thomson)	6	
38	*Fenella nigrita* Westwood	1	(1997)
39	*Aglaostigma aucupariae* (Klug)	36	
40	*Aglaostigma fulvipes* (Scopoli)	6	
41	*Tenthredopsis coquebertii* (Klug)	32	
42	*Tenthredopsis litterata* (Geoffroy in Fourcroy)	5	
43	*Tenthredopsis nassata* (L.)	4	
44	*Rhogogaster punctulata* (Klug)	1	(1989)
45	*Rhogogaster viridis* (L.)	1	(1988)
46	*Tenthredo colon* Klug (Larvae on *Epilobium hirsutum*, 1982)		
47	*Pachyprotasis rapae* (L.)	4	
48	*Cladius pectinicornis* (Geoffroy in Fourcroy)	731	
49	*Priophorus morio* (Lepeletier)	30	
50	*Priophorus pallipes* (Lepeletier)	319	
51	*Priophorus pilicornis* (Curtis)	102	
52	*Priophorus rufipes* (Lepeletier)	2	
53	*Priophorus ulmi* (L.)	1	(1997)
54	*Trichiocampus viminalis* (Fallén)	1	(1983)
55	*Dineura stilata* (Klug)	2	
56	*Hoplocampa chrysorrhoea* (Klug)	2	
57	*Hoplocampa crataegi* (Klug)	11	
58	*Hoplocampa testudinea* (Klug)	1	(1982)
59	*Stauronematus compressicornis* (Fab.)	22	
60	*Pristiphora saxsenii* (Hartig)	2	
61	*Pristiphora aquilegiae* (Vollenhoven)	57	
62	*Pristiphora armata* (Thomson)	100	
63	*Pristiphora coniceps* Lindqvist	1	(1997)
64	*Pristiphora kamchatica* Malaise	6	
65	*Pristiphora maesta* (Zaddach)	1	(1990)
66	*Pristiphora pallidiventris* (Fallén)	2	
67	*Pristiphora punctifrons* (Thomson)	26	
68	*Pristiphora rufipes* (Lepeletier)	971	
69	*Sharliphora nigella* (Foerster)	4	
70	*Amauronematus* species. (Larva on *Salix fragilis*, 1982)		
71	*Euura atra* (Jurine)	3	
72	*Phyllocolpa leucapsis* (Tischbein)	1	(1990)
73	*Phyllocolpa purpureae* (Cameron)	2	
74	*Pontania dolichura* (Thomson)	2	
75	*Croesus septentrionalis* (L.) (Larvae on *Betula pendula* in most years)		
76	*Nematus lucidus* (Panzer)	76	
77	*Nematus bergmanni* Dahlbom	1	(1990)
78	*Nematus hypoxanthus* Foerster	3	
79	*Nematus leucotrochus* Hartig	6	
80	*Nematus melanaspis* Hartig	3	
81	*Nematus myosotodis* (Fab.)	30	
82	*Nematus olfaciens* Benson	41	
83	*Nematus pavidus* Lepeletier	2	
84	*Nematus ribesii* (Scopoli)	123	
85	*Nematus salicis* (L.)	1	(1989)
86	*Nematus spiraeae* Zaddach	2	
87	*Nematus tibialis* Newman	3	
88	*Nematus viridis* Stephens	1	(1985)
89	*Pachynematus obductus* (Hartig)	1	(1987)
90	*Pachynematus trisignatus* (Foerster)	3	
91	*Pachynematus vagus* (Fab.)	2	

A sawfly, *Phymatocera aterrima* (above), whose larvae shred the leaves of Solomon's seal, *Polygonatum multiflorum*

Larvae of birch sawfly, *Croesus septentrionalis* (below), in characteristic defence posture on a birch leaf

Ametastegia carpini was the commonest species overall and in 15 of the 20 years; the proportion it formed of the total catch varied from 36.1% to 81.7%. Its larvae of feed on a variety of *Geranium* species; in the garden they reduced the leaves of *G. cinereum* to a network of veins every year.

Pristiphora rufipes was the commonest species in 1996, 1997 and 2000. The larvae of *P. rufipes* feed on leaves of red currants (*Ribes rubrum*) and gooseberries (*R. uva-crispa*), and are considered as pests. *Cladius pectinicornis* was the commonest species in 1982 and 1983; its larvae feed on both wild and cultivated roses (*Rosa* species and cultivars) and on strawberries (*Fragaria* × *ananassa*), but were never evident in the garden.

Sawflies may be compared with moths and butterflies because of the ecological similarity of their larvae. 1983 and 1992, which were good sawfly years, were only average to poor for moths; 1983 was reasonable for butterflies, but 1992 was poor, as butterfly numbers declined from their highs in the 1970s and 1980s. Evidently different factors affected sawflies and Lepidoptera.

Apart from *Ametastegia carpini*, nine species of sawflies were found breeding in the garden, four of which were not caught in the Malaise trap. Larvae of five species that were caught in the Malaise trap were found in the garden: *Aneugmenus padi* on male fern (*Dryopteris filix-mas*) in 1983, *Nematus olfaciens* on blackcurrant (*Ribes nigrum*) in 1982, *N. ribesii* on gooseberry in most years, *N. spiraeae* on goat's beard (*Aruncus dioicus*) in 1983, and *Pontania* species on *Salix fragilis* in most years. In addition to the Malaise trap list are larve of *Trichiosoma lucorum* on hawthorn (*Crataegus monogyna*) in 1977 and 1986, *Tenthredo colon* on great willowherb (*Epilobium hirsutum*) in 1982, *Amauronematus* species on crack willow (*Salix fragilis*) in 1982, and *Croesus septentrionalis* on silver birch (*Betula pendula*) in most years.

Fig. 7.2. Annual fluctuations in numbers of Psocoptera caught in Malaise trap, 1983–2001

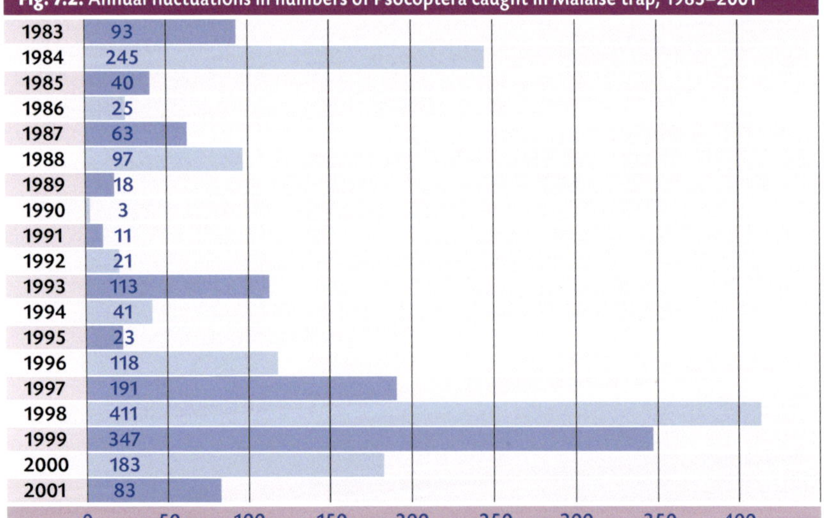

PSOCIDS

Ectopsocus briggsi, the commonest psocid in the Leicester garden in 1983 and 1984

Psocids (order Psocoptera) are small, sometimes minute, insects, frequently wingless (when they are known as booklice, dustlice or barklice, although they are not parasitic). The antennae are long and thread-like, and in most, the large compound eyes are prominent on the sides of the broad head. At rest, the relatively large wings, which in some are marbled with pigment, are held sloping, roof-like, over the body. Many species are apterous or polymorphic, with winged and wingless forms. The biting mouthparts are unusual in that part of each maxilla forms a rod-like 'pick', believed to be used for scraping food from the substrate, and the upper and lower parts of the tongue-like hypopharynx, between the maxillae, are sclerotized and apparently grind together like a mortar and pestle (Imms 1964).

Natural history of psocids

Psocids are found in most terrestrial communities, on the foliage of trees, on and beneath bark, in leaf litter, occasionally in old birds' nests or ants' nests, and also in buildings. Most outdoor species graze indiscriminately on organic debris, algae, lichens, yeasts and starchy materials. Outdoor species frequently eat their own and other insects' eggs, and psocid corpses and moulted skins, although other species are more selective (New 1971).

The indoor species feed on stored dried products, organic debris, moulds on damp books, paste on wallpaper or book-bindings, and on dried natural history specimens, such as pinned insects. The indoor species tend to be wingless, but those outdoors usually have wings at least at some stage in their life histories. In many, flight is important, as, for instance, in species such as *Ectopsocus briggsi* that make spring flights from ground litter to the foliage of trees, where they spend the summer.

Most overwinter as eggs, although nymphs and adults of some species hibernate. The arboreal species, which are generally restricted either to bark or foliage, have one, two, or occasionally three generations a year, but the domestic species breed continuously. Bark-frequenting Psocidae lay eggs in batches of up to 20 and cover them with a crust of chewed bark fragments and faecal pellets, whereas many foliage-dwellers cover their

PSOCIDS

Total Malaise trap catches, 1983–2001.
Taxonomy follows New (1974), except that *E. petersi* was first described as a new species after that publication.

Caeciliusidae		
92	*Caecilius flavidus* (Stephens)	2
93	*Caecilius burmeisteri* Brauer	24
Stenopsocidae		
94	*Graphopsocus cruciatus* (L.)	1,530
95	*Stenopsocus stigmaticus* (Imhoff & Labram)	4
96	*Stenopsocus immaculatus* (Stephens)	1
Ectopsocidae		
97	*Ectopsocus briggsi* McLachlan	229
98	*Ectopsocus petersi* Smithers	256
Peripsocidae		
99	*Peripsocus phaeopterus* (Stephens)	2
100	*Peripsocus subfasciatus* (Rambur)	1
Elipsocidae		
101	*Elipsocus hyalinus* (Stephens)	12
102	*Elipsocus mclachlani* Kimmins	17
103	*Elipsocus westwoodi* McLachlan	7
Philotarsidae		
104	*Philotarsus picicornis* (Fab.)	8
Psocidae		
105	*Amphigerontia contaminata* (Stephens)	1
106	*Trichadenotechnum sexpunctatum* (L.)	1
107	*Trichadenotechnum pearmani* Kimmins	1
108	*Trichadenotechnum variegatum* (Latreille)	3
Trichopsocidae		
109	*Trichopsocus dalii* (McLachlan)	7

eggs with silk produced from glands in the labium (lower lip), and some lay eggs singly. Most over-wintered eggs hatch in April and May, and after six nymphal instars (stages between moulting), adults are produced in June. Parthenogenesis is widespread in the order, and in some species, males are unknown (New 1971).

Diversity and abundance of psocids

About 90 species of Psocoptera are known in the British Isles, of which only about 50 live in natural surroundings out of doors (New 1974). According to observations done at Uppingham (Clark 1979), about 24km from my garden, the peak of flight activity appears to be in October, with a smaller peak in July. Although I sorted them as annual samples from the Malaise trap, most individuals were taken in the autumn. Psocids from the garden were identified by J. Sellick.

Obviously, the Malaise trap caught only winged psocids, and of those, there are many that fly little. Clark (1979) found that four species common on vegetation near to his Malaise trap were absent from trap samples. Nevertheless, 2,126 individuals of 18 species were caught in my garden Malaise trap in the period 1983–2001. Of these, 1,530 were *Graphopsocus cruciatus*, which was the commonest species in 17 of the 19 years (Tab. 7.2). The next commonest, *Ectopsocus petersi* (256) was first described as a new species in 1978, from Ireland, but is very common in Leicestershire and Rutland, and was the commonest species in the Malaise trap in 1984; *E. briggsi* was the commonest species in 1983.

Fluctuations in numbers of particular species were spectacular from year to year, which accounts for the dramatic annual fluctuations in total numbers of Psocoptera caught (Fig. 7.2 and Tab. 7.2). In general, the annual fluctuations in total numbers were caused by variations in the numbers of *Graphopsocus*

Table 7.2. Size and composition of annual and 19-year Malaise trap catches of Psocoptera in the Leicester garden, 1983–2001

Year	Individuals (N)	Species (S)	N/S	Species taken once only	Commonest species	As % N
1983	93	8	11.6	2	*Ectopsocus briggsi*	41.9
1984	245	8	30.6	-	*Ectopsocus petersi*	52.5
1985	40	10	4.0	2	*Graphopsocus cruciatus*	50.0
1986	25	6	4.2	1	*Graphopsocus cruciatus*	44.0
1987	63	9	7.0	6	*Graphopsocus cruciatus*	81.0
1988	97	11	8.8	4	*Graphopsocus cruciatus*	51.6
1989	18	5	3.6	1	*Graphopsocus cruciatus*	38.9
1990	3	2	1.5	1	*Graphopsocus cruciatus*	66.7
1991	11	3	3.7	1	*Graphopsocus cruciatus*	72.7
1992	21	2	10.5	-	*Graphopsocus cruciatus*	90.5
1993	113	6	18.8	3	*Graphopsocus cruciatus*	89.4
1994	41	7	5.9	3	*Graphopsocus cruciatus*	39.0
1995	23	3	7.7	1	*Graphopsocus cruciatus*	60.9
1996	118	4	29.5	1	*Graphopsocus cruciatus*	85.6
1997	191	4	47.8	1	*Graphopsocus cruciatus*	91.6
1998	411	6	68.5	1	*Graphopsocus cruciatus*	86.8
1999	347	7	49.6	4	*Graphopsocus cruciatus*	89.4
2000	183	7	26.1	4	*Graphopsocus cruciatus*	92.4
2001	83	4	20.8	3	*Graphopsocus cruciatus*	96.4
1983–2001	2,126	18	118.2	5	*Graphopsocus cruciatus*	72.0

cruciatus, but the peak in 1984 was caused by large numbers of *Ectopsocus petersi*. The annual variations in N shown in Fig. 7.2 and Tab. 7.2, with highs in 1984, 1998 and 1999, do not accord with variations in other groups of insects in the garden, such as butterflies, moths and hoverflies, but solitary wasps also did well in 1984 and 1999. The main similarity between the years 1984, 1998 and 1999 was that autumn, which is the major psocid flight time, was damp or wet, which would probably be more favourable for these small, fragile insects. Autumn rainfall in 1990, the poorest year for psocids, was only average, after a dry and bright spring and summer.

Six of the species captured in the Malaise trap in the garden were among the 11 species trapped at Uppingham, 1972–78, where *Ectopsocus petersi* and *Graphopsocus cruciatus* were the most abundant (Clark 1979).

BUGS

A mating pair of hawthorn shield bugs, *Acanthosoma haemorrhoidale*, one of the larger and most conspicuous heteropterans in gardens

The order Hemiptera, bugs, is a large assemblage of insects with piercing and sucking mouthparts. The stylet-like mandibles and maxillae are held firmly together in a groove on the dorsal (or anterior) side of the segmented labium or rostrum. Within the rostrum, the mandibles lie peripheral to the maxillae which are W-shaped in cross-section, so that when held tightly together they form two tubes, down one of which saliva is passed into a feeding puncture, while liquid food is sucked up the other. The mouthparts and feeding method are common to all bugs, but the order comprises two rather different groupings of insects, Heteroptera and Homoptera, which are sometimes given ordinal rank.

Heteroptera, the true bugs, are characterized by the forewings, the proximal halves of which are toughened and often pigmented, while the apical half is membranous. When the wings are closed, the forewings are held flat to the body and overlapping, covering the membranous hind wings. The head is normally held in line with the body, so that it is clearly visible from above. The beak-like rostrum arises from the front of the head, although, when not in use, it is held bent back to lie horizontally beneath the head and body, sometimes stretching back to the base of the hind legs. There are aquatic as well as terrestrial Heteroptera, and many are predatory. However, a considerable proportion of those that are primarily plant-feeders take animal food occasionally.

Homoptera include cicadas, leafhoppers of various sorts, aphids and scale insects. All have uniformly-textured forewings, although they may be toughened or membranous, and the hind wings are membranous. Moreover, at rest, the wings are closed roof-like over the body. The head is flexed downwards, and the rostrum originates from it posteriorly so that in many the beak seems to arise from between the front legs. All Homoptera are plant-feeders, and none is aquatic. Many, particularly aphids, are injurious to plants not only because of feeding damage, but also because they transmit plant viruses, and hence they are economically damaging pests.

Diversity and abundance of Heteroptera

All Heteroptera were sorted from Malaise trap catches in 1984–2001 as annual samples. In all, 1,874 individuals of at least 93 species (some females could only be identified to genus) were caught, annual totals ranging from a high of 161 in 1984 to a low of 68 in 2001. This might sound like a continuous decline during the trapping period, but as Fig. 7.3 and Tab. 7.3 show, there were peaks in 1984, 1988 and 1994, although overall numbers have probably declined. Solitary wasps, another predatory group, also did well in 1984, although not particularly well in 1988 and 1994, and other groups such as moths and hoverflies did not follow this pattern of annual abundance. 1984 was also the year when most species were trapped (33), but in 1988 and 1994 there were only 19 and 20, respectively. Fewest species (13) were caught in 1987, when the catch was average. Astonishingly, 27 of the species trapped during the 18 years were represented by a single individual, suggesting sparse populations whose members wander widely, which accords well with their partial trophic role as secondary or higher order consumers.

Common flower bug, *Anthocoris nemorum*, the most abundant heteropteran in most years in the Leicester garden

The commonest species in most years and overall was *Anthocoris nemorum*, which formed 37.8% of the total catch. This is the common flower bug, one of the commonest Heteroptera in the British Isles. Adults overwinter in such places as under bark and in litter, and move in March or April to a variety of plants, where they feed on small animals, such as aphids, and occasionally on leaves.

Table 7.3. Size and composition of annual and 18-year Malaise trap catches of Heteroptera in the Leicester garden, 1984–2001

Year	Individuals (N)	Species (S)	N/S	Species taken once only	Commonest species	As % N
1984	161	33	4.9	13	*Anthocoris nemorum*	40.4
1985	100	30	3.3	22	*Anthocoris nemorum*	51.0
1986	94	25	3.8	14	*Anthocoris nemorum*	55.3
1987	108	13	8.3	7	*Anthocoris nemorum*	82.4
1988	152	19	8.0	12	*Anthocoris nemorum*	76.3
1989	97	25	3.9	14	*Anthocoris nemorum*	45.4
1990	107	30	3.6	13	*Anthocoris nemorum*	33.6
1991	84	30	2.8	14	*Anthocoris nemorum* & *Anthocoris nemoralis*	13.1=
1992	112	28	4.0	14	*Kleidocerys resedae*	23.2
1993	117	24	4.9	12	*Anthocoris nemorum*	37.6
1994	155	20	7.8	8	*Anthocoris nemorum*	40.0
1995	95	24	4.0	11	*Anthocoris nemorum*	34.7
1996	88	22	4.0	6	*Anthocoris nemoralis*	12.5
1997	81	28	2.9	14	*Anthocoris nemoralis*	16.1
1998	76	25	3.0	14	*Anthocoris nemorum*	15.8
1999	99	21	4.7	7	*Anthocoris nemorum*	22.2
2000	80	18	4.4	9	*Acanthosoma haemorrhoidale*	31.3
2001	68	21	3.2	14	*Anthocoris nemorum*	32.4
1984–2001	1874	92	20.4	27	*Anthocoris nemorum*	37.8

Eggs, which are inserted into leaf tissue, hatch in late May and early June, and the nymphs develop into adults in late June and early July. A second generation of adults is produced by early September and, in some years, a third generation occurs (Southwood & Leston 1959).

Birch catkin bug, *Kleidocerys resedae*, was the commonest heteropteran in the Leicester garden in 1992

In the gardening context, two of the more interesting species that were caught feed on Lawson's cypress (*Chamaecyparis lawsoniana*). *Cyphostethus tristriatus* was, until recently, confined to juniper (*Juniperus communis*) and restricted to southern England, but it now feeds on exotic conifers and is spreading. *Orsillus depressus*, a recent arrival to Britain, is spreading from the southeast and feeds on Lawson's cypress.

In 1996 and 1997, *Anthocoris nemoralis*, closely related to *Anthocoris nemorum*, was the commonest species, and in 1991, *A. nemorum* and *A. nemoralis* were joint commonest, each at 13.1% of the annual catch. Like *Anthocoris nemorum, A. nemoralis* is predatory, though it primarily seeks its prey on trees and bushes, rather than low-growing plants. *Kleidocerys resedae* was much more common than usual in 1992, when it formed 23.2% of the total catch, and was the commonest heteropteran. It is a plant-feeder, primarily associated with birch, and one or two other plants, where it feeds on the catkins; in 1992, there was still a large birch tree in the garden. *Acanthosoma haemorrhoidale*, hawthorn shieldbug, a large, brightly coloured and unmistakable insect, was the commonest species in 2000, when it formed 31.3% of the catch of Heteroptera (Tab. 7.3). It too is a plant-feeder, primarily associated with hawthorn (*Crataegus*), one species of which, *C. monogyna*, is abundant in the garden.

The Malaise trap Heteroptera were identified by P. Kirby, and occasional pitfall captures were identified by D. Goddard. ►*p129*

Fig. 7.3. Annual fluctuations in numbers of Heteroptera caught in Malaise trap, 1984–2001

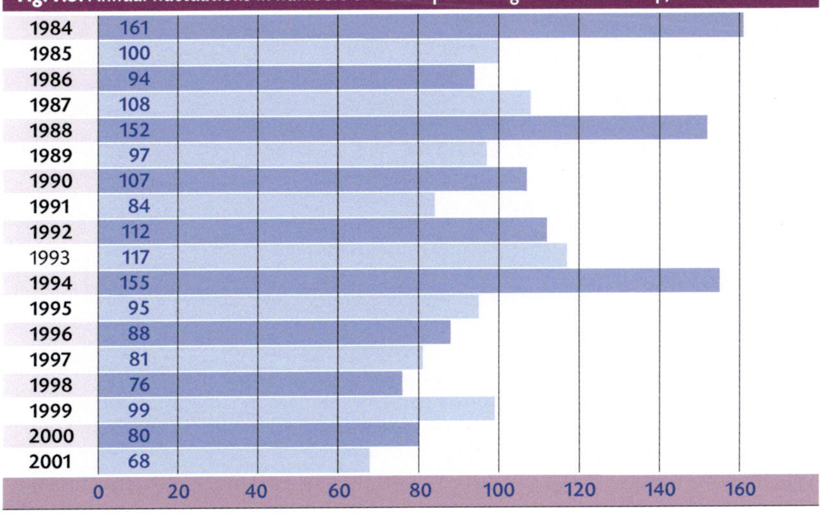

BUGS: HETEROPTERA

Total Malaise trap captures, 1984–2001.
Taxonomy and sequence follow Kloet & Hincks (1964), somewhat updated.
P indicates that the species was also caught in pitfalls.

Acanthosomatidae – shield bugs

110 Hawthorn shield bug *Acanthosoma haemorrhoidale* (L.) 66
111 Juniper shield bug *Cyphostethus tristriatus* (Fab.) 16
112 Birch shield bug *Elasmostethus interstinctus* (L.) 14
113 Parent bug *Elasmucha grisea* (L.) 8

Pentatomidae – stink bugs

114 Green shield bug *Palomena prasina* (L.) 1
115 Woundwort shield bug *Eysarcoris fabricii* (Kirkaldy) 3
116 Gorse shield bug *Piezodorus lituratus* (Fab.) 2

Coreidae – squash bugs

117 *Coriomeris denticulatus* (Scopoli) 3

Lygaeidae

118 Nettle ground bug *Heterogaster urticae* (Fab.) 2
119 European chinch bug *Ischnodemus sabuleti* (Fallén) 1
120 *Orsillus depressus* Dallas 1
121 Birch catkin bug *Kleidocerys resedae* (Panzer) 51
122 *Drymus brunneus* (Sahlberg) 1
123 *Drymus sylvaticus* (Fab.) P 26
124 *Scolopostethus affinis* (Schilling) P 4
125 *Scolopostethus thomsoni* Reuter P 1
126 *Taphropeltus contractus* (Herrich-Schäffer) 2
127 *Stygnocoris sabulosus* (Schilling) 1

Tingidae – lace bugs

128 Gorse lacebug *Dictyonota strichnocera* Fieber 1
129 *Tingis ampliata* (Herrich-Schäffer) 22
130 Spear thistle lacebug *Tingis cardui* (L.) 1
131 *Physatocheila dumetorum* (Herrich-Schäffer) 2

Nabidae – damsel bugs

132 Field damsel bug *Nabis ferus* (L.) 10

Anthocoridae – flower bugs

133 *Anthocoris confusus* Reuter 27
134 *Anthocoris nemoralis* (Fab.) 156
135 Common flower bug *Anthocoris nemorum* (L.) 708
136 *Anthocoris simulans* Reuter 4
137 *Orius vicinus* (Ribaut) 12
138 *Orius laevigatus* (Fieber) 6

Cimicidae

139 *Lyctocoris campestris* (Fab.) 3
140 *Xylocoris galactinus* (Fallén) 3

Miridae – plant bugs

141 Bracken bug *Monalocoris filicis* (L.) 5
142 *Deraeocoris lutescens* (Schilling) 106
143 *Deraeocoris ruber* (L.) 3
144 *Megalocoleus molliculus* (Fallén) 3
145 *Amblytylus nasutus* (Kirschbaum) 1
146 *Macrotylus solitarius* (Meyer-Dür) 1
147 *Harpocera thoracica* (Fallén) 23
148 *Psallus ambiguus* (Fallén) 14
149 *Psallus assimilis* Stichel 2
150 *Psallus perrisi* (Mulsant & Rey) or *Psallus wagneri* Ossiannillson 5
151 *Psallus falleni* Reuter 1
152 *Psallus flavellus* Stichel 1
153 *Psallus lepidus* (Fieber) 3
154 *Psallus diminutus* (Kirschbaum) 1
155 *Psallus varians* (Herrich-Schäffer) 15
156 *Psallus haemotodes* (Gmelin) 1
157 *Attractotomus mali* (Meyer-Dür) 3
158 *Plagiognathus arbustorum* (Fab.) 27
159 *Plagiognathus chrysanthemi* (Wolff) 4
160 *Asciodema obsoletum* (Fieber) 2
161 *Dicyphus epilobii* Reuter 5
162 *Dicyphus errans* (Wolff) 5
163 *Dicyphus pallicornis* (Fieber) 1
164 *Neodicyphus rhododendri* (Dolling) 1
165 *Campyloneura virgula* (Herrich-Schäffer) 3
166 *Pilophorus perplexus* Douglas & Scott 4
167 *Malacocoris chlorizans* (Panzer) 3
168 *Dryophilocoris flavoquadrimaculatus* (DeGeer) 2
169 *Heterocordylus tibialis* (Hahn) 3
170 *Heterotoma merioptera* (Scopoli) 11
171 *Blepharidopterus angulatus* (Fallén) 8
172 *Orthotylus marginalis* Reuter 5
173 *Orthotylus ochrotrichus* Fieber 4
174 *Orthotylus adenocarpi* (Perris) 2
175 *Orthotylus virescens* (Douglas & Scott) 11
176 *Orthotylus flavosparsus* (Sahlberg) 2
177 *Pseudoloxops coccineus* (Meyer-Dür) 1
178 *Lygus maritimus* Wagner 2
179 *Lygus rugulipennis* Poppius 7
180 *Liocoris tripustulatus* (Fab.) 50
181 *Orthops cervinus* (Herrich-Schäffer) 85
182 *Orthops campestris* (L.) 37
183 *Lygocoris pabulinus* (L.) 35
184 *Lygocoris contaminatus* (Fallén) 1
185 *Lygocoris viridis* (Fallén) 1
186 *Lygocoris lucorum* (Meyer-Dür) 4
187 *Lygocoris spinolai* (Meyer-Dür) 12
188 *Plesiocoris rugicollis* (Fallén) 2
189 *Polymerus nigritus* (Fallén) 1
190 *Calocoris norvegicus* (Gmelin) 3
191 *Phytocoris dimidiatus* Kirschbaum 1
192 *Phytocoris longipennis* Flor 1
193 *Phytocoris tiliae* (Fab.) 8
194 *Phytocoris ulmi* (L.) 1
195 *Phytocoris varipes* Boheman 1
196 *Capsus ater* (L.) 1
197 *Pantilius tunicatus* (Fab.) 1
198 *Stenodema calcaratum* (Fallén) 3
199 *Stenodema laevigatum* (L.) 24
200 *Notostira elongata* (Geoffroy in Fourcroy) 1
201 *Megaloceraea recticornis* (Geoffroy in Fourcroy) 1
202 Meadow plant bug *Leptoterna dolobrata* (L.) 3

Some examples of Homoptera: (left to right) An adult alder spittlebug, *Aphrophora alni*; adults and eggs of cabbage whitefly, *Aleyrodes proletella*; black bean aphids, *Aphis fabae*; and green leafhopper, *Cicadella viridis*

Diversity and abundance of Homoptera

The Malaise trap captured vast numbers of the smaller Homoptera, such as aphids and cicadellids, but these were not sorted systematically from the weekly catch. A number of spittlebugs (Cercopidae) and leafhoppers (Cicadellidae) were, however, sent to P. Kirby with the heteropteran sample, and he identified these also; the species list is incomplete, because some could only be identified to genus. The 69 species of cercopid (spittlebugs) and cicadellid (leafhoppers) caught in the Malaise trap in 1984–2001 form the basis of the garden list of at least 90 species of Homoptera. A number of other cicadellids were identified from Malaise trap catches by L. Jones-Walters and from pitfalls by D. Goddard. Only a few species of garden aphids were identified: those familiar to entomologists, and those whose feeding activities aroused interest and curiosity. It is not, therefore, possible to indicate any numbers for Homoptera, beyond saying that some species were very common. Furthermore, it is fully appreciated that the species list, a total of 90, is incomplete.

Three species of spittlebug (Cercopidae) were recorded in the garden, of which two feed as nymphs and adults on a wide range of plants, and one of which, *Neophilaenus lineatus*, feeds on grasses. Their feeding method is unusual in that they insert their mouthparts into the xylem vessels of plants and extract the watery xylem sap. Since this contains only very low concentrations of amino acids, from which proteins are synthesized, feeding nymphs ingest vast quantities, and continually excrete the excess water. Air is pumped into the excreted fluid by extending the posterior segments of the abdomen out of the fluid so that air enters an 'air canal' beneath the hind dorsal plate divisions, and then the body is contracted within the fluid so that a bubble of air is forced out (Weaver & King 1954). The resulting froth is the familiar cuckoo-spit that surrounds a spittlebug nymph.

Cuckoo spit, the conspicuous froth that protects of the meadow spittlebug, *Philaenus spumarius*

The meadow spittlebug (*Philaenus spumarius*) is the commonest cercopid in the garden and one of the most conspicuous Homoptera. Weaver & King (1954) listed nearly 400 species of plants on which nymphs of *P. spumarius* have been recorded, and in my garden they have been found on 143 species of 47 plant families. The extreme polyphagy of spittlebugs is doubtless a consequence of feeding on xylem sap, which is unlikely to contain secondary plant substances

BUGS: HOMOPTERA

MT indicates identification from Malaise trap samples and P identification from pitfalls.
Taxonomy and sequence of spittlebugs and leafhoppers (Auchenorrhyncha) follows Le Quesne & Payne (1981) and that of aphids follows Kloet & Hincks (1964).

Cercopidae – spittlebugs

203 Alder spittlebug *Aphrophora alni* (Fallén) MT
204 Meadow spittlebug *Philaenus spumarius* (L.) MT
205 Lined spittlebug *Neophilaenus lineatus* (L.) MT

Cicadellidae – leafhoppers

206 Green leafhopper *Cicadella viridis* (L.) MT P
207 *Rhitidodus decimusquartus* (Schrank) MT
208 *Idiocerus albicans* Kirschbaum MT
209 *Idiocerus distinguendus* Kirschbaum MT
210 *Idiocerus fulgidus* (Fab.) MT
211 *Idiocerus nitidissimus* (Herrich-Schäffer) MT
212 *Idiocerus poecilus* (Herrich-Schäffer) MT
213 *Idiocerus populi* (L.) MT
214 *Idiocerus stigmaticalis* Lewis,R. H. MT
215 *Idiocerus vitreus* (Fab.) MT
216 *Iassus lanio* (L.) MT
217 *Oncopsis flavicollis* (L.) MT
218 *Oncopsis tristis* (Zetterstedt) MT
219 *Macropsis fuscula* (Zetterstedt) MT
210 *Macropsis graminea* (Fab.) MT
211 *Macropsis scotti* Edwards MT
212 *Macropsis scutellata* (Boheman) MT
213 *Agallia consobrina* Curtis MT
214 *Aphrodes albifrons* (L.) MT
215 *Aphrodes bicinctus* (Schrank) P
216 *Aphrodes makarovi* Zakhvatkin MT
217 *Deltocephalus pulicaris* (Fallén) MT
218 *Arthaldeus pascuellus* (Fallén) MT
219 *Psammotettix sp.* MT
220 *Rhytistylus proceps* (Kirschbaum) MT
221 *Allygus mixtus* (Fab.) MT
222 *Allygus modestus* Scott MT
223 *Conosanus obsoletus* (Kirschbaum) MT
224 *Euscelis incisus* (Kirschbaum) MT
225 *Euscelis lineolatus* Brullé MT
226 *Streptanus sordidus* (Zetterstedt) MT
227 *Mocydia crocea* (Herrich-Schäffer) MT
228 *Elymana sulphurella* (Zetterstedt) MT
229 *Opsius stactogalus* Fieber MT
230 *Macrosteles sp.* MT
231 *Balclutha punctata* (Fab.) MT
232 *Alebra albostriella* (Fallén) MT
233 *Alebra wahlbergi* (Boheman) MT
234 *Empoasca decipiens* Paoli MT
235 *Empoasca vitis* (Gothe) MT
236 *Kybos betulicola* (Wagner) MT
237 *Kybos butleri* (Edwards) MT
238 *Kybos rufescens* (Melichar) MT
239 *Kybos smaragdula* (Fallén) MT
240 *Kybos virgator* (Ribaut) MT
241 *Eupteryx aurata* (L.) MT
242 *Eupteryx florida* Ribaut MT
243 Sage leafhopper *Eupteryx melissae* Curtis MT
244 *Eupteryx tenella* (Fallén) MT
245 *Eupteryx urticae* (Fab.) MT
246 *Aguriahana stellulata* (Burmeister) MT
247 *Ribautiana cruciata* (Ribaut) MT
248 *Ribautiana tenerrima* (Herrich-Schäffer) MT
249 *Ribautiana ulmi* (L.) MT
250 *Limnavuoriana sexmaculata* (Hardy) MT
251 *Ossiannilssonola callosa* (Then) MT
252 *Lindbergina aurovittata* (Douglas) MT
253 *Edwardsiana crataegi* (Douglas) MT
254 *Edwardsiana flavescens* (Fab.) MT
255 *Edwardsiana rosae* (L.) MT
256 *Alnetoidea alneti* (Dahlbom) MT
257 *Arboridia ribauti* Ossiannilsson MT
258 *Zygina angusta* (Lethierry) MT
259 *Zygina flammigera* (Geoffroy in Fourcroy) MT
260 *Zygina ordinaria* (Ribaut) MT

Delphacidae

261 *Stenocranus minutus* (Fab.) MT
262 *Javesella* species MT

Psyllidae – psyllids (jumping plant lice)

263 *Aphalara polygoni* Forster MT
264 *Psyllopsis fraxini* (L.) MT
265 *Psylla alni* (L.) MT
266 Box psyllid *Psylla buxi* (L.) MT
267 *Cacopsylla brunneipennis* Edwards, J. MT
268 *Chamaepsylla hartigii* (Flor) MT
269 *Cacopsylla melanoneura* Forster MT
270 *Cacopsylla sorbi* (L.) MT
271 *Spanioneura fonscolombii* Forster MT

Aleyrodidae – whiteflies

272 Cabbage whitefly *Aleyrodes proletella* (L.)

Callaphididae

273 Silver birch aphid *Betulaphis quadrituberculata* (Kaltenbach)

Aphididae – aphids (greenfly & blackfly)

274 Elder aphid *Aphis sambuci* L.
275 Black bean aphid *Aphis fabae* Scopoli
276 Cabbage aphid *Brevicoryne brassicae* (L.)
277 Cherry blackfly *Myzus cerasi* (Fab.)
278 Potato aphid *Macrosiphum euphorbiae* (Thomas, C. A.)

Thelaxidae

279 *Anoecia corni* (Fab.)

Pemphigidae – petiole gall aphids

280 Poplar petiole gall aphid *Pemphigus spirothecae* Passerini

Adelgidae – adelgids (gall aphids)

281 *Adelges viridis* (Ratzeburg)
282 *Adelges abietis* (L.)

and toxins, and indeed is little other than water. In contrast, most other Homoptera extract sugary phloem sap from plants; this originates in the photosynthetic tissues and is, therefore, likely to contain high concentrations of secondary plant substances, which give a plant its characteristic odour and taste.

It is because they feed on a rich mixture of chemicals that many of the phloem-feeders are restricted in which plants they can use as food. Aphids, for instance, are mostly monophagous, particularly in temperate regions, where there is a greater number of species (Dixon 1985). Amino acids occur at only low concentrations in phloem sap, so phloem-feeders tend to ingest large amounts of the sugary sap and continually defecate the excess sugar, familiar as the honeydew produced by aphids.

All the cicadellids (leafhoppers) recorded from Malaise trap catches are common and widespread, particularly in southern parts of the British Isles. Many are less than 5mm in length, but they are often brightly coloured, and many are conspicuously marked. In many years cabbage whitefly (*Aleyrodes proletella*) was abundant on brassicas, particularly curly kale.

The aphids recorded in the garden belong to several different families. Five species were abundant in all or most years: *Betulaphis quadrituberculatus* on birch (*Betula pendula*), *Aphis sambuci* on elder (*Sambucus nigra*), *Aphis fabae* on beans (*Vicia faba* and *Phaseolus coccineus*) and mock orange (*Philadelphus coronarius*), *Brevicoryne brassicae* on *Brassica*, and *Pemphigus spirothecae* on poplar (*Populus nigra* 'Italica'). In 1976, *Macrosiphum euphorbiae* was recorded from *Buddleja davidii* and *Anoecia corni* from dogwood (*Cornus alba*), and in 1978 and other years, cherry blackfly, *Myzus cerasi*, was abundant on suckers of a flowering cherry (*Prunus serrulata*). In 1980, the green 'pineapple' galls of *Adelges viridis* were abundant on Norway spruce (*Picea abies*), together with wax-producing *Adelges abietis*, and the tree's growth was much affected. An unidentified, unpigmented aphid was abundant in 1980 on roots of *Leucanthemum* × *superbum* that were growing within a nest of black ants, *Lasius niger*.

The author discovered these pineapple galls (above left), caused by a gall aphid, *Adelges viridis*, on a Norway spruce, *Picea abies*, in the Leicester garden

Section of a gall (above right) caused by a related species, *A. cooleyl*, showing the adelgids *in situ*

The author found unidentified white root aphids (below) on the roots of Shasta daisy, *Leucanthemum* × *superbum*, that were growing within a nest of garden black ants, *Lasius niger*, in 1980

LACEWINGS AND THEIR ALLIES

The sub-order Planipennia of the order Neuroptera includes the lacewings proper and a family (Coniopterygidae) of tiny insects resembling psocids, which are covered with white, waxy powder, so that they look superficially like white-flies. Some authors consider Planipennia as a separate order, Neuroptera, distinct from alderflies and snakeflies (Megaloptera).

Natural history of lacewings and their allies

Green lacewing, *Chrysoperla carnea,* a common species in the Leicester garden

Lacewings are delicate, medium-sized insects characterized by their large, flimsy wings, delicately netted with veins, which are held roof-like over the body at rest. Green lacewings (Chrysopidae) are often attracted to lighted windows at night, and some find their way into houses, particularly in winter before hibernating, when many assume a straw-coloured tinge. Brown lacewings belong to the family Hemerobiidae.

Lacewings have prominent eyes on the sides of the head, and long, slender antennae. The mouthparts are of the biting type, with strong mandibles, and although some species are rarely seen to feed, others are predatory on aphids and other small animals. They quite often drink, and are attracted by sweet fluids, such as exposed nectar on flowers or the sugar solutions used as bait by moth collectors. They fly at dusk and during the night.

Less is known about the tiny, psociform, Neuroptera (Coniopterygidae), but they, too, appear to be predatory. The wax covering the wings of coniopterygids

Fig. 7.4. Annual fluctuations in numbers of lacewings caught in Malaise trap, 1978–2001

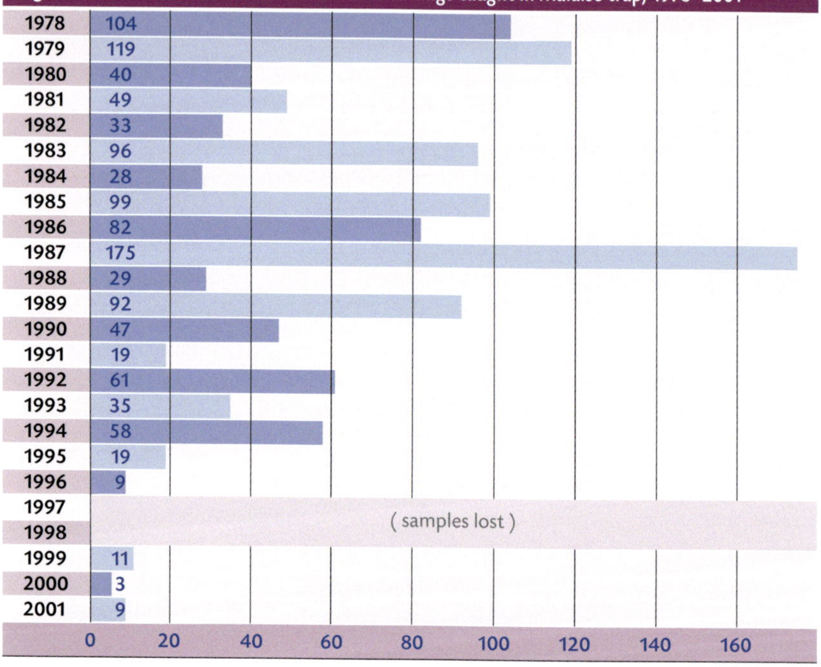

is produced from epidermal glands, mainly on the abdomen, and is spread over the wings using the hind legs (Killington 1936).

Eggs are laid singly or, more rarely, in twos or threes. Those of the Hemerobiidae and Coniopterygidae are cemented to leaves, twigs or bark, while in spongeflies (Sisyridae), they are laid in batches covered with silk on vegetation overhanging water. Egg-laying in Chrysopidae is more curious, in that just before an egg is laid on a leaf, a drop of clear, viscous fluid is extruded from the female's abdomen, which is gently raised from the substrate so that the quick-drying fluid hardens into a stalk at the end of which an egg is attached (Killington 1936). These may be laid in groups, and, in some species, the stalks of several eggs coalesce.

Waxy lacewings, *Conwentzia psociformis*, have been captured on several occasions in the Malaise trap in the Leicester garden. Wax is produced from glands on the abdomen

The eggs of Hemerobiidae, Coniopterygidae and Chrysopidae hatch into carnivorous larvae that prey on aphids, mites and other small animals, those of Sisyridae into larvae that drop into the water to feed on freshwater sponges.

Larvae have quite conspicuous, thin, sharp, tubular 'jaws', that are used to suck fluids from their prey. Each of the pair of 'jaws' is formed from grooved maxilla and mandible, which are held firmly together so that each 'jaw' forms a tube. Fluid food is sucked straight up into the pharynx, often while the struggling prey is held aloft, impaled on the 'jaws'. They are extremely slender in Sisyridae, strongly inwardly curved in Chrysopidae and Hemerobiidae, and shorter in the tiny Coniopterygidae. The larvae of some species of Chrysopidae decorate their bodies with debris, including the drained skins of their prey. The covering of debris, which is attached to hooked setae on prominent tubercles on the thorax, serves as camouflage.

The digestive tract of larval lacewings is blocked in the anterior region of the hind-gut, any waste products accumulating in the large mid-gut, to be voided after metamorphosis to the adult state. In third instar larvae, a number of the excretory Malpighian tubes that arise from the hind-gut become modified to secrete silk, which is stored in a silk reservoir, and later spun through the anus to form the cocoon in which pupation takes place. Cocoons and pupae, including those of Sisyridae, whose larvae leave the water and travel considerable distances before selecting pupation sites, are formed in crevices in bark, or other sheltered positions.

LACEWINGS & their allies

Numbers refer to total Malaise trap captures over the periods indicated

Coniopterygidae (1983–2001)		
283	*Conwentzia psociformis* Curtis	7
284	***Coniopteryx borealis*** Tjeder	3
285	***Coniopteryx pygmaea*** Enderlein	1
286	***Coniopteryx tineiformis*** Curtis	8
287	*Parasemiadalis fuscipennis* (Reuter)	2
Sisyridae (1978–2001)		
288	*Sisyra fuscata* (Fab.)	2
Hemerobiidae – brown lacewings (1978–2001)		
289	***Micromus variegatus*** (Fab.)	26
290	***Hemerobius humulinus*** L.	10
291	***Hemerobius nitidulus*** Fab.	1
292	***Hemerobius micans*** Olivier	1
293	***Hemerobius stigma*** Stephens	1
294	***Hemerobius lutescens*** Fab.	26
295	***Wesmaelius nervosus*** (Fab.)	1
296	***Wesmaelius subnebulosus*** (Stephens)	834
297	***Sympherobius pygmaeus*** (Rambur)	1
Chrysopidae (1978–2001)		
298	*Chrysopa perla* (L.)	4
299	*Chrysopidia ciliata* (Wesmael)	4
300	*Cunctochrysa albolineata* (Killington)	60
301	Green lacewing *Chrysoperla carnea* (Stephens)	227
302	*Dichochrysa flavifrons* (Brauer)	12
303	*Dichochrysa prasina* (Burmeister)	2
304	*Dichochrysa ventralis* (Curtis)	1
305	*Nineta flava* (Scopoli)	4

Diversity and abundance of lacewings and their allies

There are 67 species of Planipennia in the British Isles, 23 of which have been caught in the Malaise trap, including five psociform Neuroptera (Coniopterygidae), one Sisyridae, and 17 lacewings proper – nine Hemerobiidae and eight Chrysopidae. Lacewings were sorted from Malaise trap catches during 1978 –2001. Annual catches fluctuated considerably from year to year, the best year being 1987 (N=175) and the worst, 2000 (N=3) (Fig. 7.4 and Tab. 7.4).

By and large, lacewings did not follow the annual pattern of other garden insects caught in the Malaise trap (although hoverflies also had a good year in 1987), almost certainly because annual fluctuations in numbers were almost entirely a consequence of fluctuations in numbers of one species, *Wesmaelius subnebulosus*, of which 834 were caught altogether, out of a total lacewing catch of 1,217. This is one of the commonest and most widespread hemerobiids, which is often found in gardens. *Wesmaelius subnebulosus* is often associated with birch, and in some years the Malaise trap catch included many recently-emerged specimens, suggesting that it was breeding in the garden, presumably on the large birch tree. At the end of 1994, the birch was severely pollarded and subsequently died, and it can be no coincidence that no *W. subnebulosus* were caught in 1996 and only few after that.

The number of species of lacewing trapped varied from year to year, the most

Table 7.4. Size and composition of annual, 9-year, 15-year and 24-year Malaise trap catches of lacewings in the Leicester garden, 1978–2001

Year	Individuals (N)	Species (S)	N/S	Species taken once only	Commonest species	As % N
1978	104	6	17.3	1	*Wesmaelius subnebulosus*	73.1
1979	119	4	29.8	1	*Wesmaelius subnebulosus*	91.6
1980	40	4	10.0	-	*Chrysoperla carnea*	50.0
1981	49	4	12.3	1	*Wesmaelius subnebulosus*	85.7
1982	33	4	8.3	1	*Wesmaelius subnebulosus*	45.5
1983	96	9	10.7	6	*Wesmaelius subnebulosus*	75.0
1984	28	6	4.7	2	*Chrysoperla carnea*	57.1
1985	99	7	14.1	1	*Wesmaelius subnebulosus*	67.7
1986	82	6	13.7	3	*Wesmaelius subnebulosus*	74.4
1987	175	8	21.9	5	*Wesmaelius subnebulosus*	73.1
1988	29	3	9.7	2	*Chrysoperla carnea*	69.0
1989	92	2	46.0	-	*Wesmaelius subnebulosus*	95.7
1990	47	2	23.5	1	*Wesmaelius subnebulosus*	97.9
1991	19	2	9.5	-	*Wesmaelius subnebulosus*	57.9
1992	61	6	10.2	2	*Chrysoperla carnea*	65.6
1993	35	5	7.0	1	*Wesmaelius subnebulosus*	40.0
1994	58	7	8.3	1	*Wesmaelius subnebulosus*	62.1
1995	19	2	9.5	-	*Wesmaelius subnebulosus*	73.7
1996	9	2	4.5	1	*Chrysoperla carnea*	88.9
		(1997, 1998 – samples lost)				
1999	11	4	2.8	2	*Wesmaelius subnebulosus*	63.6
2000	3	3	1.0	3	*Wesmaelius subnebulosus* *Micromus variegatus* *Diochrysa prasina*	33.3=
2001	9	4	2.3	3	*Wesmaelius subnebulosus*	68.5
1978–1986	650	13	50.0	2	*Wesmaelius subnebulosus*	70.9
1987–2001	576	16	35.4	5	*Wesmaelius subnebulosus*	65.8
1978–2001	1217	23	52.9	7	*Wesmaelius subnebulosus*	68.5

diverse being 1983 when nine species were caught and 1987 when eight were caught, but in five years there were only two species. 1987, when numbers peaked, and diversity was high, was a cool, damp, dull summer after a mild spring, when rainfall and sunshine were average. The overall impression was that lacewings declined in numbers, particularly during the 1990s, as with other groups of garden insects; no Coniopterygidae was caught after 1988.

The single individual of *Sisyra fuscata* caught in 1983 was the first adult spongefly to be recorded in the county of Leicestershire, but another individual turned up in 2001. Larvae of spongeflies are common in the county, but there are three species that cannot at present be separated.

Coniopterygidae were identified by J. Sellick, and other Planipennia by D. Goddard, until his untimely death, after which they were identified by C. Plant, although the collections for 1997 and 1998 were lost. In the list on p133, taxonomy and sequence follow Plant (1994) updated by C. Plant, and numbers refer to total Malaise trap captures (1983–2001 for Coniopterygidae, and 1978–2001 for the others).

A brown lacewing, *Wesmaelius subnebulosus* (above left), associated with birch, once common in the Leicester garden, that became rarer after the large birch tree died

A spongefly, *Sisyra fuscata* (above right), twice recorded in the Leicester garden, that in 1983 being the first Leicestershire record

8 : Beetles

A ground beetle (Carabidae), *Notiophilus biguttatus* (opposite), regularly caught in pitfall traps in the Leicester garden

Beetles are not usually conspicuous insects in gardens, but the gardener may be aware of several different types. Shiny, black ground beetles lurk beneath vegetation and under stones, and scurry away when disturbed; small flea beetles, yellowish-brown or black and yellow, are often abundant, chewing holes in the leaves of many sorts of plants; tiny, shiny or metallic, dark beetles are common in flowers; bright ladybirds attract attention as they crawl from the shelter of vegetation in the first warm days of spring, and they and their splay-legged, prehistoric-looking larvae are often to be found amongst aggregations of aphids.

There are more different species of beetles in the world than of any other type of animal, and they are to be found just about everywhere, in all sorts of habitats, feeding on all sorts of foods. They are not, however, as amenable to trapping as some other sorts of insects, and the list of 422 species I have recorded in my garden probably represents only a fraction of the garden fauna. Furthermore, beetles form an order of insects, and so are a larger taxonomic group than most of the other types of insects considered in Chapters 3–7.

NATURAL HISTORY OF BEETLES

There is usually little difficulty in recognising a beetle. With their hard, often rounded, outer covering, they are one of the most distinctive types of insects. A detailed, authoritative account of their biology is given by Crowson (1981) and has been referred to for this section. Beetles constitute the insect order Coleoptera, characterized, particularly, by the modification of the front pair of wings into hard, protective elytra, or wing-cases, which shield the functional, yet typically flimsy, hind wings. In their development, they undergo complete metamorphosis, from larval forms looking quite unlike beetles, through an immobile pupal stage to adult beetles. Typically, adults have biting mouthparts, as do the active larvae characteristic of many families.

Elytra probably evolved as an adaptation for crawling into crevices or beneath soil, stones, leaves, bark or litter without damaging the functional wings. In evolutionary terms, this modification was evidently successful, for of the 1.82 million described species of plants and animals, half are insects and a quarter are beetles (Stork 1988). It is accepted, however, that there are many more undescribed species than are known, giving a total of perhaps 30 million species of animals and plants in the world (May 1987) or even more, for Stork estimates that there are up to 80 million arthropod species in the world's tropical forests alone. On the higher estimate, and extrapolating from known species, there could be in excess of 20 million species of beetle.

The vast assemblage of beetles that are known calls for detailed classification, and taxonomists are by no means agreed on how this should be achieved. In this chapter, I follow the Kloet & Hincks checklist of British insects (Pope 1977). The order is divided into three suborders, Adephaga (including six British families, such as Carabidae), Myxophaga (with one family), and Polyphaga (which incorporates 88 British families). The checklist recognizes 3,729 species as British, with an additional 116 species either extinct or of doubtful status.

Adult beetles range in size from about 0.25mm in the featherwing beetles (Ptilidae) to 100mm in longhorn beetles (Cerambycidae), but considering the diversity in how and where they live, their appearance is remarkably uniform. Most have a very toughened external skeleton, which in some is reputed to have the mechanical strength of mild steel; the soft exoskeleton of cantharid beetles is exceptional. The body surface of terrestrial beetles is normally hydrofuge, and that of water beetles more hydrophilic, although areas where air is held for respiration while submerged, often beneath the elytra, are hydrofuge.

The majority of beetles are black, but others are brightly coloured, as in ladybirds (Coccinellidae), where carotenoid pigments are responsible, or leaf beetles (Chrysomelidae), where the colours are structural, resulting from light interference in different layers of the cuticle. Many species, including black beetles, are iridescent because of micro-sculpturing of the cuticle. Patterning in cryptic colours, as in many weevils (Curculionidae) is often achieved by a dense covering of short, recumbent setae (hairs) that may be scale-like in structure.

Anatomy

Although elytra are characteristic of beetles, in some, particularly the rove beetles (Staphylinidae), they are abbreviated, exposing the abdomen. Furthermore, staphylinids tend to be elongate and sinuous, although they still generally have fully functional wings, which are folded in a complicated fashion beneath the short elytra. In some instances, such as in some ground beetles (Carabidae) and weevils (Curculionidae), the elytra are fused along the mid-line into a solid carapace, and the beetles cannot fly. In female glow-worms (Lampyridae), elytra and functional wings are lacking. Many species, such as ladybirds, hold the elytra raised and separated while in flight, but others, such as the rose chafer, *Cetonia aurata* (Scarabaeidae), fly without extending the elytra.

Antennae are variable, being very short in whirligig beetles (Gyrinidae) and ladybirds, and sometimes very long, more than twice as long as the body in Cerambycidae. In others, the individual segments are expanded, giving saw-like (some click beetles, Elateridae), comb-like (some Elateridae) or lamellate (cock-chafers, Scarabaeidae) antennae, and in weevils the antennae are elbowed.

Beetles' legs reflect the way of life of the insect: those of active, predatory ground beetles (Carabidae) are long and slender; the hind legs of flea beetles, such as *Phyllotreta* species (Chrysomelidae) have enlarged femora for jumping; those of dung beetles, such as *Geotrupes* species (Geotrupidae), are expanded and strengthened for digging; the tarsi of leaf beetles (Chrysomelidae) have adhesive lobes for clinging to vegetation; and the hind and sometimes the middle legs of many aquatic beetles are expanded and fringed with hairs, forming paddles for swimming.

Although the vast majority are recognisably beetles, their larvae are very different. Active predators on the ground, such as carabid larvae, on plants (ladybird larvae) or in water, such as diving beetle larvae (Dytiscidae), are more or less elongate with well-developed legs; herbivores that live openly on plants, such as chrysomelid larvae, also have fully functional legs. In contrast, the many beetle larvae that live in their food, whether dung, decaying vegetation, wood, or some other substrate, have much reduced legs or even lack them altogether, so that they resemble maggots.

Larval appearance reflects the need for activity and mobility; newly hatched oil beetle (Meloidae) larvae, for instance, which climb vegetation and attach themselves to solitary bees, have well-developed legs ending in claws, but after being inadvertently carried by a bee back to its nest, where they feed on the stored pollen and honey, they become grub-like, with much reduced legs.

Typical life cycles of beetles

Life histories may be illustrated by ground beetles (Carabidae) and ladybirds (Coccinellidae). Some carabids breed in late summer or autumn and the resulting larvae hibernate; others breed in spring or early summer then die, and the new generation of adults, which appears in autumn, then hibernates. Others vary the time of breeding dependent on the weather, and some diapause as adults before reproduction. They usually lay eggs singly, either depositing them on the soil surface, or using the tip of the abdomen to dig a small hole. Most have three larval instars, and the active larvae are mainly carnivorous, whereas adults are often omnivorous or even phytophagous.

The carabid pupa is usually formed in a compacted cell in the soil, and lies on its back supported by specialized, long dorsal spines. The pupal cuticle is soft and white, but segments and head are recognizable and all the appendages are visible and free although adpressed to the ventral side. Adults of *Nebria brevicollis*, common in gardens, are active for only two or three weeks after emergence in early summer, feeding rapidly and building up fat reserves before entering diapause aggregated under stones or logs until the beginning of the breeding season, which lasts from mid-October until the following April (Forsythe 1987).

In contrast to ground beetles, most ladybirds lay their eggs in early summer in batches on vegetation, often among aggregations of aphids, the larval and adult prey of most species. Eggs hatch into active predatory larvae, which range freely over plants, and, when food is short, turn cannibal. There are four larval instars, and the larvae pupate on leaf surfaces in a characteristic hunched position so that they appear rather featureless.

The time taken from laying of an egg to emergence of an adult varies from 10 to 61 days, depending on temperature, in 7-spot ladybird, *Coccinella septempunctata* (Majerus & Kearns 1989), and most adults emerge in mid- to late summer. There may occasionally, in a few species, such as 2-spot (*Adalia bipunctata*) and 14-spot (*Propylea quattuordecimpunctata*) ladybirds, be two generations a year. Ladybirds pass the winter as adults, either hibernating fully, as in the 2-spot, so that they are immobile all winter with a reduced metabolic rate, or merely slowing to immobility when it is cold but stirring on warm winter days, as in the 7-spot.

Water beetles

Aquatic adaptations of beetles are usually most evident in the locomotory and respiratory systems. Adult water beetles can be loosely divided into open-water swimmers, such as *Dytiscus* species, and crawlers on vegetation, such as *Helophorus* species (Hydrophilidae). In most swimmers, the tibiae and tarsi of the hind legs are wide and hair-fringed, whereas in crawlers, the claws are enlarged. Many water beetles, such as *Dytiscus* species, use for respiration a bubble of air trapped beneath the elytra; such beetles are good swimmers, making repeated trips to the surface to replenish the air bubble, and typically occur in water of 2m in depth, or less. In addition, in many smaller dytiscids the air bubble behaves as a physical gill into which dissolved oxygen diffuses to replace that used by the beetle. Other water beetles, such as the non-swimming Elmidae, trap air bubbles in coatings of hydrofuge hairs or scales on the ventral side of the thorax and abdomen, and use this for respiration; although this so-called plastron to some extent acts as a physical gill, they can also use the oxygen bubbles produced by photosynthesizing algae, and so remain submerged in deep water for weeks at a time.

Feeding and defence

The mouthparts of most beetles are of a fairly typical biting type, but a few groups show extreme specialization. In some flower-feeding species, such as Rhipiphoridae for instance, each maxilla has developed one or more long, hairy lobes. In Meloidae, the maxillae fit together to form an elongate suctorial proboscis. In some small species, such as Cerylonidae, which live under bark, in fungi or in decomposing vegetable matter, mandibles, maxillae and labium are sharp and pointed for piercing and sucking. In many carnivorous beetle larvae, digestion is extra-oral, in that enzymes from the mid-gut are injected into the prey by the slender, grooved mandibles and the liquefied prey is then sucked into the narrow, muscular oesophagus.

Many beetles live in or on living or dead trees, often in association with fungi, which convert the nutrients in wood into a form the beetles can eat. Adult *Scolytus* beetles, for instance, when they feed on young shoots of elm, infect the trees with the sticky spores of Dutch elm disease, *Ophiostoma novi-ulmi*; as the fungus spreads, branches are killed, the beetles make their egg burrows under the bark of dying or recently-dead fungus-infested branches, and the larvae tunnel and feed in the cambium, being apparently dependent for some elements of their nutrition on the presence of the fungus. Ambrosia beetles (also Scolytidae) burrow more extensively into trees than *Scolytus* species, and have a closer relationship with a fungus. Adults carry fungal spores in special cavities on their body, and these develop within the burrows into carpets of fungal growth on which the larvae feed. In other beetles, fungal spores are transported internally in the alimentary canal, infection of the tree being by defecation.

Not all tree-feeding beetles, however, have symbiotic relationships with fungi, and they eat the actual tissues of the tree, including wood. In many cases, such beetles are known to have symbiotic bacteria in the gut, which presumably aid digestion. Some, particularly Scolytidae, have a number of adaptations appropriate for burrowing: the body is markedly cylindrical, the prothorax is humped or hooded, the head is more or less bent down, the legs are short, and

The distinctive pattern of tunnels left under the bark of elm (*Ulmus* species) by larvae of *Scolytus scolytus* beetles. The wood is killed, softened and made more palatable by Dutch elm disease, a fungus transmitted by the beetles themselves

the elytra bend sharply down at the back, so that they block the end of a burrow, sometimes also being modified to scoop 'sawdust' backwards.

Beetles of many different families associate with social insects, particularly ants and termites, living in their nests, sometimes as scavengers, often as parasites or predators. In the majority of cases, those in ant nests, particularly Paussidae (not found in the UK) and many Staphylinidae, are tolerated because glands on the surface of their bodies produce sweet secretions that the hosts lick; many are tended, fed and even carried by the ants, which grasp the beetles' characteristically enlarged and thickened antennae. Their successful association with ants is dependent, in the evolutionary sense, on the beetles having broken into the ants' code of chemical and tactile signals; since they smell, taste and feel like ants, they are treated as such.

Beetles are themselves subject to predation, and among the defences they use are chemical ones. Most Adephaga, for instance, have pygidial glands, opening dorsally in the region of the ninth abdominal tergite, which secrete an aqueous solution containing *inter alia* a toxin. The most extraordinary development of the pygidial glands is in certain carabids, such as the bombadier beetle, *Brachinus crepitans*; a mixture of hydroquinones and hydrogen peroxide is exploded by enzymes in a 'combustion' chamber so that a hot, irritant liquid is forcibly ejected from the gland openings. As the abdomen is markedly flexible, this boiling jet can be discharged directly at the predator.

Some *Stenus* species (Staphylinidae) use their pygidial glands to move on the surface film of water; the glandular secretion lowers the surface tension of the water behind the beetle so that it is rapidly propelled forwards by the pull of surface tension in front of it.

Beetles fill a great diversity of ecological niches, as predators, herbivores, scavengers, and even parasites, such as louse-like *Platypsyllus castoris*, which parasitizes beavers. They abound on land and in aquatic habitats, and some have become adapted to the most physiologically stressful of habitats where water is almost totally lacking, such as hot deserts and stores of grain and dried foodstuffs. Adaptations to such difficult conditions demonstrate the diversity and evolutionary plasticity of beetles.

BEETLES (from pitfall traps)

In the following list of species, total numbers caught in pitfalls (including during the winter of 1985–1986) are given. MT indicates that the species was also caught in the Malaise trap; MV that it was also caught in the mercury-vapour light trap.

Carabidae

1	*Carabus violaceus* L.	40	
2	*Leistus fulvibarbis* Dejean	1	
3	*Leistus rufomarginatus* (Duftschmid)	1	
4	*Leistus spinibarbis* (Fab.)	67	MT
5	*Nebria brevicollis* (Fab.)	2,824	MT
6	*Notiophilus biguttatus* (Fab.)	53	MT
7	*Loricera pilicornis* (Fab.)	3	MT
8	*Trechus quadristriatus* (Schrank)	22	MT MV
9	*Asaphidion flavipes* (L.)	2	MT
10	*Bembidion aeneum* Germar	1	
11	*Bembidion guttula* (Fab.)	2	MT
12	*Pterostichus madidus* (Fab.)	8,644	
13	*Calathus fuscipes* (Goeze)	1	
14	*Synuchus nivalis* (Panzer)	9	
15	*Amara aenea* (Degeer)	1	MT MV
16	*Amara familiaris* (Duftschmid)	2	MT
17	*Amara plebeja* (Gyllenhal)	1	MT
18	*Amara similata* (Gyllenhal)	1	MT
19	*Harpalus rufipes* (Degeer)	6	
20	*Harpalus rufibarbis* (Fab.)	1	MT MV
21	*Harpalus affinis* (Schrank)	4	
22	*Badister bipustulatus* (Fab.)	1	

Dytiscidae

23	*Ilybius fuliginosus* (Fab.)	1	MV

Leiodidae

24	*Ptomophagus subvillosus* (Goeze)	1	MT
25	*Catops nigricans* (Spence)	3	MT

Staphylinidae

26	*Omalium excavatum* Stephens	2	MT
27	*Anotylus inustus* (Gravenhorst)	1	MT
28	*Anotylus sculpturatus* (Gravenhorst)	1	MT
29	*Lathrobium brunnipes* (Fab.)	1	
30	*Lathrobium fulvipenne* (Gravenhorst)	1	
31	*Lathrobium multipunctum* Gravenhorst	1	
32	*Rugilus orbiculatus* (Paykull)	1	MT
33	*Gyrohypnus angustatus* Stephens	4	MT
34	*Xantholinus linearis* (Olivier)	12	MT
35	*Philonthus cognatus* Stephens	2	MT
36	*Philonthus marginatus* (Ström)	1	MT
37	*Philonthus politus* (L.)	1	MT
38	*Philonthus tenuicornis* Mulsant & Rey	2	MT
39	*Philonthus varius* (Gyllenhal)	2	MT
40	*Staphylinus (Tasgius) ater* Gravenhorst	161	MT
41	*Staphylinus olens* Müller	5	
42	*Quedius boops* (Gravenhorst)	1	MT
43	*Quedius curtipennis* Bernhauer	4	
44	*Quedius molochinus* (Gravenhorst)	1	MT
45	*Quedius tristis* (Gravenhorst)	13	MT
46	*Tachinus signatus* Gravenhorst	5	MT

Strawberry beetle, *Pterostichus madidus*, feeding on strawberry fruit (*Fragaria* x *ananassa* cultivar). It was by far the commonest carabid caught in pitfall traps in the Leicester garden

Scarabaeidae

47	*Aphodius rufipes* (L.)	3	MV

Byrrhidae

48	*Simplocaria semistriata* (Fab.)	1	MT

Elateridae

49	*Melanotus erythropus* (Gmelin in L.)	1	
50	*Agriotes pallidulus* (Illiger)	1	MT

Nitidulidae

51	*Meligethes aeneus* (Fab.)	1	MT
52	*Meligethes rotundicollis* Brisout	1	
53	*Glischrochilus hortensis* (Fourcroy)	1	

Coccinellidae

54	*Adalia bipunctata* (L.)	1	MT
55	*Coccinella septempunctata* L.	1	MT
56	*Propylea quattuordecimpunctata* (L.)	1	MT

Lathridiidae

57	*Aridius nodifer* (Westwood)	1	MT

Apionidae

58	*Apion (Aspidapion) radiolus* Marsham	1	MT

Curculionidae

59	*Otiorhynchus porcatus* (Herbst)	51	
60	*Otiorhynchus rugostriatus* (Goeze)	97	MT
61	*Otiorhynchus singularis* (L.)	211	MT
62	*Otiorhynchus sulcatus* (Fab.)	33	MT MV
63	*Phyllobius maculicornis* Germar	1	
64	*Barypeithes pellucidus* (Boheman)	209	MT
65	*Sciaphilus asperatus* (Bonsdorff)	4	MV
66	*Barynotus obscurus* (Fab.)	92	MV

DIVERSITY AND ABUNDANCE OF BEETLES

Pitfall traps were operated between 1 April and 31 October from 1979 to 1990 inclusive. Eight were sited in four different areas of the garden, and from 1980 a further three were positioned below the Malaise trap (see Chapter 2). In addition, the pitfalls were operated during winter 1985/86, from 31 October 1985 to 1 April 1986. The rationale for using pitfalls below the Malaise trap is that it is generally assumed that many beetles that fly into the open sides of the trap drop to the ground on striking the central baffle and walk away. Beetles that have thus evaded capture in the Malaise trap might be expected to fall into pitfalls beneath it. Beetles from the pitfall and Malaise traps were identified by D. Goddard until his death, and subsequently by D. Lott.

Beetles from pitfall traps

The size of the catch of Carabidae in a pitfall depends on the density of vegetation and litter around the trap, more being caught where there is little to impede the movement of beetles near to the trap. Consequently, pitfall trapping cannot be used for assessing population sizes, other than by saturation trapping and removal from a confined area, or for comparing the beetle fauna in different habitats (Greenslade 1964). In my garden, however, pitfall traps were used annually in identical sites, and can legitimately be used to compare the size of the ground beetle community in different years, as well as for compiling a species list.

Nebria brevicollis, a carabid beetle, often caught in pitfall traps in the Leicester garden

Formalin, which was used as killing and preservative agent in pitfalls in the garden, appears to attract Staphylinidae, Curculionidae and Carabidae, particularly *Pterostichus madidus* (Luff 1968), militating against its use in population or community studies. Pitfalls with formalin are, however, very effective as traps, and the arguments against their use do not apply to pitfall trapping as conducted in the garden, nor compromise the way that the trapping data are used to compare years. Luff (1975) also found that plastic pitfalls, as used in the Leicester garden, were less effective than glass, and use of plastic drinking beakers in the garden may to some extent counter the effects of using formalin. The catch was removed from traps at regular intervals, at least every two weeks, and the formalin changed when it became dirty or diluted.

Beetles were identified as annual samples by D. Goddard, and nomenclature, sequence and classification follows Pope (1977). During the 12-year period 1979–90, 12,620 individuals of 66 species of beetle were caught in pitfall traps, including 22 species of Carabidae, 21 species of Staphylinidae and eight species of Curculionidae.

The annual catches of beetles from pitfall traps rose more or less constantly from 1979 to 1986, remained high for a further two years, then began to drop off, being only 385 in 1990 compared with 1,877 in 1986 (Tab. 8.1, Fig. 8.1). The pattern of rise in numbers and subsequent decline was accounted for by changes in the abundance of Carabidae, which formed 92.6% of the total catch. The commonest beetle in pitfalls was *Pterostichus madidus*, which formed 68.5% of the total catch and was the commonest in every year in traps set in the open garden, i.e. away from the Malaise trap. The percentage of the catch formed by *P. madidus* was lower in years when *Nebria brevicollis* was common, particularly from 1984 to 1987.

The reasons for the increasing numbers of Carabidae and other ground-dwelling beetles during the 1980s are not entirely clear, even allowing for the apparent attractiveness of formalin. *Pterostichus madidus* is predatory as a larva and omnivorous as an adult, and *Nebria brevicollis* preys on Collembola as a larva and as an adult. It is probable that during the 1980s the garden became lusher, with more dead and decomposing vegetation on the soil surface; this may have improved conditions for small soil animals, such as Collembola, on which both species of beetle feed.

The number of beetles caught on average in each of the three pitfalls below the Malaise trap was consistently smaller than the catch per trap in the open garden, and fewer species were caught. There is, however, some evidence, although inconclusive, that these pitfalls were catching beetles 'lost' from the Malaise trap, such as a few ladybirds. Beetles that would not normally be caught in pitfalls were caught below the Malaise trap, and *Nebria brevicollis*, rather than *Pterostichus madidus* was the commonest beetle in some years in these traps.

Fig. 8.1. Annual fluctuations in numbers of beetles caught in pitfall traps, 1979–1990

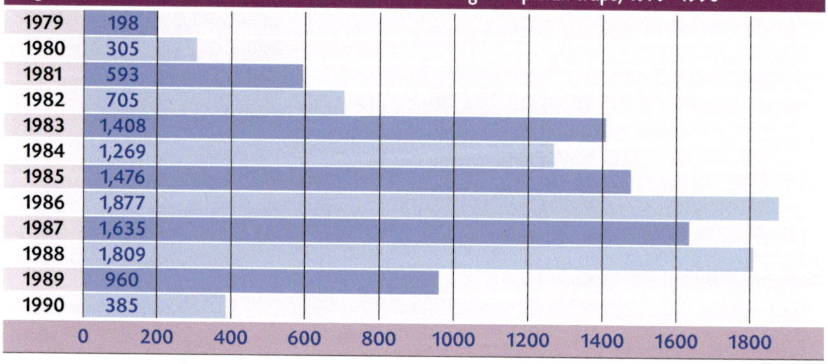

Table 8.1. Size and composition of annual and 12-year catches of beetles in pitfall traps in the Leicester garden, 1979–1990

Year	Individuals (N)	Species (S)	N/S	Species taken once only	Commonest species	As % N
1979	198	20	9.9	8	*Pterostichus madidus*	72.73
1980	305	19	16.05	5	*Pterostichus madidus*	74.43
1981	593	17	34.88	5	*Pterostichus madidus*	66.95
1982	705	21	33.57	8	*Pterostichus madidus*	75.74
1983	1,408	22	64.00	9	*Pterostichus madidus*	73.01
1984	1,269	23	55.17	15	*Pterostichus madidus*	52.56
1985	1,476	20	73.80	8	*Pterostichus madidus*	61.38
1986	1,877	25	75.08	12	*Pterostichus madidus*	54.98
1987	1,635	15	109.00	3	*Pterostichus madidus*	67.77
1988	1,809	21	86.14	6	*Pterostichus madidus*	81.15
1989	960	17	56.47	6	*Pterostichus madidus*	88.54
1990	385	17	22.65	6	*Pterostichus madidus*	73.51
1979–1990	12,620	66	191.21	33	*Pterostichus madidus*	68.49

There is a pronounced similarity of the garden beetle fauna as recorded from pitfall traps to that of open ground and wasteland sites, and most of the recorded species are xerophilous (liking dry places) or eurytopic (range widely). This accords with the widely held ecological view of gardens as open, disturbed sites. There are also typical woodland species, such as *Leistus fulvibarbis*, and hydrophilic species, such as *Bembidion guttula*, the former caught below the Malaise trap, the latter in the open garden. *Pterostichus madidus* is common in open country, often on cultivated soil and in gardens, and *Nebria brevicollis* is extremely eurytopic. Of the other Carabidae that were most frequently captured, *Notiophilus biguttatus* is common in somewhat shady but dry places, and *Leistus spinibarbis* is widespread and fairly eurytopic (Lindroth 1974).

The 51 garden records of a curculionid, *Otiorhynchus porcatus*, were the first for the county, although, as all four species of *Otiorhynchus* captured are common garden pests, it may well have been under-recorded. Of the other two abundant weevils, *Barypeithes pellucidus* is locally common, and *Barynotus obscurus* is widespread and common.

Beetles, other than ladybirds, from the Malaise trap

A total of 14,483 beetles, other than ladybirds, of at least 398 species in 41 families were caught in the Malaise trap during the period 1981–2001. Ladybirds (*Coccinellidae*) are considered separately from other beetles caught in the Malaise trap for three reasons: ladybirds are conspicuous and familiar to most people; unlike other beetles they were sorted from trap samples, identified and tabulated from 1972 to 2001; and they differ from other beetles in that there were large-scale influxes to the garden of certain species in some years.

The annual catch of individuals and species of beetle, other than ladybirds, in the Malaise trap fluctuated somewhat, with 1981 and 1987 being particularly poor years (Tab. 8.2 and Fig. 8.2), but the striking thing was the marked ►p150

Fig. 8.2. Annual fluctuations in numbers of beetles other than ladybirds caught in Malaise trap, 1981–2001

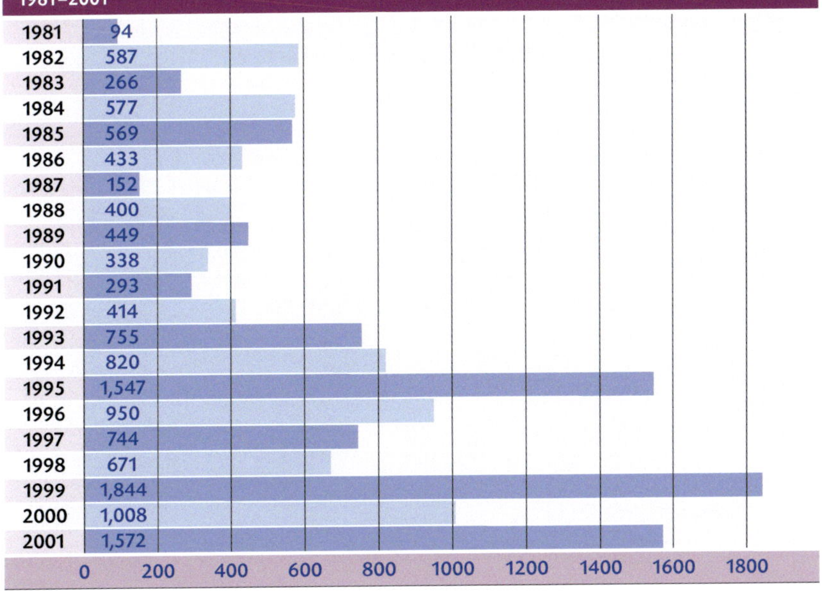

BEETLES (from Malaise trap)

In the following list, species also caught in pitfalls are given the same number (in parentheses) as in the previous section. Numbers refer to total Malaise trap catch, 1981–2001. For ladybirds, see p153.
Species also caught in the mercury-vapour light trap are indicated by MV.
Taxonomy and sequence follow Pope (1977).

Carabidae – ground beetles

(4)	*Leistus spinibarbis* (Fab.)	1	
(5)	*Nebria brevicollis* (Fab.)	1	
(6)	*Notiophilus biguttatus* (Fab.)	26	
67	*Notiophilus substriatus* Waterhouse	1	
(7)	*Loricera pilicornis* (Fab.)	5	
(8)	*Trechus quadristriatus* (Schrank)	47	
68	*Asaphidion curtum* (Heyden)	2	
(9)	*Asaphidion flavipes* (L.)	1	
69	*Bembidion quadrimaculatum* (L.)	6	
(11)	*Bembidion guttula* (Fab.)	19	
70	*Bembidion lunulatum* (Fourcroy)	1	
71	*Tachys parvulus* Dejean	1	
(15)	*Amara aenea* (Degeer)	1	
72	*Amara apricaria* (Paykull)	1	MV
(16)	*Amara familiaris* (Duftschmid)	3	
(17)	*Amara plebeja* (Gyllenhal)	1	
(18)	*Amara similata* (Gyllenhal)	2	
(20)	*Harpalus rufibarbis* (Fab.)	1	
73	*Bradycellus verbasci* (Duftschmid)	8	MV
74	*Stenolophus mixtus* (Herbst)	1	
75	*Demetrias atricapillus* (L.)	5	
76	*Dromius agilis* (Fab.)	3	
77	*Dromius meridionalis* Dejean	4	

Hydrophilidae

78	*Helophorus brevipalpis* Bedel	116
79	*Helophorus minutus* Fab.	1
80	*Sphaeridium scarabaeoides* (L.)	4
81	*Cercyon atomarius* (Fab.)	2
82	*Cercyon haemorrhoidalis* (Fab.)	3
83	*Cercyon lateralis* (Marsham)	5
84	*Cercyon quisquilius* (L.)	2
85	*Cryptopleurum minutum* (Fab.)	2
86	*Anacaena globulus* (Paykull)	1
87	*Anacaena limbata* (Fab.)	1

Histeridae

88	*Acritus nigricornis* (Hoffman)	1

Leiodidae

89	*Leiodes badia* (Sturm)	2
90	*Cyrtusa pauxilla* (Schmidt)	1
91	*Ptomophagus medius* Rey	9
(24)	*Ptomophagus subvillosus* (Goeze)	53
92	*Choleva agilis* (Illiger)	1
93	*Choleva angustata* (Fab.)	12
(25)	*Catops nigricans* (Spence)	2

Scydmaenidae

94	*Euconnus* species	2

Staphylinidae

95	*Metopsia retusa* (Stephens)	2	
96	*Megarthrus affinis* Miller	1	
97	*Megarthrus depressus* (Paykull)	3	
98	*Proteinus brachypterus* (Fab.)	4	
99	*Proteinus ovalis* Stephens	5	
100	*Anthobium atrocephalum* (Gyllenhal)	1	
101	*Lesteva heeri* Fauvel	1	
102	*Lesteva longoelytrata* (Goeze)	2	
103	*Lesteva punctata* Erichson	1	
104	*Eusphalerum primulae* (Stephens)	1	
105	*Phyllodrepa floralis* (Paykull)	2	
106	*Omalium caesum* Gravenhorst	1	
(26)	*Omalium excavatum* Stephens	3	
107	*Omalium rivulare* (Paykull)	1	
108	*Carpelimus pusillus* (Gravenhorst)	1	
109	*Platystethus arenarius* (Fourcroy)	8	
110	*Anotylus complanatus* (Erichson)	28	
(27)	*Anotylus inustus* (Gravenhorst)	3	
111	*Anotylus rugosus* (Fab.)	49	
(28)	*Anotylus sculpturatus* (Gravenhorst)	92	
112	*Anotylus tetracarinatus* (Block)	6	
113	*Oxytelus laqueatus* (Marsham)	6	
114	*Stenus clavicornis* (Scopoli)	4	
115	*Stenus nitidiusculus* Stephens	1	
116	*Stenus ossium* Stephens	3	
117	*Stenus rogeri* Kraatz	1	
118	*Stenus similis* (Herbst)	2	
119	*Achenium humile* (Nicolai)	1	
120	*Sunius propinquus* (Brisout)	3	
121	*Lithocharis obscurella* Erichson	1	
(32)	*Rugilus orbiculatus* (Paykull)	6	
122	*Rugilus rufipes* Germar	3	
123	*Phacophallus parumpunctatus* (Gyllenhal)	1	
(33)	*Gyrohypnus angustatus* Stephens	1	
124	*Gyrohypnus fracticornis* (Müller)	5	
125	*Gyrohypnus punctulatus* (Paykull)	2	
126	*Xantholinus glabratus* (Gravenhorst)	4	
(34)	*Xantholinus linearis* (Olivier)	14	
127	*Xantholinus longiventris* Heer	12	
128	*Philonthus cephalotes* (Gravenhorst)	1	
(35)	*Philonthus cognatus* Stephens	83	
129	*Philonthus decorus* (Gravenhorst)	14	
130	*Philonthus discoideus* (Gravenhorst)	1	
131	*Philonthus (Bisnius) fimetarius* (Gravenhorst)	40	MV
132	*Philonthus intermedius* (Boisduval & Lacordaire)	1	
133	*Philonthus jurgans* Tottenham	14	
134	*Philonthus laminatus* (Creutzer)	5	
(36)	*Philonthus marginatus* (Ström)	23	
(37)	*Philonthus politus* (L.)	2	
135	*Philonthus sanguinolentus* (Gravenhorst)	4	
136	*Philonthus sordidus* (Gravenhorst)	2	
137	*Philonthus splendens* (Fab.)	3	
138	*Philonthus succicola* Thomson	2	
(38)	*Philonthus tenuicornis* Mulsant & Rey	21	
139	*Philonthus umbratilis* (Gravenhorst)	1	
140	*Philonthus varians* (Paykull)	363	
(39)	*Philonthus varius* (Gyllenhal)	94	
141	*Gabrius subnigritulus* (Reitter)	1	
142	*Gabrius* species	1	

(40)	***Staphylinus* (*Tasgius*) *ater*** Gravenhorst	8
143	***Ontholestes murinus*** (L.)	1
(42)	***Quedius boops*** (Gravenhorst)	38
144	***Quedius brevis*** Erichson	1
145	***Quedius cinctus*** (Paykull)	9
146	***Quedius cruentus***(Olivier)	3
147	***Quedius lateralis*** (Gravenhorst)	3
148	***Quedius mesomelinus*** (Marsham)	5
(44)	***Quedius molochinus*** (Gravenhorst)	2
149	***Quedius nemoralis*** Baudi	1
150	***Quedius picipes*** (Mannerheim)	1
151	***Quedius scintillans*** (Gravenhorst)	1
152	***Quedius semiaeneus***(Stephens)	1
153	***Quedius semiobscurus*** (Marsham)	5
(45)	***Quedius tristis*** (Gravenhorst)	2
154	***Habrocerus capillaricornis*** (Gravenhorst)	1
155	***Mycetoporus clavicornis*** (Stephens)	1
156	***Mycetoporus lepidus*** (Gravenhorst)	11
157	***Mycetoporus nigricollis*** Stephens	19
158	***Mycetoporus punctus*** (Gravenhorst)	1
159	***Mycetoporus* (*Ischnosoma*) *splendidus*** (Gravenhorst)	7
160	***Lordithon thoracicus*** (Fab.)	3
161	***Bolitobius analis*** (Fab.)	3
162	***Bolitobius cingulatus*** (Mannerheim)	1
163	***Sepedophilus marshami*** (Stephens)	2
164	***Tachyporus chrysomelinus*** (L.)	138
165	***Tachyporus hypnorum*** (Fab.)	594
166	***Tachyporus nitidulus*** (Fab.)	168
167	***Tachyporus obtusus*** (L.)	53
168	***Tachyporus pusillus*** Gravenhorst	2
169	***Tachyporus solutus*** Erichson	11
170	***Tachinus humeralis*** Gravenhorst	5
171	***Tachinus marginellus*** (Fab.)	1
172	***Tachinus pallipes*** (Gravenhorst)	1
(46)	***Tachinus signatus*** Gravenhorst	103
173	***Cypha longicornis*** (Paykull)	5
174	***Autalia rivularis*** (Gravenhorst)	6
175	***Falagria caesa*** Erichson	4
176	***Tachyusa*** species	2
177	***Aloconota gregaria*** (Erichson)	51
178	***Amischa analis*** (Gravenhorst)	52
179	***Amischa decipiens*** (Sharp)	15
180	***Amischa soror*** (Kraatz)	23
181	***Nehemitropia sordida*** (Marsham)	3
182	***Dinaraea angustula*** (Gyllenhal)	2
183	***Atheta* (*Philhygra*) *elongatula*** (Gravenhorst)	6
184	***Atheta malleus*** Joy	1
185	***Atheta palustris*** (Kiesenwetter)	8
186	***Atheta indubia*** (Sharp)	1
187	***Atheta* (*Microdota*)** species	2
188	***Atheta trinotata*** (Kraatz)	4
189	***Atheta* (*Mocyta*) *fungi*** (Gravenhorst)	311
190	***Atheta aterrima*** (Gravenhorst)	2
191	***Atheta muscorum*** (Brisout)	6
192	***Atheta parvula*** (Mannerheim)	1
193	***Atheta* (*Datomicra*) *nigra*** (Kraatz)	12
194	***Atheta sordidula*** (Erichson)	1
195	***Atheta graminicola*** (Gravenhorst)	1
196	***Atheta triangulum*** (Kraatz)	2
197	***Atheta xanthopus*** (Thomson)	6
198	***Atheta coriaria*** (Kraatz)	1
199	***Atheta crassicornis*** (Fab.)	24
200	***Atheta* (*Mycetota*) *laticollis*** (Stephens)	48
201	***Atheta ravilla*** (Erichson)	2
202	***Atheta* (*Dimetrota*) *atramentaria*** (Gyllenhal)	3
203	***Atheta setigera*** (Sharp)	1
204	***Atheta* (*Chaetida*) *longicornis*** (Gravenhorst)	8
205	***Chiloporata longitarsis*** (Erichson)	2
206	***Oxypoda haemorrhoa*** (Mannerheim)	1
207	***Oxypoda induta*** Mulsant & Rey	1
208	***Oxypoda opaca*** (Gravenhorst)	3
209	***Oxypoda umbrata*** (Gyllenhal)	9
210	***Haploglossa pulla*** (Gyllenhal)	5
211	***Tinotus morion*** (Gravenhorst)	2
212	***Aleochara bilineata*** Gyllenhal	9
213	***Aleochara bipustulata*** (L.)	4
214	***Aleochara curtula*** (Goeze)	25
215	***Aleochara inconspicua*** Aubé	1
216	***Aleochara lanuginosa*** Gravenhorst	25

Pselaphidae

217	***Bryaxis bulbifer*** (Reichenbach)	1

Scarabaeidae

218	***Aphodius contaminatus*** (Herbst)	15
219	***Aphodius prodromus*** (Brahm)	1

Scirtidae

220	***Microcara testacea*** (L.)	3
221	***Cyphon variabilis*** Thunberg	1

Byrrhidae

(48)	***Simplocaria semistriata*** (Fab.)	1

Elateridae

222	***Athous haemorrhoidalis*** (Fab.)	1
223	***Agriotes acuminatus*** (Stephens)	1
(50)	***Agriotes pallidulus*** (Illiger)	2

Cantharidae

224	***Cantharis decipiens*** Baudi	1
225	***Cantharis lateralis*** L.	7
226	***Cantharis rufa*** L.	2
227	***Rhagonycha femoralis*** (Brullé)	1
228	***Rhagonycha fulva*** (Scopoli)	1
229	***Rhagonycha testacea*** (L.)	1
230	***Malthinus flaveolus*** (Herbst)	1
231	***Malthodes marginatus*** (Latreille)	2

Dermestidae

232	***Anthrenus fuscus*** Olivier	38
233	***Anthrenus museorum*** (L.)	8
234	***Anthrenus verbasci*** (L.)	61

Anobiidae

235	***Ptinomorphus imperialis*** (L.)	1
236	***Ochina ptinoides*** (Marsham)	1
237	***Stegobium paniceum*** (L.)	1

238	***Anobium punctatum*** (Degeer)	11
239	***Ptilinus pectinicornis*** (L.)	1
240	***Dorcatoma serra*** Panzer	1

Ptinidae

241	***Ptinus sexpunctatus*** Panzer	8

Melyridae

242	***Dasytes aeratus*** Stephens	3
243	***Dasytes plumbeus*** (Müller)	3

Nitidulidae

244	***Kateretes bipustulatus*** (Paykull)	2
245	***Kateretes pedicularis*** (L.)	1
246	***Brachypterus glaber*** (Stephens)	117
247	***Brachypterus urticae*** (Fab.)	40
248	***Pria dulcamarae*** (Scopoli)	1
(51)	***Meligethes aeneus*** (Fab.)	2,730
249	***Meligethes difficilis*** (Heer)	1
250	***Meligethes morosus*** Erichson	207
251	***Meligethes nigrescens*** Stephens	374
252	***Meligethes ovatus*** Sturm	24
253	***Epuraea aestiva*** (L.)	3
254	***Epuraea deleta*** Sturm	2
255	***Epuraea pusilla*** (Illiger)	1
256	***Soronia grisea*** (L.)	2

Rhizophagidae

257	***Monotoma picipes*** Herbst	14

Cryptophagidae

258	***Cryptophagus dentatus*** (Herbst)	29
259	***Cryptophagus laticollis*** Lucas	2
260	***Cryptophagus pseudodentatus*** Bruce	1
261	***Cryptophagus pubescens*** Sturm	2
262	***Cryptophagus scanicus*** L.	18
263	***Atomaria atricapilla*** Stephens	1
264	***Atomaria fuscata*** (Schoenherr)	4
265	***Atomaria lewisi*** Reitter	1
266	***Atomaria nitidula*** (Marsham)	7
267	***Atomaria fimetarii*** (Fab.)	1
268	***Ephistemus globulus*** (Paykull)	1

Byturidae

269	***Byturus ochraceus*** (Scriba)	1
270	***Byturus tomentosus*** (Degeer)	2

Erotylidae

271	***Dacne bipustulata*** (Thunberg)	2

Phalacridae

272	***Olibrus aeneus*** (Fab.)	5
273	***Olibrus liquidus*** Erichson	2
274	***Stilbus oblongus*** (Erichson)	1
275	***Stilbus testaceus*** (Panzer)	112

Coccinellidae

276	***Coccidula rufa*** (Herbst)	6
277	***Rhizobius litura*** (Fab.)	12
278	***Stethorus punctillum*** Weise	14
279	***Scymnus nigrinus*** Kugelann	3
280	***Scymnus limbatus*** Stephens	2
281	***Scymnus suturalis*** Thunberg	1

Endomychidae

282	***Sphaerosoma piliferum*** (Müller)	3

Lathridiidae

283	***Stephostethus lardarius*** (Degeer)	44
284	***Aridius bifasciatus*** (Reitter)	79
(57)	***Aridius nodifer*** (Westwood)	28
285	***Lathridius minutus*** (L.)	1
286	***Enicmus transversus*** (Olivier)	67
287	***Corticaria punctulata*** Marsham	8
288	***Corticarina fuscula*** (Gyllenhal)	162
289	***Cortinicara gibbosa*** (Herbst)	795

Cisidae

290	***Cis boleti*** (Scopoli)	5

Mycetophagidae

291	***Typhaea stercorea*** (L.)	1

Tenebrionidae

292	***Lagria hirta*** (L.)	3

Salpingidae

293	***Rhinosimus planirostris*** (Fab.)	18

Melandryidae

294	***Anisoxya fuscula*** (Illiger)	5

Scraptiidae

295	***Anaspis humeralis*** (Fab.)	31
296	***Anaspis lurida*** Stephens	1
297	***Anaspis maculata*** Fourcroy	1,248
298	***Anaspis regimbarti*** Schilsky	229
299	***Anaspis rufilabris*** (Gyllenhal)	12
300	***Anaspis thoracica*** (L.)	3

Rhipiphoridae

301	***Metoecus paradoxus*** (L.)	2

Oedemeridae

302	***Oedemera lurida*** (Marsham)	12

Anthicidae

303	***Anthicus antherinus*** (L.)	1
304	***Anthicus floralis*** (L.)	6
305	***Anthicus (Omonadus) formicarius*** (Goeze)	7

Aderidae

306	***Aderus populneus*** (Creutzer in Panzer)	1

Cerambycidae

307	***Grammoptera ruficornis*** (Fab.)	90

Bruchidae

308	***Bruchus loti*** Paykull	2
309	***Bruchus rufimanus*** Boheman	8
310	***Bruchidius ater*** (Marsham)	1
311	***Bruchidius varius*** (Olivier)	1
312	***Bruchidius villosus*** (Fab.)	6

Chrysomelidae		
313	*Oulema lichenis* Voet	2
314	*Oulema melanopa* (L.)	39
315	*Oulema rufocyanea* (Suffrien)	13
316	*Phaedon tumidulus* (Germar)	1
317	*Plagiodera versicolora* (Laicharting)	1
318	*Lochmaea crataegi* (Forster)	3
319	*Phyllotreta aerea* Allard	1
320	*Phyllotreta atra* (Fab.)	43
321	*Phyllotreta consobrina* (Curtis)	2
322	*Phyllotreta cruciferae* (Goeze)	40
323	*Phyllotreta nigripes* (Fab.)	97
324	*Phyllotreta ochripes* (Curtis)	4
325	*Phyllotreta undulata* Kutschera	65
326	*Phyllotreta vittula* Redtenbacher	62
327	*Aphthona atrocaerulea* (Stephens)	83
328	*Aphthona euphorbiae* (Schrank)	899
329	*Aphthona herbigrada* (Curtis)	62
330	*Aphthona melancholica* Weise	202
331	*Longitarsus atricillus* (L.)	1
332	*Longitarsus exoletus* (L.)	18
333	*Longitarsus ganglbaurei* Heikertinger	1
334	*Longitarsus jacobaeae* (Waterhouse)	13
335	*Longitarsus luridus* (Scopoli)	133
336	*Longitarsus melanocephalus* (Degeer)	5
337	*Longitarsus parvulus* (Paykull)	814
338	*Longitarsus rubiginosus* (Foudras)	2
339	*Longitarsus succineus* (Foudras)	1
340	*Longitarsus suturalis* (Marsham)	26
341	*Longitarsus suturellus* (Duftschmid)	2
342	*Altica brevicollis* Foudras	4
343	*Altica lythri* Aube	1
344	*Altica oleracea* (L.)	4
345	*Altica palustris* Weise	2
346	*Chalcoides* (*Crepidodera*) *aurata* (Marsham)	15
347	*Chalcoides aurea* (Fourcroy)	4
348	*Chalcoides fulvicornis* (Fab.)	8
349	*Epitrix pubescens* (Koch)	1
350	*Chaetocnema concinna* (Marsham)	109
351	*Chaetocnema hortensis* (Fourcroy)	29
352	*Sphaeroderma testaceum* (Fab.)	2
353	*Psylliodes affinis* (Paykull)	1
354	*Psylliodes chalcomera* (Illiger)	1
355	*Psylliodes chrysocephala* (L.)	97
356	*Psylliodes napi* (Fab.)	10
357	*Cassida viridis* L.	5

Attelabidae		
358	*Rhynchites* (*Neocoenorrhinus*) *aequatus* (L.)	4

Apionidae		
359	*Apion* (*Perapion*) *curtirostre* Germar	27
360	*Apion hydrolapathi* (Marsham)	5
361	*Apion* (*Malvapion*) *malvae* (Fab.)	33
362	*Apion* (*Pseudapion*) *rufirostre* (Fab.)	8
363	*Apion* (*Aspidapion*) *aeneum* (Fab.)	1
(58)	*Apion radiolus* (Marsham)	87
364	*Apion* (*Exapion*) *ulicis* (Forster)	2
365	*Apion miniatum* Germar	2
366	*Apion* (*Thymapion*) *atomarium* Kirby	1
367	*Apion* (*Catapion*) *seniculus* Kirby	1
368	*Apion* (*Ceratapion*) *carduorum* Kirby	2
369	*Apion* (*Oxystoma*) *craccae* (L.)	1
370	*Apion* (*Protapion*) *assimile* Kirby	2
371	*Apion dichroum* Bedel	3
372	*Apion nigritarse* Kirby	6
373	*Apion trifolii* (L.)	6
374	*Nanophyes marmoratus* (Goeze)	21

Curculionidae		
(60)	*Otiorhynchus rugostriatus* (Goeze)	6
(61)	*Otiorhynchus singularis* (L.)	1
(62)	*Otiorhynchus sulcatus* (Fab.)	15
375	*Phyllobius pyri* (L.)	1
(64)	*Barypeithes pellucidus* (Boheman)	1
376	*Sitona cambricus* Stephens	1
377	*Sitona cylindricollis* (Fahraeus)	1
378	*Sitona griseus* (Fab.)	1
379	*Sitona hispidulus* (Fab.)	17
380	*Sitona humeralis* Stephens	3
381	*Sitona lepidus* Gyllenhal	28
382	*Sitona lineatus* (L.)	416
383	*Sitona puncticollis* Stephens	26
384	*Sitona suturalis* Stephens	8
385	*Hypera postica* (Gyllenhal)	1
386	*Magdalis ruficornis* (L.)	6
387	*Euophryum confine* (Brown)	4
388	*Cossonus parallelepipedus* (Herbst)	1
389	*Cidnorhinus quadrimaculatus* (L.)	4
390	*Ceuthorrynchidius* (*Trichosirocolodea*) *troglodytes* (Fab.)	11
391	*Ceutorhynchus assimilis* (Paykull)	231
392	*Ceutorhynchus campestris* Gyllenhal	3
393	*Ceutorhynchus erysimi* (Fab.)	7
394	*Ceutorhynchus floralis* (Paykull)	27
395	*Ceutorhynchus quadridens* (Panzer)	165
396	*Ceutorhynchus rugulosus* (Herbst)	19
397	*Ceutorhynchus triangulum* Boheman	4
398	*Rhinoncus pericarpius* (L.)	2
399	*Rhinoncus perpendicularis* (Reich)	2
400	*Amalus scortillum* (Herbst)	3
401	*Anthonomus rubi* (Herbst)	26
402	*Anthonomus pedicularius* (L.)	10
403	*Amalorrhynchus melanarius* (Stephens)	17
404	*Curculio pyrrhoceras* Marsham	6
405	*Curculio nucum* L.	1
406	*Curculio salicivorus* Paykull	4
407	*Tychius* species	1
408	*Miccotrogus* (*Tychius*) *picirostris* (Fab.)	2
409	*Gymnetron pascuorum* (Gyllenhal)	7
410	*Rhynchaenus* (*Orchestes*) *fagi* (L.)	9

Scolytidae		
411	*Scolytus intricatus* Ratzeberg)	11
412	*Scolytus mali* (Bechstein)	3
413	*Scolytus multistriatus* (Marsham)	4
414	*Scolytus rugulosus* (Müller)	17
415	*Scolytus scolytus* (Fab.)	2
416	*Hylesinus oleiperda* (Fab.)	5
417	*Leperisinus varius* (Fab.)	14
418	*Acrantus vittatus* (Fab.)	2
419	*Xylechinus pilosus* (Ratzeberg)	1
420	*Hylastinus obscurus* (Marsham)	1
421	*Pityogenes* species	1

improvement after 1992, with 1995, 1999 and 2001 being particularly good years. The average number of individuals per species followed much the same pattern, indicating that in 1981 and 1987 the relatively few species present were each less common than in other years, whereas in 1995, 1999 and 2001, the high number of species were each represented by many individuals. Furthermore, in the years of maximum total catch, the commonest species formed a high proportion of the catch, 25.9% (*Meligethes aeneus*) in 1995, 32.4% (*Aphthona*

Table 8.2. Size and composition of annual and 21-year catch of beetles, other than ladybirds, in Malaise trap in the Leicester garden, 1981–2001

Year	Individuals (N)	Species (S)	N/S	Species taken once only	Commonest species	As % N
1981	94	33	2.9	17	*Anaspis maculata*	18.1
1982	587	86	6.8	40	*Ceutorhynchus assimilis*	18.1
1983	266	55	4.8	28	*Philonthus varians*	26.7
1984	577	86	6.7	42	*Stilbus testaceus*	7.8
1985	569	101	5.6	42	*Tachyporus hypnorum*	19.3
1986	433	78	5.6	35	*Meligethes aeneus*	25.5
1987	152	43	3.5	24	*Meligethes aeneus*	26.3
1988	400	64	6.3	27	*Meligethes aeneus*	12.0
1989	449	56	8.0	25	*Atheta fungi*	16.0
1990	338	59	5.7	17	*Anaspis maculata*	11.0
1991	293	37	7.9	10	*Meligethes aeneus*	17.1
1992	414	66	6.3	23	*Meligethes aeneus*	14.5
1993	755	91	8.3	43	*Anaspis maculata*	25.2
1994	820	66	12.4	26	*Meligethes aeneus*	24.4
1995	1,547	95	16.3	36	*Meligethes aeneus*	25.9
1996	950	91	10.4	37	*Meligethes aeneus*	31.7
1997	744	99	7.5	37	*Anaspis maculata*	18.7
1998	671	97	6.9	49	*Meligethes aeneus*	29.8
1999	1,844	106	17.4	47	*Aphthona euphorbiae*	32.4
2000	1,008	107	9.4	51	*Longitarsus parvulus*	32.4
2001	1,572	114	13.8	60	*Meligethes aeneus*	59.4
1981–2001	14,483	395	36.7	128	*Meligethes aeneus*	18.9

Fig. 8.3. Annual fluctuations in numbers of all beetles caught in the Malaise trap, 1981–2001

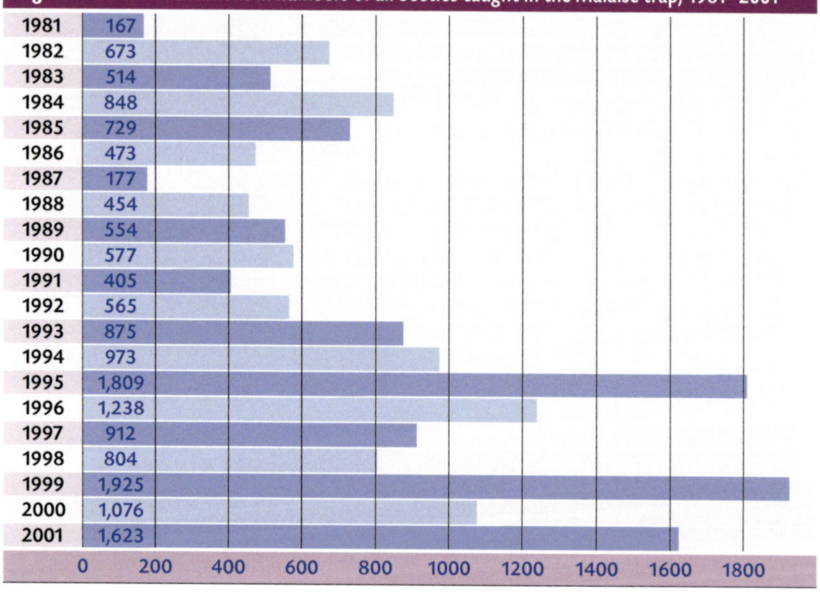

euphorbiae) in 1999 and 59.4% (*M. aeneus*) in 2001. Before 1991, the commonest species tended to change from year to year, but subsequently, *M. aeneus* dominated, and it was the commonest species overall, 1981–2001, making up more than 18% of the total catch.

Pollen beetles, *Meligethes* species, on a flower of Shasta daisy, *Leucanthemum × superbum*. *Meligethes aeneus* is one of the commonest beetles in the Leicester garden

Of the commonest species each year, flower-feeding *M. aeneus* (Nitidulidae), a pollen beetle, sometimes so common as to be considered a pest, is non-specific as to flowers, but in some years was particularly abundant on sweet peas in my garden; adults mainly eat pollen, but larvae chew their way indiscriminately through flower buds of their host plants. *Anaspis maculata*, a scraptiid beetle, is also a flower-feeder, the larvae developing in rotten wood. *Ceutorrhynchus assimilis*, a weevil (Curculionidae), is often a pest on cabbages and other cruciferous plants and, like *M. aeneus*, has become more common with widespread cultivation of oilseed rape.

Philonthus varians, a staphylinid, feeds on dung, carrion and compost, but *Tachyporus hypnorum*, another staphylinid, is a predator in leaf litter and decaying vegetation on the ground, and both species occur in compost heaps. *Atheta fungi*, another common and widespread staphylinid, also abounds in leaf litter, feeding on decomposing plant material. *Stilbus testaceus* (Phalacridae) is said to live mainly on dry grass and is relatively common, although little

Table 8.3. Size and composition of total annual and 21-year catches of all beetles in Malaise trap in the Leicester garden, 1981–2001

Year	Individuals (N)	Species (S)	N/S	Species taken once only	Commonest species	As % N
1981	167	37	4.5	17	*Adalia bipunctata*	31.7
1982	673	91	7.4	42	*Ceutorhynchus assimilis*	15.8
1983	514	60	8.6	28	*Propylea quattuordecimpunctata*	26.7
1984	848	92	9.2	42	*Adalia bipunctata*	13.6
1985	729	106	6.9	42	*Tachyporus hypnorum*	15.1
1986	473	82	5.8	36	*Meligethes aeneus*	23.3
1987	177	47	3.8	24	*Meligethes aeneus*	22.6
1988	454	70	6.5	28	*Meligethes aeneus*	10.6
1989	554	63	8.8	26	*Atheta fungi*	13.0
1990	577	66	8.7	17	*Adalia bipunctata*	17.5
1991	405	43	9.4	10	*Meligethes aeneus*	12.6
1992	565	73	7.7	23	*Meligethes aeneus* *Adalia bipunctata*	10.6=
1993	875	96	9.1	43	*Anaspis maculata*	21.7
1994	973	73	13.3	26	*Meligethes aeneus*	20.6
1995	1,809	100	18.1	37	*Meligethes aeneus*	22.1
1996	1,238	99	12.5	39	*Meligethes aeneus*	24.3
1997	912	106	8.6	39	*Anaspis maculata*	15.2
1998	804	104	7.7	51	*Meligethes aeneus*	24.9
1999	1,925	113	17.0	48	*Aphthona euphorbiae*	31.0
2000	1,076	114	9.4	53	*Longitarsus parvulus*	30.4
2001	1,623	120	13.5	61	*Meligethes aeneus*	57.5
1981–2001	17,371	422	41.2	128	*Meligethes aeneus*	15.7

is known of its development. Both *Aphthona euphorbiae* and *Longitarsus parvulus* are flea beetles, Chrysomelidae, and are pests of flax (*Linum usitatissimum*), although the former is more commonly found on spurges, as the specific name implies.

Ladybirds (Coccinellidae) are considered separately in the next section, but in assessing annual fluctuations in numbers of beetles caught in the Malaise trap, it may be better to include them here. When this is done (Tab. 8.3 and Fig. 8.3), it becomes clear that 1984 was a good year, particularly for the 1980s. Moreover, in three years in the 1980s, ladybirds were the commonest species of beetle, *Adalia bipunctata* in 1981 and 1984, and *Propylea quattuordecimpunctata* in 1983, and in 1992 *Adalia bipunctata* was equally common with *Meligethes aeneus*, each making up 10.6% of the year's catch.

For most groups of insects considered in earlier chapters, there seems to have been a decline in abundance, as judged by Malaise trapping, in the 1990s and on to 2001. Beetles, however, showed a striking increase in abundance during this period, a trend only shown otherwise by solitary wasps and, perhaps, psocids and sawflies. There is no obvious similarity between the ecologically diverse group of beetles and the predatory wasps to explain this, particularly as the commonest beetle species in each year of the 1990s were herbivores. Both may have been responding to the changing weather conditions, as the English Midlands, as represented by Leicester, became slightly warmer, sunnier and wetter.

Fig. 8.4. Annual fluctuations in numbers of ladybirds caught in the Malaise trap, 1972–2001

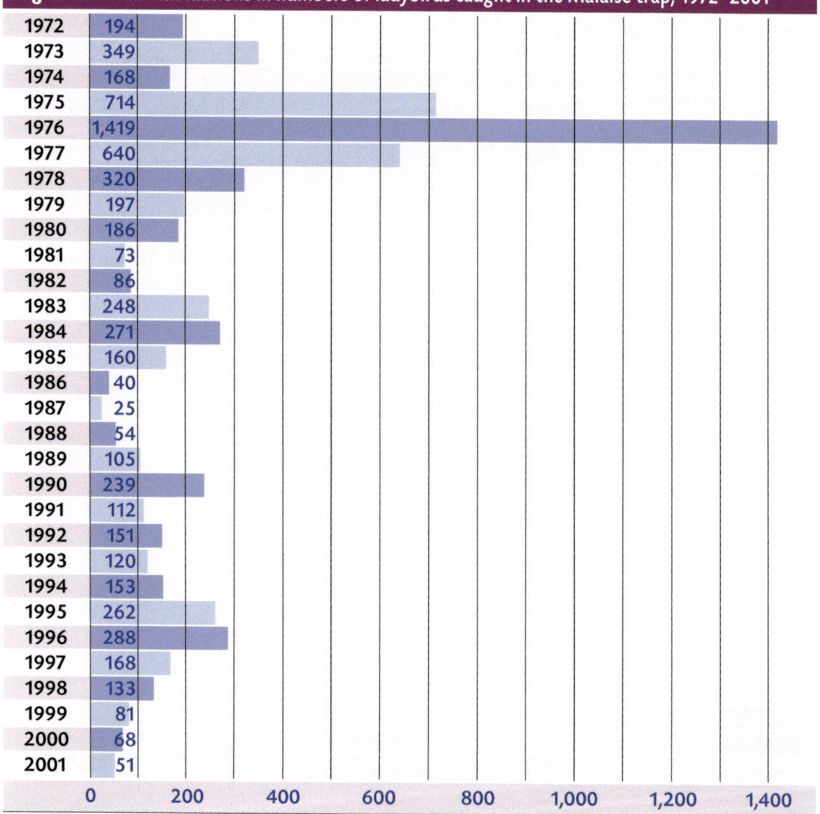

LADYBIRDS

7-spot ladybird, *Coccinella septempunctata*, is a familiar species in gardens. There were influxes of large numbers of this spcies to the Leicester garden in 1975 and 1976

Forty-two species of coccinellid are British, and 24 of these are generally considered as ladybirds (Majerus & Kearns 1989). This section deals only with ladybirds caught in the garden, the other coccinellids being included in the previous section. In the 30-year period 1972–2001, 7,075 ladybirds of 12 species were caught in the Malaise trap and one additional species in the mercury-vapour light trap. They were identified by D.F. Owen until 1996, and subsequently by myself; some early results were published (Owen, D.F. 1982/83).

It is frequently observed that numbers of ladybirds fluctuate markedly from year to year. Annual fluctuations in the total Malaise trap catch are shown in Tab. 8.4 and Fig. 8.4. These show that 1976 was a particularly good year for ladybirds ,when 1,419 of the 30-year total of 7,075 were trapped, and 1987 was a particularly poor year, when only 25 were trapped, but numbers of the six more abundant species fluctuated in different ways.

The 2-spot ladybird (*Adalia bipunctata*) is a common, resident garden ladybird, whose larvae feed on aphids on woody shrubs and trees. It was the commonest ladybird in Malaise trap catches in 17 of the 30 years, and joint commonest in another year. Furthermore, it was the commonest ladybird in the first fifteen years, 1972–1986, and commonest overall for the entire 30-year period 1972–2001 (Tab. 8.4). In 1981, 1984 and 1990 it was the commonest beetle in the garden (Tab. 8.3), and joint commonest in 1992. Numbers peaked in 1975, but from 1981 onwards, it was much less common, although 1984 and 1990 were reasonable years; it is possible that in 1975 and 1976 (the drought year) some 2-spots moved into the garden from the surrounding countryside.

The summers of 1975 and 1976 were hot, dry and sunny, and in the countryside at large there was a population explosion of aphids, followed by a

LADYBIRDS

Taxonomy and classification follow Pope (1977) in the following list, which gives the numbers caught in the Malaise trap. Species also caught in the mercury vapour trap are marked MV

422	**Kidney-spot ladybird** *Chilocorus renipustulatus* (Scriba)	4	
423	**Larch ladybird** *Aphidecta obliterata* (L.)	1	
424	**16-spot ladybird** *Tytthaspis sedecimpunctata* (L.)	1	
(54)	**2-spot ladybird** *Adalia bipunctata* (L.)	3,099	
425	**10-spot ladybird** *Adalia. decempunctata* (L.)	217	
(55)	**7-spot ladybird** *Coccinella septempunctata* L.	1,392	
426	**11-spot ladybird** *Coccinella undecimpunctata* L.	228	
427	**Cream-streaked ladybird** *Harmonia quadripunctata* (Pontoppidan)	3	
(56)	**14-spot ladybird** *Propylea quattuordecimpunctata* (L.)	1,938	
428	**Eyed ladybird** *Anatis ocellata* (L.)	(1 in MV trap only)	
429	**Cream-spot ladybird** *Calvia quattuordecimguttata* (L.)	25	MV
430	**Orange ladybird** *Halyzia sedecimguttata* (L.)	8	
431	**22-spot ladybird** *Thea vigintoduopunctata* (L.)	258	

22-spot ladybird, *Thea vigintiduopunctata* (right), was usually present in the Leicester garden but never common

A larva of 22-spot ladybird (far right). This species feeds on mildew, an unusual food, as other ladybirds and their larvae eat aphids

Table 8.4. Size and composition of annual, 15-year and 30-year catches of ladybirds in Malaise trap in the Leicester garden, 1972–2001

Year	Individuals (N)	Species (S)	N/S	Species taken once only	Commonest species	Commonest as % N
1972	194	6	32.3	1	*Adalia bipunctata*	86.6
1973	349	4	87.3	1	*Adalia bipunctata*	92.8
1974	168	4	42.0	1	*Adalia bipunctata*	92.3
1975	714	6	119.0	-	*Adalia bipunctata*	59.4
1976	1,419	7	202.7	1	*Coccinella septempunctata*	39.8
1977	640	6	106.7	-	*Adalia bipunctata*	56.1
1978	320	5	64.0	-	*Adalia bipunctata*	74.7
1979	197	4	49.3	1	*Propylea quattuordecimpunctata*	75.6
1980	186	5	37.2	1	*Adalia bipunctata*	69.4
1981	73	4	18.3	-	*Adalia bipunctata*	72.6
1982	86	5	17.2	2	*Adalia bipunctata*	65.1
1983	248	5	49.6	-	*Propylea quattuordecimpunctata*	55.2
1984	271	6	45.2	-	*Adalia bipunctata*	42.4
1985	160	5	32.0	-	*Adalia bipunctata*	48.1
1986	40	4	10.0	1	*Adalia bipunctata*	65.0
1987	25	4	6.3	-	*Adalia bipunctata* *Propylea quattuordecimpunctata*	40.0=
1988	54	6	9.0	1	*Propylea quattuordecimpunctata*	68.2
1989	105	7	15.0	1	*Adalia bipunctata*	29.5
1990	239	7	34.1	-	*Adalia bipunctata*	42.3
1991	112	6	18.7	-	*Adalia bipunctata*	42.00
1992	151	7	21.6	-	*Adalia bipunctata*	39.7
1993	120	5	24.0	-	*Propylea quattuordecimpunctata*	45.0
1994	153	7	21.9	-	*Coccinella septempunctata*	36.6
1995	262	5	52.4	1	*Coccinella septempunctata*	58.8
1996	288	8	36.0	2	*Propylea quattuordecimpunctata*	42.0
1997	168	7	24.0	2	*Propylea quattuordecimpunctata*	53.0
1998	133	7	19.0	1	*Propylea quattuordecimpunctata*	51.1
1999	81	7	11.6	1	*Propylea quattuordecimpunctata*	49.4
2000	68	7	9.7	2	*Propylea quattuordecimpunctata*	61.8
2001	51	6	8.5	1	*Propylea quattuordecimpunctata*	39.2
1972–1986	5,065	8	633.1	2	*Adalia bipunctata*	51.1
1987–2001	2,010	12	167.5	1	*Propylea quattuordecimpunctata*	35.9
1972–2001	7,075	12	589.6	1	*Adalia bipunctata*	43.8

rise in numbers of predatory ladybirds (and, in 1975, of hoverflies). As their aphid food became depleted, either because of predators, or when crops were harvested (or, in 1976, when the drought took hold and withered vegetation), large numbers of ladybirds (and in 1975, hoverflies) took to the wing and moved, often ending up in gardens.

The 10-spot ladybird (*Adalia decempunctata*) was never common, although trapped in every year, numbers reaching a maximum of 25 in 1976. Its ecological requirements are similar to those of *A. bipunctata*, and it is suspected to have bred in the garden, although this is not known. Adults were often seen feeding on aggregations of aphids with *A. bipunctata*, and a few have hibernated in dense vegetation in the garden.

14-spot ladybird, *Propylea quattuordecimpunctata* (above), in eleven years the commonest ladybird in the Leicester garden, and in 1983, the commonest beetle

The 7-spot ladybird (*Coccinella septempunctata*), a common, large ladybird, is often reported as migratory, although more probably it periodically irrupts and undertakes mass movements. It was not caught in the Malaise trap in 1973, 1974, 1986 or 1987, but in other years numbers trapped varied from one to a spectacular 564 in 1976. There was a subsequent marked increase in 1995 to 154, and this was sustained to 1996, when 95 were caught. The drought year, 1976, has been called 'The year of the ladybird', for large numbers of 7-spot ladybirds appeared in many areas, especially in eastern counties of England and on the east coast. This species, like most ladybirds, is aphidivorous, and is a countryside ladybird, associated with aphids feeding on low-growing vegetation, particularly field crops. In 1976, and to a lesser extent in 1975, numbers built up on the aphid population explosion in the countryside and then moved away in vast numbers as the vegetation became parched and brown in the drought. 1995 and 1996 were also hot, dry and bright summers, and although they were not designated as drought years, it appears that the 7-spot ladybird responded in the same way.

The 11-spot ladybird (*Coccinella undecimpunctata*) is probably not a characteristic garden ladybird in central England (Owen, D. F. 1982/83), none having been trapped in 1973, 1979–1983, 1985–1988, 1990–1993 or 1995–2001, although 108 were caught in 1975 and 98 in 1976. There was a movement into the garden in the first two weeks of August 1975, followed by successful over-wintering, and another influx in July 1976. Although 11-spots have over-wintered in the garden, there has been no evidence of breeding, and they seem to be subject to irruptive movements similar to those of *C. septempunctata*, although on a smaller scale.

Although especially associated with deciduous woodland, the 14-spot ladybird (*Propylea quattuordecimpunctata*), another aphid-feeder, was trapped every year in the garden in numbers varying from eight to 346 (in 1976, when many invaded the garden in July and August). Numbers were also high in 1977, 1979, 1983, 1996 and 1997, when there were evidently further invasions to the garden. The 14-spot was never found breeding in the garden, but frequently

2-spot ladybird, *Adalia bipunctata*, is a common resident in the Leicester garden, and in some years was the commonest beetle trapped

over-wintered in dense vegetation. In ten years the 14-spot was the commonest ladybird caught in the Malaise trap, and in the 15-year period 1987–2001, it was overall the commonest (Tab. 8.4). In 1983, it was the commonest beetle caught in the Malaise trap (Tab. 8.3). The 1976 invasion may have been a response to drought, and 1983 and 1996 were hot, dry and bright summers, but other invasions have no particular association with weather.

The 22-spot ladybird (*Thea vigintiduopunctata*) is reputedly a mildew-feeder, and was trapped in 25 of the 30 years 1972–2001, the best year being 1976, with a total of 39. There was no evidence that it over-wintered in the garden, but in August 1985, a pupa was found on a willowherb leaf and produced an adult in September.

Undoubtedly, the garden has provided a good habitat for ladybirds. There is a resident population of 2-spots, which became large in some years, and there have been periodic large influxes of 7-spots, 11-spots and 14-spots, with the 7-spot occasionally breeding. The years 1975 and 1976 were evidently special for ladybirds, particularly for mass movements, and there was no comparable repetition despite several good summers. It does seem that overall ladybirds declined over the 30-year period 1972–2001.

OTHER BEETLES

Additional species, captured at night in the mercury-vapour light trap (MV) or seen and collected (C).

Dytiscidae			
432	*Ilybius ater* (Degeer)	1 MV	
Silphidae			
433	*Nicrophorus humator* (Gleditsch)	1 MV	
434	*Nicrophorus investigator* Zetterstedt	2 MV	
435	*Necrodes littoralis* (L.)	1 MV	1 C
Staphylinidae			
436	*Tachinus subterraneus* (L.)	1 MV	
Geotrupidae			
437	*Geotrupes stercorarius* (L.)	2 MV	
Scarabaeidae			
438	*Aphodius depressus* (Kugelann)	3 MV	
439	*Aphodius rufus* (Moll)	9 MV	
Cantharidae			
440	*Cantharis livida* L.		1 C
Cleridae			
441	*Thanasimus formicarius* (L.)		1 C
Chrysomelidae			
442	*Phyllodecta vulgatissima* (L.)		1 C

OTHER BEETLES

Although the vast majority of information about the garden's beetles comes from Malaise and pitfall trapping, a number of additional species were captured at night in the mercury-vapour light trap or seen and collected. These were variously identified by D. Goddard and K. Porter.

Eight species, including all the Silphidae recorded and the only geotrupid, are thus known for the garden only from light-trapping at night. Three species, including ant beetle (*Thanasimus formicarius*), the only clerid to have been caught, are known only from observation and collection. Pitfall and Malaise trapping, useful as they are for describing a beetle fauna, do not give the complete picture, and it is necessary to employ other methods, particularly to record nocturnal species.

BEETLES IN LEICESTERSHIRE

In all, a total of 442 species of beetle from the Leicester garden have been identified, at least to genus, with the possibility of additional species among undetermined Aleocharinae (Staphylinidae) and *Meligethes* (Nitidulidae). A few species, such as *Pterostichus madidus*, a carabid, and *Adalia bipunctata*, a coccinellid, were common, but the majority were rare, and represented by only one or two individuals.

Nicrophorus investigator, one of three species of sexton or burying beetle attracted to the mercury-vapour light in the Leicester garden. The adults excavate soil around dead mammals and birds and, once they are buried, lay their eggs on the corpse which the beetle larvae eat

In contrast to most groups of insects studied in the garden, beetles became more abundant in the 1990s and at the turn of the century. The 442 species recorded in the garden form only about 12% of the British list. They are, I assume, only a fraction of those really present or at least passing through, for A.A. Allen recorded more than 700 species in his southeast London garden between 1926 and 1973 (Davis 1978). Many beetles are fairly sedentary, others do not fly much, and only some walk about on the soil surface. The Malaise trap, pitfalls and light trap used in the garden could consequently catch only a proportion, perhaps a small proportion, of those present.

The county of Leicestershire is poor for beetles, its central position making it too far north for southern species and too far south for northern species. Furthermore, the destruction of ancient woodland in the last 80 years has meant the increasing rarity or disappearance from the county of many species associated with dead wood. The carabid and staphylinid fauna of the garden has a similarity to the fauna of open ground and of wasteland sites, which is perhaps what might be expected, as the majority are ground-dwellers and a garden is by definition a disturbed site. There is, however, an additional woodland element among the ground-frequenting beetles, reflecting, perhaps, the densely planted aspect of the garden. However, the entire beetle list, including the Malaise trap captures, is much more diverse, and includes many associated with dead wood, with dung or with compost, and also a good number of fungus- and mould-feeders. Furthermore, Malaise trapping in the garden yielded beetle species either very rare in Leicestershire, or new to the county, for example a bruchid beetle, *Bruchidius varius* (Lott & Owen, J. 2003).

Worldwide, Coleoptera are a varied and large group, and it is perhaps to be expected that the garden list would be long. A few garden species are abundant, but most are rare, and species diversity is higher than in any other group investigated.

9: Other insects

A mating pair of common blue damselflies, *Enallagma cyathigerum* (opposite), not usually garden visitors but sometimes appearing in summer

Other insect groups are less well known in the Leicester garden than those allocated separate chapters. Groups such as Odonata (dragonflies) are not typical of gardens and occur only occasionally. Others, such as Dermaptera (earwigs), are represented by only a single species, and Ichneumonidae were only sorted from Malaise trap samples in three years. Many, such as most groups of Diptera (flies), were not systematically collected. Some of these groups, however, have proved to be abundant and diverse, and collectively they augment the garden list considerably.

My knowledge of the insects dealt with in this chapter is patchy, and the fact that there is more information, for instance, about Ichneumonidae is because a world specialist was willing and able to identify them. Although there is more information about some groups than others, 'other insects' are discussed here in taxonomic sequence, rather than separating off the better-described groups, and named species are numbered in sequence.

WINGLESS INSECTS

A globular springtail (Collembola). Springtails are common among leaf litter in gardens

Those insects that are generally assumed to be primitively wingless, and to represent ancestral insects before the evolution of wings and the ability to fly, are included in the subclass Apterygota, although there is some question as to whether all are really insects. They are small to very small, many are difficult to identify, and few entomologists specialize in them or even collect them. I have no detailed knowledge of apterygotes in the garden, although springtails (Collembola) were common throughout the study, many of the larger ones falling into pitfall traps. I occasionally saw a shiny, sinuous silverfish (Thysanura) scuttling over the floor of the kitchen before it was modernized. The majority of apterygotes, however, live in leaf litter, beneath bark, in soil, or in similarly humid places.

MAYFLIES

Delicate, weak-flying, long-tailed mayflies, often called 'fishing flies', constitute the insect order Ephemeroptera. The nymphs are fully aquatic, breathing by means of plate-like abdominal gills, and the adult life is short, less than a day in many species. They are usually found close to water, and are not a constituent part of the insect fauna of suburban gardens, but they may get carried considerable distances when the wind catches their relatively large wings, and in most years occasional individuals were caught in the Malaise trap or at night in the mercury-vapour light trap, although none was identified to species.

DRAGONFLIES AND DAMSELFLIES

The long, slender body, prominent compound eyes, long wings, and rapid, agile flight of dragonflies make them one of the most easily recognized orders of insects. The nymphs are aquatic, and so dragonflies are particularly associated with water, although being strong fliers, they are not infrequently encountered away from it. The adults are efficient predators of insects (including other dragonflies), which they catch on the wing, and the mouthparts include powerful, toothed mandibles, hence the name of the order, Odonata, which means 'toothed'. Dragonflies, and their more fragile relatives, damselflies, fly too swiftly and strongly to be captured in the Malaise trap, but occasional individuals are seen in the garden. A common blue damselfly (*Enallagma cyathigerum*) was caught with a net in July 1983. Large, powerful *Aeshna* were occasionally seen hawking over the lawn, particularly in late July and August, and in July 1974, one was caught and eaten by a starling; in August 2001, one flew into one of the downstairs windows of the house, but appeared to recover.

The dragonflies sometimes flew close to me, and a certain identification was made of *Aeshna cyanea* in 1995, and a probable identification of *A. juncea* in 1992. Occasionally hawking dragonflies alighted briefly on the sweet pea sticks beside the lawn, and consequently, at various times, it was possible to identify a further four species. Odonata were identified using Hammond (1977).

DRAGONFLIES & DAMSELFLIES

1 **Common blue damselfly *Enallagma cyathigerum*** (Charpentier)
2 **Azure hawker *Aeshna cyanea*** (Müller)
3 **Common hawker *Aeshna juncea*** (L.)
4 **Keeled skimmer *Orthetrum coerulescens*** (Fab.)
5 **Four-spotted chaser *Libellula quadrimaculata*** L.
6 **Broad-bodied chaser *Libellula depressa*** L.
7 **Common darter *Sympetrum striolatum*** (Charpentier)

Grasshoppers cannot be expected in large numbers in gardens where the lawn is closely mown, but field grasshoppers, *Chorthippus brunneus*, were caught in the Malaise trap in the Leicester garden in most years

GRASSHOPPERS AND CRICKETS

Easily recognized by their robust body and head, long hind legs modified for jumping, and their propensity for chirping, grasshoppers and crickets, forming the order Orthoptera, are familiar to most people. They have biting mouthparts, and grasshoppers are vegetarians, feeding mainly on grass, although crickets are omnivorous, taking some animal food. House crickets (*Acheta domestica*) are not native, but are well established, particularly in association with heated buildings and warm rubbish dumps. The species was noisily conspicuous in the garden in the warm August of 1975, although at no other time. Grasshoppers are characteristic of meadowland, and cannot be expected in large numbers in a suburban garden where the lawn is kept closely mown, although one or two were caught in the Malaise trap in most

GRASSHOPPERS & CRICKETS

8 **House cricket *Acheta domestica*** (L.)
9 **Meadow grasshopper *Chorthippus parallelus*** (Zetterstedt)
10 **Field grasshopper *Chorthippus brunneus*** (Thunberg)
11 **Oak bush cricket *Meconema thalassinum*** (Degeer)

years. Most of these captures were of common field grasshopper (*Chorthippus brunneus*), although a nymph of meadow grasshopper (*C. parallelus*) was caught in 1984, and another was seen on the lawn, suggesting that it might have bred in the garden. Oak bush cricket (*Meconema thalassinum*) was caught in the Malaise trap several times between 1996 and 2001, and another, unidentified species of bush cricket was captured in 2000. Orthoptera were identified by J. F. Burton.

EARWIGS

Earwigs, *Forficula auricularia*

Despite the fearsome-looking pincers at the hind end of their bodies, earwigs (order Dermaptera) are harmless insects. Although mainly scavengers, they often chew flower petals and occasionally prey on living insects. Their most endearing habit is that of maternal care, the female tending her young until they are well grown; family groups are a common sight in spring. The only species recorded in the garden, *Forficula auricularia*, is ubiquitous and common in the British Isles.

EARWIGS

12 ***Forficula auricularia*** L.

CADDISFLIES

CADDISFLIES

13 ***Phrygaena grandis*** L.
14 ***Phrygaena varia*** Fab.
15 ***Limnephilus marmoratus*** Curtis
16 ***Limnephilus rhombicus*** (L.)
17 ***Stenophylax*** species

Caddisflies (Trichoptera) are closely associated with running water and so are not really garden insects, but larger species that fly more strongly are sometimes caught in light traps away from water. The five species recorded in the garden were caught in the mercury-vapour light trap. The smaller ones may be mistaken for small moths, but caddisfly wings are covered with tiny hairs, and they all have biting mouthparts, though these are poorly developed and rarely used.

To most people, especially children who go stream-dipping, the aquatic larvae of caddisflies are more familiar. This is because most build protective tubular cases of sand grains, plant material or other fragments on a foundation of sticky silk, and these are often found in freshwater habitats. Head and legs protrude from the case, which is open at both ends, and body movements maintain a through-current of water over the feathery gills along the sides of the abdomen. Most larvae are omnivorous, but some tend to be carnivorous, and a few species that live in running water make no cases, but trap food in silken nets spun among vegetation. Those caught in the garden were identified by K. Porter. Nomenclature follows Kloet & Hincks (1964).

A species of caddisfly, *Limnephilus marmoratus*, was occasionally caught in the mercury-vapour light trap in the Leicester garden

FLIES

Many true, two-winged flies (Diptera), other than hoverflies, have been recorded from the Leicester garden. The majority have been captured in the Malaise trap, but others have been seen and collected, or attracted to special baits. By no means all flies captured in the Malaise trap have been identified, however; inclusion in the garden list being dependent on the readiness of a specialist to identify them. Nomenclature and sequence of Diptera in this chapter follow Chandler (1998).

A cranefly, *Tipula paludosa*, whose larvae are common garden leatherjackets

CRANE FLIES

18 *Tipula paludosa* Meigen

Crane flies (daddy-long-legs) (Tipulidae), with their fragile, gangling legs, were present in the Malaise trap during summer and autumn, and in some years were conspicuous in the garden in late summer. A female caught laying eggs in the lawn in September 1980 was identified by A. E. Stubbs as *Tipula paludosa*, which as a larva is the common garden leatherjacket. Each female lays several hundred eggs during the summer, inserting her ovipositor well into the ground. The eggs hatch into grubs that, as in other crane flies, grow into tough-skinned, greyish leatherjackets, which feed with their biting mouthparts on the roots and lower stems of an enormous range of plants, and are the bane of gardeners, farmers and groundsmen. They feed throughout the winter and spring, and pupate in early summer, still in the soil, the adult emerging about a fortnight later, after the pupa has worked its way to the soil surface using its downwardly-directed spines (Colyer & Hammond 1968).

OWL MIDGES (MOTH FLIES)

19 *Pericoma pilularia* Tonnoir
20 *Pericoma trivialis* Eaton
21 *Psychoda albipennis* Zetterstedt
22 *Psychoda cinerea* Banks
23 *Psychoda grisescens* Tonnoir
24 *Psychoda minuta* Banks
25 *Psychoda phalaenoides* (L.)
26 *Psychoda setigera* Tonnoir
27 *Psychoda surcoufi* Tonnoir
28 *Tinearia alternata* (Say)

Owl midges (moth flies) (Psychodidae) are tiny flies whose body, legs and wings are clothed in long hairs, and which hold their wings roof-like over the body at rest. They may abound in damp, humid locations, and sometimes occur in sinks in houses or around drains, where they are recognizable by the jerky way they run and circle about. The larvae develop in dung, mud, rotting vegetation or any wet, decaying matter. (Blood-sucking sand flies, which transmit several diseases, are psychodids, although absent from Britain.)

Psychoda surcoufi, one of the species of owl midges or moth flies captured in the Malaise trap in the Leicester garden

From 1983 to 1986, Psychodidae were sorted from Malaise trap catches and sent to P. Withers, who identified 10 species, mostly in the genus *Psychoda*. The larvae of *Psychoda* species occur in a variety of decaying organic material, and *P. albipennis, P. cinerea* and the closely related *Tinearia alternata* have colonized the bacteria beds of sewage purification works. A number of species occur on farmland, the larvae developing in cow dung, the most important numerically being *Psychoda phalaenoides* (Satchell 1947). There are a number of suitable habitats for them in gardens: *P. albipennis*, for instance, breeds in kitchen waste and *P. minuta* in rotting lawn clippings. As the adults fly weakly and do not disperse far, it can be assumed that those caught in the Malaise trap are garden residents.

Soldier flies (Xylomyidae and Stratiomyidae) are small to fairly large, somewhat flattened flies, often with a colourful, metallic lustre. They are rather sluggish in behaviour, and are often found resting on vegetation. They vary considerably in shape, *Microchrysa* species, for instance, being short and broad, and *Sargus* species elongate. Adults are fluid-feeders and occasionally take exposed nectar from flowers, whereas larvae occur in a variety of sites, feeding on decomposing organic material. Some larvae are aquatic, but all the species recorded in the garden have terrestrial larvae that feed on and in dung, rotting wood or organic

debris of one sort or another. Soldier flies were sorted from Malaise trap catches in the years 1980–1986 and identified by J. Mousley.

Total annual catches for the seven-year period were 56, 66, 95, 264+, 99, 40 and 97, respectively. In other words, 1983 was far and away the best year and 1985 the worst. 1983 was a hot, dry, sunny summer, when many insect groups were abundant, although most, unlike soldier flies, fared even better in 1984; the summer of 1985 was cool and damp, with only average hours of sunshine, and insect numbers in general were down. The most abundant soldier fly overall was *Microchrysa polita*, a small, shining-green fly, which is generally common.

SOLDIER FLIES

Numbers in the following list refer to Malaise trap captures 1980–1986.

Xylomyidae

29	*Solva marginata* (Meigen)	7

Stratiomyidae

30	*Beris chalybeata* (Forster)	47
31	*Beris clavipes* (L.)	36
32	*Beris morrisii* Dale	1
33	*Beris vallata* (Forster)	10
34	*Pachygaster leachii* Stephens in Curtis	10
35	*Chloromyia formosa* (Scopoli)	7
36	*Microchrysa cyaneiventris* (Zetterstedt)	10
37	*Microchrysa flavicornis* (Meigen)	7
38	*Microchrysa polita* (L.)	550+
39	*Sargus bipunctatus* (Scopoli)	1 (in 1987)
40	*Sargus flavipes* Meigen	28
41	*Sargus iridatus* (Scopoli)	4

Microchrysa polita (above left), the most abundant soldier fly in the Leicester garden

Robber flies (Asilidae) are medium-sized to large, narrow-bodied and rather bristly predators of other insects. They have horny and more or less rigid, piercing mouthparts with which they suck the fluids from their prey. Only one individual was captured in the Malaise trap during the period 1980–1986, and that was *Dioctria rufipes*, identified by J. Mousley.

ROBBER FLIES

42 *Dioctria rufipes* (Degeer)

A mating pair of robber flies, *Dioctria rufipes*

Many asilids lie in wait for suitable prey to approach, but *D. rufipes* is one of the species that take their prey on the wing, with the front legs extended like a scoop. *Dioctria* species feed mainly on Hymenoptera, especially ichneumonids. Asilid larvae feed on and in decomposing vegetable material such as rotting wood or cow dung, but are found in drier situations than many fly larvae (Colyer & Hammond 1968).

Thick-headed flies (Conopidae) are medium-sized to fairly large flies, many of which (*Conopinae*) bear a striking resemblance to solitary wasps and other Hymenoptera. They have large heads with conspicuous eyes, the antennae are long, and the brightly yellow- and brown- or black-banded abdomen is constricted at the base, especially in males. Species of Myopinae look more like typical flies, although the head and eyes are large. Conopid larvae are solitary internal parasites of aculeate Hymenoptera and occasionally of Orthoptera. Eggs are laid directly into the host's abdomen in flight, and develop into larvae that feed on the host's body fluids, while tapping the host's tracheal system for respiration.

THICK-HEADED FLIES

43 ***Conops ceriaeformis*** Meigen
44 ***Conops flavipes*** L.
45 ***Conops quadrifasciata*** Degeer
46 ***Physocephala rufipes*** (Fab.)

A species of thick-headed fly, *Conops quadrifasciata*, which was caught in most years in the Leicester garden

Pupation takes place within the host's abdomen, which ruptures when the adult conopid emerges. The diurnal adults are often found at flowers, where they feed on nectar (Colyer & Hammond 1968, Smith 1969).

Four species of Conopinae have been caught in the Malaise trap; early captures were identified by K. G. V. Smith, and subsequent ones by myself. *Conops quadrifasciata* was caught in most years, and was also collected at flowers of goldenrod (*Solidago canadensis*). *Conops flavipes* is parasitic on *Bombus lapidarius*, *Osmia* species and *Vespula rufa*, *C. quadrifasciata* on *Bombus lapidarius*, and *Physocephala rufipes* on several species of *Bombus* and on *V. rufa*, but the host or hosts of *Conops ceriaeformis* are unknown (Smith 1969).

NEST FLIES

47 **Nest fly *Neottiophilum praeustum*** (Meigen)

Seventy larvae of **nest fly**, *Neottiophilum praeustum* (Piophilidae), were found in July 1976 in a blackbird's nest from which four young had successfully fledged. Larvae of this species suck the blood of a wide variety of birds.

TEPHRITID FLIES

Tephritidae

48 **Celery fly *Euleia heraclei*** (L.)

Larvae of **celery fly**, *Euleia heraclei* (Tephritidae), were found in the leaves of lovage (*Levisticum officinale*) in every year.

LEAF MINERS

Agromyzidae

49 ***Aulagromyza hendeliana*** (Hering)
50 ***Phytomyza ilicis*** Curtis

A blister on a holly leaf formed by the leaf-mining activities of a larva of a small fly, *Phytomyza ilicis*

Leaf miners (Agromyzidae) are very small flies whose tiny larvae bore in plant stems or mine the tissue between the upper and lower epidermises of leaves, making characteristic patterns. Moreover, most species are restricted to one or a few closely related species of plant, and so can be identified from their mines. Female *Phytomyza ilicis* lay eggs in June at the base of the midrib beneath holly leaves. The young larva eats its way forwards along the midrib until moving out into the leaf tissue in autumn, where it forms a characteristic blister. The larvae pupate within the mine in March and April, and new adults emerge in May and June.

Holly trees (*Ilex* species and cultivars) in my garden and most of the neighbourhood are conspicuously affected. Holly trees tend to shed mined leaves in the period January–May, whereas their normal leaf-fall is in June and July (Owen, D.F. 1978). Characteristic meandering leaf-mines every year on leaves of honeysuckle (*Lonicera periclymenum*) were assumed to be those of *Aulagromyza hendeliana*.

The **muscid flies** (Muscidae) include species, such as *Musca domestica*, that commonly enter houses, green-bottles, flies that cluster around the heads and bodies of cattle and horses, and also biting flies such as *Stomomyx calcitrans* and *Haematobia irritans*. *Mesembrina meridiana*, a large black fly with conspicuously yellow wing-bases, attracted attention in the garden, because it is not only handsome, but also common. Adults of this species feed on nectar from umbellifers and other accessible flowers and bask on tree trunks and on the ground. The female lays large eggs, singly, each of which hatches immediately, in cattle or horse dung (Colyer & Hammond 1968).

The **flesh flies** (Sarcophagidae) recorded in the garden are all common, except *Amobia signatus*, whose larvae are found in nests of solitary and social bees and wasps. The larvae of *Sarcophaga* species feed on decaying animal material, and those of other species of Sarcophagidae are found in bee and wasp nests, feeding on stored food or on the host larvae. The larvae of *Brachicoma devia* occur in nests of bumblebees and wasps (*Bombus* species and *Vespula* species), those of *Sarcophaga africa* and *S. subvicina* in decaying organic matter, and those of *S. incisolobata* in excrement.

Of the **blow flies** (Calliphoridae) in the garden, most are carrion-feeders in the larval stage, although *Lucilia ampullacea*, a scarce species generally found in woods, is an occasional internal parasite of toads, and *Protophormia terraenovae* feeds and breeds in decaying organic matter. Most are generally common, although the garden record of *Calliphora subalpina*, common in Scotland, was the most southerly record for England, *C. vomitaria* is not found in numbers in cities, *Lucilia caesar* is a common woodland species, and *L. richardsi* is a species of open grassland. Most blow-flies frequent decaying animal matter and excrement, and females lay eggs on meat and carcasses or in open wounds and sores on live animals; the larvae of other species are external parasites of nestling birds, or parasites of other invertebrates, such as earthworms.

In a campaign that was well publicised on radio, dipterists at the British Museum (Natural History),

MUSCID FLIES

51 *Mesembrina meridiana* (L.)

Mesembrina meridiana, a common fly in gardens, often to be seen basking in the sun

FLESH FLIES

MT signifies capture in the Malaise trap; F indictes capture in trap baited with fish heads, skins and offal.

Sarcophagidae

52 *Amobia signata* (Meigen) MT
53 *Brachicoma devia* (Fallén) MT
54 *Sarcophaga africa* (Wiedermann) MT, F
55 *Sarcophaga carnaria* (L.) MT
56 *Sarcophaga subvicina* Rohdendorf MT, F
57 *Sarcophaga incisolobata* Pandellé MT, F

A flesh fly, *Sarcophaga* species (above left)

A blow fly, *Calliphora vicina* (above right); its larvae feed on carrion

BLOW FLIES

MT signifies capture in the Malaise trap; F indictes capture in trap baited with fish heads, skins and offal.

Calliphoridae

58 *Calliphora subalpina* (Ringdahl) MT
59 *Calliphora vicina* Robineau-Desvoidy MT, F
60 *Calliphora vomitaria* (L.) MT, F
61 *Protophormia terraenovae* (Robineau-Desvoidy) F
62 *Lucilia ampullacea* Villeneuve MT
63 *Lucilia caesar* (L.) Maestrus MT, F
64 *Lucilia illustris* (Meigen) MT, F
65 *Lucilia richardsi* Collin in Richards MT
66 *Lucilia sericata* (Meigen) MT, F

keen to record as fully as possible the status and distribution of blow-flies (Calliphoridae) in Britain, organized a number of National Fish Skin Weeks in 1977, 1978 and 1979. Anyone interested was asked to put out rotting carcasses and fish skins, collect any flies attracted to the bait, and send them to the British Museum for identification. A trap in the garden was duly baited with mackerel heads, skins and offal in July 1978 and blow-flies collected and sent to the British Museum, together with those sorted from 1977 and 1978 Malaise trap catches.

Flesh flies (Sarcophagidae) were inadvertently included in the samples, but these as well as the Calliphoridae were identified by J.P. Dear, who also provided additional information about the various species. The samples were so large that the offer of further material in 1979 was politely declined, as it was felt that the locality had been sufficiently covered. The National Fish Skin Weeks were not simply academic exercises, for the distribution of calliphorids is important for veterinary and public health work, and often for forensic medicine.

PARASITE FLIES

67 *Voria ruralis* (Fallén)
68 *Epicampocera succinata* (Meigen)

Two species of **parasite fly** (Tachinidae) have been bred from parasitized Lepidoptera found in the garden. The larvae of all tachinids are internal parasites of arthropods, mainly insects, either pupating in the host's skin, or boring out to pupate in the ground. The host lives at least until the tachinid pupates, but eventually dies, so the tachinids are best considered as parasitoids, similar in host relationships to Ichneumonidae.

A parasite fly, *Voria ruralis*; this species was bred from larvae of burnished brass moths, *Diachrysia chrysitis*, found in the Leicester garden

Voria ruralis was bred on several occasions from caterpillars of burnished brass (*Diachrysia chrysitis*), a moth that feeds on *Origanum majorana*; *Epicampocera succinata* was bred from a caterpillar of small white butterfly (*Pieris rapae*) feeding on *Hesperis matronalis*. Adult parasite flies are black, grey or brown, and rather bristly; they were identified by R. Belshaw.

ICHNEUMON WASPS (ICHNEUMONIDAE)

The Ichneumonidae comprises a vast and varied assemblage of parasitic wasps. They are slender, wasp-waisted insects, often with the abdomen rather compressed laterally. Some are quite large, up to 40mm long, most are small, and many are tiny, down to 2mm. They have a characteristic jerky walk over leaves, with the long, slender antennae in constant, quivering motion. The juvenile stages of ichneumonids are parasites of insect larvae or pupae, i.e. of the developmental stages of insects with complete metamorphosis, particularly Lepidoptera, sawflies and wood-boring beetles, or, less frequently, spiders or spider eggs. The majority develop inside the host, i.e. as endoparasites,

Table 9.1. Size and composition of annual Malaise trap catch and 2-year catch of ichneumon wasps in the Leicester garden, 1972–1973

Year	Individuals (N)	Species (S)	N/S	Species taken once only	Commonest species	As % N
1972	2,495	326	7.7	118	*Helictes* species	3.2
1973	3,950	355	11.1	117	*Charitopes* species	6.6
1972–1973	6,445	455	14.2	141	*Syrphoctonus pictus*	5.0

and, because the host is eventually killed, they are more akin to predators than parasites, so are often called parasitoids.

It was once assumed that adults do not feed, and only drink from dew, but many larger species are now known to feed on nectar, pollen, honeydew and plant sap. Ichneumonids are active and wide-ranging, and adult females are very particular about where they seek hosts, being niche-specific rather than host-specific.

There are 2,028 recorded species of ichneumonid in the British Isles, although the estimated fauna is 2,900. In 1972 and 1973, Ichneumonidae were sorted from Malaise trap samples, identified by Henry Townes, and analysed in detail, and in 1974, additional species in Malaise trap samples were recorded. This yielded an astonishing total of 529 species, and a further four species were bred from parasitized larvae. *Rhinotorus atratus* was bred from larvae of a sawfly, *Croesus septentrionalis*, on birch, as was a species of *Olesicampe*; *Promethes bridgmani* emerged from a hibernating hoverfly larva, and *Hypomechus quadriannulatus* came from a larva of a geometrid moth, *Chloroclysta truncata*, feeding on bramble, *Rubus fruticosus*. The size and composition of the 1972 and 1973 catches of ichneumonids are shown in Tab. 9.1. The commonest species in 1972, a species of *Helictes*, parasitizes flies that live in rotting vegetation, and the commonest in 1973, a species of *Charitopes*, parasitizes brown lacewings (Hemerobiidae), whereas *Syrphoctonus pictus*, overall commonest, is a parasite of hoverflies.

A striking feature of the ichneumonid sample from the garden, compared with other groups, such as hoverflies (Syrphidae) (see Tab. 5.1), is its extraordinary diversity. In 1972 and 1973, 6,445 individuals of 455 species were caught. Of these, 141 species were represented by only one individual in the two years, and the commonest species, *Syrphoctonus pictus*, formed only 5% of the two-year catch. In other words, most of the garden ichneumonids were rare, and none is particularly common. The average number of individuals per species in 1972 and 1973 was 7.7 and 11.1, respectively.

To emphasize how relatively rare, by comparison, ichneumonids are, the average number of individuals per species of hoverfly captured in the Malaise trap was 28.5 in 1972 and 51.0 in 1973. Evidently ichneumonid populations are relatively small. However, the ichneumonid fauna of northern Europe as a whole is rich and varied, and the cool, damp climate of England is particularly favourable for them. The garden ichneumonids are described in detail in Owen, J.

ICHNEUMON WASPS

69 *Piogaster albina* Perkins
70 *Netelia inedita* Kokujev
71 *Oedomopsis scabriculus* (Gravenhorst)
72 *Lysibia nana* (Gravenhorst)
73 *Dichrogaster aestivalis* (Gravenhorst)
74 *Maestrus longulus* Horst.
75 *Maestrus mandibularis* Horst.
76 *Maestrus varicoxis* Tasch.
77 *Odontoneura punctiscutum* Horst.
78 *Megacara rusticellae* Bridgman
79 *Megacara vagans* Gravenhorst
80 *Lissonota stigmator* Aubert
81 *Banchus volutatorius* (L.)
82 *Rhinotorus atratus* (Hufnagel)
83 *Dusona terebrator* (Fab.)
84 *Hyposoter tricolor* (Ratzeburg)
85 *Apolophus borealis* Townes
86 *Syrphoctonus fissorius* Gravenhorst
87 *Syrphoctonus pictus* Gravenhorst
88 *Syrphoctonus tarsatorius* Panzer
89 *Enizemum ornatum* (Gravenhorst)
90 *Syrphophilus tricinctus* (Ashmead)
91 *Tymmophorus graculus* (Gravenhorst)
92 *Diplazon laetatorius* (Fab.)
93 *Diplazon tetragonus* (Thunberg)
94 *Promethes bridgmani* Fitton
95 *Promethes sulcator* (Gravenhorst)
96 *Platylabus iridipennis* (Gravenhorst)
97 *Hypomechus quadriannulatus* (Gravenhorst)

NB For details of species numbers **98–601**, see Owen, J., Townes & Townes (1981) and Owen, J. (1991)

An ichneumonid, *Diplazon laetatorius*, a parasitoid of hoverfly larvae, which was seen feeding at honeydew excreted by aphids, *Brevicoryne brassicae*, on cabbage

(1991), and in Owen, J., Townes and Townes (1981). In these publications, species are accorded a number within genera, but a partial list of garden species has resulted from rearing records and other identifications by Henry Townes. Not all of these species appear in the British checklist (Fitton *et al.*, 1978).

A further 504 species of Ichneumonid were recorded in the Malaise trap catches of 1972–1974; identified to genus, they are listed in Owen, J., Townes and Townes (1981) and Owen, J. (1991).

A tiny serphid wasp, *Exallonyx* species (below), a parasite of insects found in moist soil or rotting vegetation

OTHER PARASITIC WASPS

There are 40 families of parasitic British Hymenoptera (Fitton et al. 1978), including nearly 5,500 species; the Ichneumonidae and the two families described here, Serphidae and Heloridae, represent less than 40% of those described (and presumably, therefore, in my garden), the notable exclusions being Braconidae and the entire superfamily Chalcidoidea. Serphidae and Heloridae (superfamily Serphoidea =Proctotrupoidea) that were inadvertently included in trap samples of Ichneumonidae sent to Townes were also identified and so are included here.

Serphidae are small, black wasps, the front wing measuring 1.6–7.4mm. They have weakly compressed bodies, and reduced wing venation with a large stigma on the forewing. They are internal parasites of beetles, particularly staphylinids and carabids, and of mycetophilid and sciarid flies, seeking their hosts in moist soil or rotting plant material; most species occur in damp habitats.

Thirty-six species are included in the checklist of British Hymenoptera (Fitton *et al.* 1978), but the garden list (recorded between 1972 and 1974) is an astonishing 22 species, five of them new to Britain (Owen, J., Townes & Townes 1981). It also includes eight species that appear not to be in Fitton *et al.* (1978), even allowing for synonymy.

The **Heloridae** are a small group comprising one genus of seven species. They are parasites of Chrysopidae (green lacewings), laying their eggs into the host larva, which is killed after it has spun a cocoon, within which the parasite pupates. Two species of *Helorus* have been recorded from the garden.

More information on Serphidae and Heloridae is given in Owen, J. (1991) and Owen, J., Townes and Townes (1981).

OTHER PARASITIC WASPS

This list, compiled by Henry Townes, uses the nomenclature of Townes & Townes (1981), which differs from that of Fitton *et al.* (1978).

Serphidae

602 *Brachyserphus parvulus* Nees
603 *Codrus niger* Panzer
604 *Codrus picicornis* Foerster
605 *Cryptoserphus aculeator* Haliday
606 *Disogmus areolator* Haliday
607 *Phaenoserphus chittii* Morley
608 *Phaenoserphus viator* Haliday
609 *Phaneroserphus calcar* Haliday
610 *Serphus gravidator* L.
611 *Tretoserphus laricis* Haliday
612 *Exallonyx ater* Gravenhorst
613 *Exallonyx brevicornis* Haliday
614 *Exallonyx brevimala* Townes
615 *Exallonyx ligatus* Nees
616 *Exallonyx longicornis* Nees
617 *Exallonyx microcerus* Kieffer
618 *Exallonyx nixoni* Townes
619 *Exallonyx pallidistigma* Morley
620 *Exallonyx quadriceps* Ashmead
621 *Exallonyx subserratus* Kieffer
622 *Exallonyx trichomus* Townes
623 *Exallonyx trifoveatus* Kieffer

Heloridae

624 *Helorus nigripes* Foerster
625 *Helorus ruficornis* Foerster

ANTS

There is just one large family (Formicidae) of ants, placed in a separate superfamily, Formicoidea of the Hymenoptera. All are social insects living in colonies, which may be very large. Colonies consist of an egg-laying queen and many female workers, which are usually infertile, although some may lay eggs. Winged males and reproductive females are produced at certain seasons, and these leave the colonies in great swarms under particular atmospheric conditions. Flights leave nests of the same species in a particular area in synchrony, increasing the chances of out-breeding; mating occurs in the air, and when they land, males soon die and females break off their own wings and initiate new nests. Once the first batch of workers has been produced, females devote themselves to egg-laying, and workers take over the tasks of tending the brood, feeding the larvae, and extending and defending the nest.

There are just two species of ant in my garden. Garden black ants (*Lasius niger*) are the most abundant, with many nests. On sultry afternoons in July, swarms of winged males and females rise into the air like plumes of smoke from all the gardens in the neighbourhood. In the 1970s and early 1980s, there was just one small colony of yellow meadow ant (*L. flavus*); this led to new colonies being established in different parts of the garden.

ANTS

626 **Garden black ant *Lasius niger*** (L.)
627 **Yellow meadow ant *Lasius flavus*** (Fab.)

Winged male and female garden black ants, *Lasius niger*, attended by non-breeding workers, prepare to embark on a mating flight

SUMMARY

In total, 627 species of 'other insects' were recorded in the garden, Ichneumonidae being the most diverse. It is important, however, to emphasize that identification of different groups was very patchy and uneven, perhaps best illustrated by the fact that house fly (*Musca domestica*) is not on the garden list, although I am sure that it occurred.

10 : Other invertebrates

As has been indicated, the dominant group in my garden, and indeed in all gardens, is the insects. They are the most conspicuous group of invertebrates, although many other invertebrate groups occur, and some of those in which individual size is small or microscopic, such as mites, are undoubtedly very abundant. Several types of larger invertebrates, particularly spiders, were collected and identified, providing species lists. Furthermore, myriapoda (Chilopoda and Diplopoda) and harvestmen (Arachnida – Opiliones) that fell into pitfall traps, were identified, providing comparable annual samples for the eleven years 1980–1990. Most non-insect invertebrates, however, such as spiders and molluscs, are not amenable to trapping, and so there are no data for these groups comparable with that obtained by Malaise trapping of insects.

Zebra spider, *Salticus scenicus* (opposite). These attractive little spiders are often encountered in gardens and occasionally in houses. Also known as jumping spiders, they stalk their prey and, when close enough, jump on it

As with the insects, regular identification and tabulation of samples of other invertebrates depends upon a specialist being willing and able to deal with them. A visit to the garden by A.J. Rundle, as part of a survey of Leicester's terrestrial invertebrates that he undertook for the Leicestershire Museums, Arts and Records Service, resulted in many additions to the garden species list. Over the years, an extensive species list of invertebrates, other than insects, was compiled. Much of the systematics and nomenclature used in this chapter is from an unpublished systematic list of selected invertebrate groups prepared by A.J. Rundle in 1986, a copy of which is held by Leicestershire Museums.

FLATWORMS

The flattened, simple worms of the phylum Platyhelminthes, which lack a body cavity and circulatory system and have only one opening to the gut, are known as flatworms. Only the class Turbellaria are free-living, the other classes being parasitic, and the only suborder represented in the garden is the Tricladida, relatively large planarians, with a three-branched intestine. They are carnivorous, using their protrusible pharynx to catch small creatures, but will also eat freshly dead animals. Prey too large to be ingested whole is partially digested externally by enzymes poured onto it from the pharynx, and small particles are then sucked in.

Flatworms are covered with cilia which, together with an investing film of mucus, enable them to

FLATWORMS	
Planariidae	
1	*Polycelis nigra* (Müller)
Dendrocoelidae	
2	*Dendrocoelum lacteum* (Müller)
Rhynchodemidae	
3	*Microplana terrestris* (Müller)

glide over surfaces, including in the case of aquatic forms, the underside of the surface film of water. *Polycelis nigra* is superabundant in the pond, and *Dendrocoelum lacteum* is common. Several individuals of *Microplana terrestris*, a terrestrial flatworm, were found under damp stones. Flatworms were identified by U. Bowen and A.J. Rundle.

SNAILS AND SLUGS

Snails and slugs are molluscs of the class Gastropoda; most, including all those found in the garden, belong to the subclass Pulmonata, in which the mantle cavity (beneath the shell if there is one) has become a vascular lung for air-breathing. Snails have a hard, spirally coiled shell into which they can withdraw at times of adversity; in most slugs, the shell is reduced to a thin, internal plate or a few calcareous granules, but others have a small shell perched on the hind end of the body. The outer skin is soft and wet, making slugs and snails prone to desiccation, and consequently their activity is normally confined to night-time or to wet days. The skin is coated with mucus, which also lubricates the sole of the muscular creeping foot, so that they leave a characteristic slime trail. Slugs, in particular, can burrow into soil, moving down more than a metre in dry weather.

Pulmonate slugs and snails are hermaphrodite, but cross-fertilization, after a more or less elaborate courtship and mating ritual, is the general rule. Clutches of eggs, often pearl-like or translucent, but sometimes in a hard, calcareous shell, are laid in soil or under logs or stones. The small newly hatched young are vulnerable, and mortality is high, but the larger species may be long-lived, up to ten years or longer for some individuals.

The aspect of snails and, in particular, slugs that most frequently draws them to the attention of a gardener is their feeding habits. Most feed on rotting vegetation, on fungi, algae and lichens, or on underground storage organs, such as radishes and potatoes, and do not attack healthy green plants. Many species also eat carrion, and some are carnivorous. A few species that normally eat decaying matter are, however, pests, preferentially eating cultivated plants, which tend to be softer and more nutritious than wild ones.

The feeding equipment of slugs and snails is unusual in that they have a single horny jaw in the roof of the mouth, and a strip of continuously growing horny tissue, the radula, bearing numerous rows of tough, sharp teeth, in the floor of the mouth. Jaw and radula are scraped and rasped over food, tearing off particles which become entangled in mucus and move back into the oesophagus.

SNAILS & SLUGS

Lymnaeidae	
4	*Lymnaea stagnalis* (L.)
Planorbidae	
5	*Planorbis planorbis* (L.)
Cochlicopidae	
6	*Cochlicopa lubrica* (Müller)
Vallonidae	
7	*Vallonia costata* (Müller)
Arionidae	
8	*Arion ater* (L.) agg.
9	*Arion hortensis* Ferussac
10	*Arion distinctus* Mabille
11	*Arion fasciatus* (Nilsson)
Zonitidae	
12	*Vitrea contracta* (Westerlund)
13	*Aegopinella nitidula* (Draparnaud)
14	*Oxychilus draparnaudi* (Beck)
15	*Oxychilus alliarius* (Miller)
Milacidae	
16	*Tandonia budapestensis* (Hazay)
Agriolimacidae	
17	*Deroceras reticulatum* (Müller)
18	*Deroceras caruanae* (Pollonera)
Helicidae	
19	*Trichia striolata* (Pfeiffer)
20	*Helix aspersa* (Müller)

Great pond snail (*Lymnaea stagnalis*) and ramshorn (*Planorbis planorbis*) have been common in the pond since its re-filling in 1984, when they were almost certainly reintroduced with aquatic vegetation. Seven species of slug occur in the garden, and all are common; the four *Arion* species are common and widespread in the British Isles; *Tandonia budapestensis* and the two *Deroceras* species are common in association with man and his cultivation, and *T. budapestensis* and *D. reticulum* are generally considered as pests. Five species of snail are common in the garden; tiny *Cochlicopa lubrica* with its tall shell, tiny *Vallonia costata* with a flat, spiral shell; small *Oxychilus draparnaudi* and *Trichia striolata* with flat, spiral shells; and *Helix aspersa*, common garden snail, with its intricately patterned shell. Tiny *Vitrea contracta* and small *Aegopinella nitidula* and *Oxychilus alliarius*, all with flat, spiral shells, were occasionally collected.

All eight species of snail found in the garden are common and widespread in many parts of the UK, and *Oxychilus draparnaudi, Trichia striolata* and *Helix aspersa* are particularly associated with gardens. Interestingly, some of the recorded snail species, such as *Cochlicopa lubrica* and *Aegopinella nitidula*, usually occur in moderately damp habitats, whereas others, such as *Vallonia costata* and *Vitrea contracta*, favour drier, more open places (Kerney & Cameron 1979). Garden snails and slugs, 17 species in all, were variously identified by A J. Rundle and D. Whiteley.

A black slug, *Arion ater* (top). Although they normally eat rotting vegetation or underground storage organs, such as potatoes, slugs are the bane of gardeners' lives when they turn to cultivated plants

Common garden snails, *Helix aspersa*. In winter they hibernate, often beneath stones, by retreating into the shell and secreting a thick mucus seal across the opening

EARTHWORMS

Worms of the family Lumbricidae (Annelida – Oligochaeta) are the familiar earthworms, so abundant in garden soil. The annulated appearance arises because the body consists of many, serially repeated segments, in each of which bristle-like external chaetae, muscles, excretory organs and nerve branches are the same. Earthworms require damp surroundings, and quickly desiccate if exposed to dry air. The unprotected skin must remain moist for respiration, water is necessary for excretion, and movement is achieved by a hydrostatic skeleton based on the pressure of fluids in the cavities of the segments, which are separated by septa. Earthworms are hermaphrodite, the reproductive organs being restricted to a few segments. Cross-fertilization happens when two worms overlap their front ends and bring them together, bound by mucus, and there is mutual exchange of sperm. After they have separated the prominent saddle or clitellum secretes the cocoon which forms a ring around the worm. The worm then backs out of the ring and injects its own eggs and the other worm's sperm into it. As the worm slips out, the ends of the cocoon seal.

It has long been recognized that earthworms enhance the organic content of soil and improve its texture and quality. They do this because of the way they

EARTHWORMS

Lumbricidae

21 *Allolobophora chlorotica* (Savigny)
22 *Aporrectodea longa* (Ude)
23 *Dendrilus rubidus* (Savigny)
24 *Eisenia fetida* (Savigny)
25 *Lumbricus terrestris* L.

Lumbricus terrestris is the largest garden earthworm. They may burrow down several metres in adverse conditions, an activity that helps drain and aerate the soil

move and eat. In loose soil they may burrow by alternately contracting and extending, then, as the body is anchored in position by the chaetae, expanding and shortening. In compact soil, however, many species excavate a tunnel by eating the soil. Soil is sucked in at the mouth, the particles are cleaned by digestion of all organic matter, and then defecated.

Some species defecate in their burrows, but others back up to the soil surface, where they deposit characteristic worm casts. Charles Darwin calculated that this amounts annually to a 5mm layer of fine soil deposited on the surface. Some species, such as the familiar, large *Lumbricus terrestris*, which may burrow to depths of several metres to escape unfavourable surface conditions, drag leaves and other vegetable material into the entrances of their burrows, and eat it when it is decomposed. Earthworms promote the overall soil-forming process in a number of ways. They influence soil pH by the alkalinity of their faeces, they break down the organic content of soil and enhance microbial activity by promoting humus formation. They also improve soil texture as organo-mineral complexes or 'crumbs' are formed in the gut, and by mixing soil and improving aeration and drainage. The soil is enriched by conversion of minerals to forms that are available to plants. The presence of earthworms in garden soil is undoubtedly beneficial to the gardener. Five species were found in the garden, brandling (*Eisenia fetida*) particularly in the compost heap; most were identified by A.J. Rundle.

LEECHES

One species of leech, *Glossiphonia heteroclita*, was found in the pond and identified by A.J. Rundle. Leeches (Hirudinea), like earthworms, are annelids, but have a sucker at each end. Most are blood-suckers, using suckers to attach, usually temporarily, to their prey, and may be regarded either as specialized predators or as ectoparasites. *Glossiphonia* species feed on aquatic snails, sucking out the body fluids and soft parts using a proboscis they can turn inside-out.

LEECHES

Glossiphonidae

26 *Glossiphonia heteroclita* (L.)

WOODLICE

The crustacean order Isopoda, members of which are dorso-ventrally flattened, with seven pairs of similar walking legs on the thorax, includes the terrestrial woodlice. On the head they have a pair of long antennae, with a second, rudimentary, pair between them, and usually, at the sides, a pair of eyes, which in most species are compound. The mouthparts, which resemble biting and chewing insect mouthparts, include a pair of stout mandibles, and two pairs of maxillae and a pair of maxillipeds which manipulate the food. Behind the thorax or pereion, with its seven pairs of walking legs, is the abdomen or pleon, which, on all but the last segment, bears biramous appendages (pleopods) modified as gills. On the last segment is a pair of flattened appendages known as uropods. The first two pairs of pleopods in males are modified for sperm transfer, fertilization being internal, as is typical for land animals. Females retain eggs and newly hatched young in a brood pouch that develops beneath the thorax from overlapping growths of cuticle.

Woodlice, being Crustacea with gills for respiration, are particularly susceptible to desiccation. Only in the species most highly adapted to terrestrial life is there a waterproofing layer in the cuticle, but this is not as efficient as the waxy covering that insects have. There are consequently continual risks of water loss, which is avoided by behavioural adaptations. Woodlice move along humidity, light and temperature gradients into damp, dark, cool places, and they also respond positively to contact, including contact with other woodlice. They tend, consequently, to aggregate in cool, dark, confined places, which on balance will also be humid, and are usually only active at night. The permeable body covering facilitates water absorption by osmosis as well as water loss, so they are also subject to waterlogging, and move away from a very wet place. They, to some extent, make good the water loss from their food by drinking, by absorbing water through the body surface (particularly the pleopods), and by anal drinking, achieved by capillary movement of water along the closely adpressed uropods to the anus. Excess water is lost by transpiration through the cuticle, with the faeces, and through the uropods by a reverse of the capillary action used in anal drinking (Sutton 1972).

Woodlice vary in the extent to which they have become independent of water, showing increasing specialization for life on land in the series Trichoniscidae, Oniscidae, Porcellionidae and Armadillidiidae. In those most highly adapted to terrestrial existence, there is a mass of minute air-channels or pseudotracheae in the outer part of the pleopods, connecting to the outside through a small pore. The Armadillidiidae includes *Armadillidium vulgare*, a common pill-bug, which is often seen in the sun, and is restricted to calcareous soils where it can take in enough calcium to build its substantial exoskeleton.

Woodlice generally feed on dead plant material, but in some circumstances living plants may be eaten, particularly by *A. vulgare*. They may also feed

WOODLICE

Trichoniscidae

27 *Androniscus dentiger* Verhoeff
28 *Trichoniscus pusillus* Brandt
29 *Trichoniscus pygmaeus* Sars

Oniscidae

30 *Oniscus asellus* L.

Philosciidae

31 *Philoscia muscorum* (Scopoli)

Platyarthridae

32 *Platyarthrus hoffmannseggi* Brandt

Armadillidiidae

33 *Armadillidium vulgare* (Latreille)

Porcellionidae

34 *Porcellio scaber* Latreille

Oniscus asellus, a common garden woodlouse, with two young. Newly hatched young remain in a brood pouch beneath the thorax until large enough to venture out

extensively on animal remains and dung, and die if they are prevented from eating their own faeces. The reason for coprophagy is the need to maintain high levels of copper, which is a constituent of the respiratory pigment haemocyanin. Copper compounds are present in most foods, but copper is not available for absorption until released by bacterial activity in the faeces.

Woodlice abound in my garden, and lifting any stone reveals a seething mass of little bodies struggling to escape the light. *Oniscus asellus, Porcellio scaber, Trichoniscus pusillus* and *T. pygmaeus* are common, *Philoscia muscorum* and *Armadillidium vulgare* are occasionally found, and A.J. Rundle found one *Androniscus dentiger* beneath a loose paving slab by the pond, and several individuals of *Platyarthrus hoffmannseggi*, a white, blind species, in a nest of black ants, *Lasius niger*. Garden woodlice were variously identified by A.J. Rundle and S.L. Sutton. Taxonomy and sequence follows Harding & Sutton (1985).

The most conspicuous garden centipede, *Lithobius forficatus*. They are predatory, with a pair of poison claws on the first body segment

CENTIPEDES

Centipedes form the class Chilopoda, one of four classes of terrestrial arthropods, with a distinct head bearing jaws, possessing more than eight pairs of legs, and are generally known as myriapods. Centipedes are distinguished from the other three groups by being flattened in form, and being predatory, with a pair of poison claws on the first body segment; in contrast to millipedes, there is only one pair of legs on each body segment. Three orders of centipedes, differing in form and structure, occur in Britain and are represented in the Leicester garden.

The British orders of centipede

Geophilomorph centipedes are long, thin, blind, burrowing animals, with up to 101 pairs of short legs. The large number of intersegmental joints is augmented by a subsidiary joint in each segment, giving great flexibility, such that movement of three segments can produce a body flexion of 180 degrees. The numerous joints also enable both shortening of the body, as the segmental sclerites (hardened plates) override each other at the joints, and also a thickening as the loosely articulated sides of the body expand; the converse produces lengthening and narrowing. Consequently geophilomorphs can burrow, as do earthworms, by alternately extending and contracting the body, their numerous legs acting as anchors in the soil.

Lithobiomorph centipedes, on the other hand, are shorter and more compact, with only 15 pairs of powerful legs, and eyes on both sides of the head; these are the familiar, active, brown centipedes. They have no subsidiary segmental joints, and, moreover, lateral flexion is limited because long and short tergites (dorsal sclerites) alternate along the body, so that inter-tergal joints do not lie immediately above inter-sternal joints.

Scolopendromorph centipedes are in many ways intermediate to the other two orders, being blind, with subsidiary joints in the sternites (ventral plates), although not in the tergites, which show some tendency to alternate in length along the body. They are small, active animals, with 21 pairs of legs, and although their legs are not so long, they can run almost as fast as Lithobiomorpha (Eason 1964).

CENTIPEDES

Numbers indicate total pitfall captures, 1980–1990.

ORDER: GEOPHILOMORPHA		
Himantariidae		
35	*Haplophilus subterraneus* (Shaw)	
Schendylidae		
36	*Schendyla nemorensis* (Koch)	8
Geophilidae		
37	*Geophilus insculptus* Attems	3
38	*Necrophloeophagus longicornis* (Leach)	3
ORDER: SCOLOPENDROMORPHA		
Cryptopsiidae		
39	*Cryptops hortensis* Leach	
ORDER: LITHOBIOMORPHA		
Lithobiidae		
40	*Lithobius forficatus* (L.)	729
41	*Lithobius microps* Meinert	117

Natural history of centipedes

All types of centipede have long antennae, which are assumed to have a tactile function. The mouthparts consist of a pair of mandibles and two pairs of maxillae, which move food towards the mandibles. Immediately behind the head, but curving forwards in front of the mouth, are the poison claws with which prey is grasped and paralysed or killed; these are the terminal part of the much-modified appendages of the segment immediately behind the head, and move and grasp in a horizontal plane. A poison gland opens just behind the tip of each claw, and in some tropical centipedes, the claws can inflict a venomous and painful bite to humans. Centipedes breathe by means of a ramifying tracheal system, they have a dorsal tubular heart, and the reproductive openings of both males and females are at the hind end of the body.

Reproductive processes are poorly known, but it is believed that there is copulation and internal fertilization in some species, whereas in others, males deposit spermatophores on specially constructed webs from which females pick them up with the appendages of the genital segments. Female Geophilomorpha lay 30–40 eggs in an excavated cavity in the soil and remain coiled around eggs and young for some time. If the female is removed, the eggs become infected with fungus, which she normally seems to keep at bay with fungicidal secretions from her mouth. Newly hatched geophilomorph young have the adult number of legs, although these are not fully developed, but they lack such structures as tracheae. Lithobiomorph females, however, lay a number of eggs separately, coating each with mucus and soil particles. The newly hatched young are active and independent, able to feed and breathe, but have fewer than the adult number of segments and legs, which are progressively added at each moult.

Centipedes require moist surroundings, having no waxy, waterproof covering. They also avoid light, and tend to creep into crevices and crannies where both dorsal and ventral surfaces are in contact with something firm. Consequently they are found beneath stones, beneath the bark of decayed logs, or in similar situations, and are most active at night, when they do their hunting.

Centipedes in the Leicester garden

There are 32 species of centipede native to the British Isles. Five species fell into pitfall traps in the garden, and were identified as annual samples by D.T. Richardson. The annual catches in the eleven years 1980–1990 were 73, 90, 94, 111, 130, 82, 86, 39, 50, 49 and 55, consisting largely of *Lithobius* species, particularly *L. forficatus*. 1984 was thus the year of maximum capture (130); this was a year when summer was hot, dry and bright, and many insect groups also did well in the garden.

Schendyla nemorensis is not one of the commonest British species, but is found in a range of microhabitats, including under bark and leaf litter, usually in woodland, but also in gardens. Of the other species, several *Haplophilus subterraneus* were found by A.J. Rundle under stones, and he also found one *Cryptops hortensis* under a paving slab. *Haplophilus subterraneus* is the only British centipede which sometimes turns vegetarian and damages root crops.

MILLIPEDES

Millipedes, like centipedes, are generally known as myriapods, but they form a separate class, Diplopoda, characterized by two pairs of legs on each of the body segments, each of which is in reality a double segment. They are vegetarians, eating a wide range of plant material, especially when it is soft and decaying, although they sometimes eat dead animals, such as worms and insects. The mouthparts include mandibles and behind them, a broad plate, the gnathochilarium, formed from a pair of appendages equivalent to maxillae. Spotted snake millipede, *Blaniulus guttulatus*, is frequently a pest of arable crops, but it is doubtful whether its weak jaws can effect entry to potatoes, for example, in which they often occur in numbers, and their attack almost certainly follows damage by fungus or another animal.

Most millipedes, but not *B. guttulatus*, have ocelli (eyes) on the head; they have only short antennae, but these constantly tap the substrate ahead as they move. Many millipedes are cylindrical in shape, the tergites much larger than the sternites, so that the many little legs seem to originate centrally beneath the body. The backstroke of the legs is of longer duration than the forward stroke, so that many legs are involved in a metachronal wave of movement, and they have a marked ability to push forwards using the motive force of the legs. They are essentially animals of the soil, and the rounded head and massive tergite (collum) behind it help the animal push into soil or litter. Other species have lateral or dorso-lateral keels on the tergites, or are fringed with bristles so that they look almost chiton-like; some of these species are very small.

Spotted snake millipede, *Blaniulus guttulatus*, frequently a pest of arable crops such as potatoes

MILLIPEDES		
Polydesmidae		
42	*Polydesmus gallicus* Latzel	33
Macrosternodesmidae		
43	*Macrosternodesmus palicola* Brolemann	
Blaniulidae		
44	*Blaniulus guttulatus* (Fab.)	
45	*Archiboreoiulus pallidus* (Brade-Birks)	
Julidae		
46	*Ophyiulus pilosus* (Newport)	16

Protective measures

Many species have a pair of repugnatorial glands on each segment, and in *Blaniulus guttulatus* the contents of these appear as a row of bright red spots down each side. The glands produce a noxious mixture of substances, chiefly hydrocyanic acid, iodine, quinine and chlorine, the latter often giving a characteristic odour (Cloudesley-Thompson 1968). The larger tropical species seem able to walk about with impunity, being presumably distasteful, although many predators do eat millipedes. Many species are also protected by a tough cuticle, impregnated with calcium as well as phenolic tannins. The pill-millipedes, family

Glomeridae, have substantial calcareous exoskeletons and roll into protective balls so that they resemble pill-bugs (a type of woodlouse). Most millipedes, however, are prone to desiccation; they are usually found in damp places, avoid light, and are more active at night.

Reproduction

The genital apertures are on the third segment, near the front end of the animals. Males of most species have special appendages, called gonopods, on the seventh segment, which they charge with sperm and use as intromittent organs. In other species, the hindermost legs are used to introduce sperm into females. Females of some species coat their eggs with soil or excrement and leave them in soil crevices, but others make nests of soil particles moistened with saliva, and line them with excrement. In some species, the female stays tightly coiled around the nest for several days, and, as in centipedes, her actions seem to protect the eggs from fungal infections. In some millipedes of the family Julidae, in cold climates, males revert in summer to forms without gonopods and in which the testes are inactive; this prolongs their lives to two or more years. Millipedes moult their skin several times as they grow, adding segments and legs each time, and many species construct special moulting chambers of soil or, in some cases, silk for this purpose. The cast skin is eaten, thus restoring lost calcium supplies, and further development seems to depend on this.

Millipedes are not especially common in my garden, but two species were taken in pitfalls during the period 1 April–31 October 1980–1990, the numbers trapped being given in the species list. One female *Archiboreoiulus pallidus* was caught in a pitfall during the winter 1985/86; this is one of the less-common British species, but had already been recorded for Leicestershire. A.J. Rundle found *Macrosternodesmus palicola* commonly under stones, and found a single *Blaniulus guttulatus* beneath a paving slab by the compost heap. Millipedes trapped in pitfalls were identified by D.T. Richardson.

SYMPHYLA

Scutigerella immaculata, a symphylan. Symphylans eat delicate root systems and, although tiny, are so abundant that they are serious economic pests

Symphyla are minute, pale, many-legged animals, with long antennae. They were once included in the group Myriapoda, but are now recognized as a separate class, and are perhaps closer to ancestral insects than other myriapods. Superficially they look like tiny centipedes, the largest species being only about 8mm long, but they lack poison claws. They have 11 or 12 pairs of legs, but many more tergites, up to 24. Symphylans live in soil, in moss, under bark and in organic debris in a variety of habitats, but are a poorly known group. They are herbivorous, feeding mainly on soft or decaying plant material, but often attacking delicate root systems, and are sometimes so abundant as to be serious economic pests, densities of 22 million per acre (roughly 54 million per hectare) outdoors and 90 million per acre (roughly 222 million per hectare) in glasshouses being reported (Cloudesley-Thompson, 1968). A.J. Rundle collected two symphylans from the garden, but these have not as yet been identified.

FALSE-SCORPIONS

Tiny, retiring false-scorpions, with their relatively enormous pincers, or pedipalps, form the Pseudoscorpiones, one of the orders of the class Arachnida, all of whose members have four pairs of walking legs. Other orders of arachnids include scorpions, harvestmen and spiders. False-scorpions look somewhat like scorpions on a vastly reduced scale, the largest British species being less than 4mm long, but they lack the tapering tail and stinging organ, the body instead ending bluntly. They are predators of other tiny animals for which they lie in wait, pedipalps extended, grabbing with them whenever the long tactile setae covering the appendages are stimulated. The prey is injected with toxins from glands that open on the pedipalp claws, and is transferred by them to the jaws, or chelicerae. After these make a wound, digestive enzymes are introduced, and fluid food sucked out.

FALSE-SCORPIONS	
Chernetidae	
47	***Lamprochernes nodosus*** (Schrank)

A tiny false-scorpion, *Lamprochernes nodosus*, hitching a ride by clinging to the leg of a harvestman

False-scorpions produce silk from glands that open on the chelicerae, and this is used to spin chambers for moulting, for overwintering and for brooding the young. They have elaborate courtships, involving male and female grasping each other's pedipalps and walking to and fro. In some species, the male displays with the 'ram's-horn organ', a sac that can be turned inside-out, between the second and third sternites of the abdomen, close to the genital aperture; the male then produces a spermatophore over which he guides the female while she takes it up. Females lay eggs into a membranous sac extruded from the oviduct; the eggs hatch in this, and the young feed on 'milk' produced from the mother's degenerating ovaries. When, after four or five weeks, they leave the sac, the young remain for some time clinging to the mother's abdomen (Cloudesley-Thompson 1968). Dispersal seems to be by phoresy (hanging on to another animal), as false-scorpions are frequently found clinging to flies and other insects, harvestmen or other animals. They live in such places as leaf litter, in moss, and beneath bark, and A.J. Rundle found two individuals of *Lamprochernes nodosus* in the drier top layers of the compost heap.

HARVESTMEN

The familiar harvestmen or harvest-spiders (order Opiliones) are arachnids characterized, among other features, by the uniting of cephalothorax and abdomen (so that the body appears as a single, often rounded, entity), by conspicuously long, often gangling, legs, especially the second pair, and by the absence (in contrast to spiders) of silk glands. Most have a pair of simple eyes, usually placed on either side of a prominent tubercle, the ocularium, near the middle of the cephalothorax. Near the attachment of the first pair of legs are a pair of odoriferous glands, whose secretion makes harvestmen distasteful to most invertebrate predators. The chelicerae are pincer-like, and there is a pair of short, leg-like pedipalps, which are chiefly sensory and help in grasping food.

Harvestmen, although primarily carnivorous, are omnivores, often scavenging from bird droppings and dead animal material, and some species feed on snails. The sensitive second legs are particularly important in locating and recognizing prey, which is held down by the legs and palps, and torn to pieces by the pincers on the chelicerae. Harvestmen regularly clean legs and pedipalps by pulling them through the chelicerae, cleanliness of all appendages being essential to their hunting and feeding behaviour. In adversity, harvestmen will shed a leg to escape, four-legged individuals being able to survive as long as they retain one of the long, sensitive second legs; without them, they presumably cannot feed. Harvestmen need to drink at least every two days, although they can fast for two or more weeks. As there is no lipid layer in the cuticle, they require humid surroundings, and they are normally nocturnal in their activity.

Mating is a casual, brief affair, often repeated with other individuals, and involves copulation, the male having a relatively long penis which, like the opening of the female's oviduct, is on the first sternite of the abdomen, beneath the genital plate. Females use their long ovipositors to lay eggs in crevices in soil, under stones, beneath bark, and in other moist places, the number varying from a few to several hundred in *Phalangium opilio*. Newly hatched young, each about one millimetre in length, resemble the adults in most features, but moult several times as they grow (Cloudesley-Thompson 1968).

All British harvestmen belong to the suborder Palpatores, but not all have the familiar long legs and round bodies. Trogulidae, for instance, which feed on snails, are flattened, slow-moving species with shorter legs and a longish body often covered with earth particles. Some species are active all year round, and take more than one year to complete their life cycle, a few are annuals overwintering as immatures, and many, such as *Mitopus morio*, are annuals that overwinter as eggs (Williams 1963).

Todd (1948) found the greatest seasonal density of harvestmen, about 50 per square metre of ground litter, in deciduous woodland in June, but these were mostly immature individuals, and the biomass was at its greatest in July. She found that many woodland species hatch on the ground, but then move vertically up to the field layer and then to the tree trunks, even in some species up to the branches, as they get older. Different species typically live at different levels in accord with their humidity preferences as tested in the laboratory: *Nematostoma lugubre* is found on the ground, where the relative humidity (RH) is 85–100%,

HARVESTMEN

Nomenclature and sequence follow Sankey (1988), numbers refer to total pitfall captures 1 April–31 October 1980–1990, and MT indicates that the species was also caught in the Malaise trap.

Phalangidae			
48	***Oligolophus tridens*** (Koch)	3	
49	***Paroligolophus agrestis*** (Meade)	9	MT
50	***Odiellus spinosus*** (Bosc)	567	MT
51	***Mitopus morio*** (Fab.)	1	
52	***Phalangium opilio*** L	11	MT
53	***Opilio parietinus*** (Degeer)	3	
54	***Opilio saxatilis*** Koch	93	MT
Leiobunidae			
55	***Dicranopalpus ramosus*** (Simon)		MT
56	***Leiobunum rotundum*** (Latreille)	10	MT
57	***Leiobunum blackwalli*** Meade	84	MT
58	***Nelima gothica*** Lohmander	4	

Leiobunum rotundum, a species of harvestman, sometimes captured in pitfalls and in the Malaise trap in the Leicester garden

Oligolophus tridens in the field layer (RH 70–80%), *Leiobunum rotundum* on tree trunks (RH 60–75%), and *Paroligolophus agrestis* on tree branches (RH 50–60%). She also found that *Odiellus spinosus*, which often occurs in gardens, has a much higher range of temperature preference than woodland species.

Harvestmen in the Leicester garden

Ten species of harvestmen were trapped in pitfalls between 1 April and 31 October in the eleven years 1980–1990, and identified by J.H.P. Sankey. Many harvestmen are not available for capture on the ground, because they are higher up in the vegetation, and it is significant that five of the species captured in pitfalls, and an additional species, *Dicranopalpus ramosus*, were also taken in the Malaise trap, which would have entailed a climb of nearly 2m; all are species that are usually found well above the ground. Four of them, *Paroligolophus agrestis, Phalangium opilio, Dicranopalpus ramosus* and *Leiobunum blackwalli* were also beaten from vegetation, and several females of *L. rotundum* were found by A.J. Rundle in the compost heap. *Odiellus spinosus, Opilio saxatalis* and *Leiobunum blackwalli* were also trapped in pitfalls during winter 1985–1986. The most interesting aspect of the pitfall captures was the large catch of *Odiellus spinosus*, the largest British harvestman and one of the less common species, which has a southeastern distribution in the British Isles and inhabits relatively dry and warm places, especially on walls, fences and under window ledges in urban situations (Sankey 1988).

The annual total pitfall captures of harvestmen in the eleven years 1980–1990 were 44, 50, 18, 50, 38, 70, 75, 118, 104, 154 and 145; thus, the best year was 1989, a hot, dry, bright summer, and the poorest 1982, a mild, damp, dull summer. The later Malaise trap identifications were made by J. Daws.

SPIDERS

Spiders (order Araneae) are distinguished from other arachnids by the possession of silk glands that open through spinnerets at the end of the abdomen. They are an abundant and diverse group, found in an astonishing range of terrestrial habitats. More than 600 species are found in the British Isles, and representatives can be used to illustrate the diversity of the group.

The cephalothorax is separated by a narrow pedicel from the abdomen, which may be tubular, as in *Harpactea* species, or spherical as in Theridiidae. *Segestria florentina* females are up to 22mm long, but some Linyphiidae are less than 2mm. The eyes, always simple ocelli, number six or, more usually, eight, and are sometimes on tubercles, as in crab spiders, Thomisidae. They have four pairs of legs usually of similar length, although in Thomisidae the first and second pair are very long and robust for prey capture. The tarsi usually end in two toothed claws, occasionally in three, and the orb-web builders, Araneidae, have claw-like hairs at the end of the tarsi for gripping the silk of their webs. Many hunting spiders also have dense tufts of hair, scopulae, under their claws, that help maintain a grip on vertical or overhanging surfaces. Most spiders are nocturnal in their activity, but some conspicuous species, such as Salticidae, are diurnal.

Spiders have one or, in Atypidae (part of suborder Mygalomorpha), two pairs of spiracles ventrally on the forepart of the abdomen leading to book lungs, consisting of multiple horizontal air pockets; there is also one posterior spiracle (occasionally four), leading to tubular tracheae which branch throughout the abdomen and, in some species, also through the cephalothorax. The body cavity is filled with haemolymph containing the copper-based respiratory pigment haemocyanin, and there is an open-ended, tubular dorsal heart. The air pockets in the book lungs are in contact with haemolymph-filled lamellae, and most of the blood goes through them on its way back to the heart. The tracheae are all open-ended, rather than contacting a cell, so haemolymph is always responsible for the final delivery of oxygen. Mygalomorphs have only book lungs, but the faster-moving spiders have only tracheae, presumably the most efficient way of oxygenating the tissues (Foelix 1982).

The familiar orb-web spider of gardens, *Araneus diadematus*, on her web, awaiting the entangling of an insect as food. This is a female, males being smaller and more mobile

Abdominal silk glands open on many little 'spigots' on three pairs of mobile spinnerets situated just below the anus. Silk, a proteinaceous substance, is extruded as a liquid but solidifies as it is stretched and exposed to the air. Hunting spiders have four types of silk glands producing different types of silk for attachment discs, for drag lines, for trapping prey and making a sperm web or outer egg sac, and for making the inner layer of egg sacs. Orb-weavers have two additional types, producing silk for making the spiral threads of their orb-webs, and for producing the viscous liquid that coats the sticky spiral that ensnares prey (Jones 1983). An oily coating to their legs prevents them from becoming caught themselves.

Webs

Many spiders are wanderers and do not make webs, but more sedentary species construct webs of various designs for catching their prey. For instance, those of Agelenidae are slightly concave mats, suspended by vertical threads, with a funnel-shaped retreat at one end, Linyphiidae build convex sheets also with vertical trip-lines, Theridiidae make irregular meshes, Araneidae construct intricate orb-webs, and Amaurobiidae weave a tubular retreat from which radiate signal lines and catching threads. The catching threads of *Amaurobius* species and other cribellates are flocculent, bluish ribbons which are made by combing with the calamistrum (a comb-like series of curved spines along the metatarsi of the fourth legs) the fine silk that emerges from a sieve-like plate (the cribellum) crowded with spigots in front of the spinnerets, so that it forms a dense ribbon around the web threads.

The body and limbs of all spiders are covered with hairs and bristles, and on the legs are fine bristles (trichobothria) that are sensitive to air currents and vibrations. These play a part in alerting spiders to prey, usually insects, although hunters, such as Lycosidae, rely heavily on sight. Almost all spiders have poison glands opening just behind the tip of the fangs that form the second segment of the chelicerae. The venom is a mixture composed mostly of neurotoxic

SPIDERS

Taxonomy, nomenclature and sequence follows Crocker & Daws (2001). MT refers to capture in the Malaise trap, and P to capture in pitfalls.

Amaurobiidae

59	*Amaurobius fenestralis* (Stroem)		
60	*Amaurobius similis* (Blackwall)		
61	*Amaurobius ferox* (Walckenaer)		P

Dictynidae

62	*Dictyna arundinacea* (L.)		
63	*Lathys humilis* (Blackwall)		

Oonopidae

64	*Oonops pulcher* Templeton		
65	*Oonops domesticus* Dalmas		

Dysderidae

66	*Harpactea hombergi* (Scopoli)		

Gnathosidae

67	*Drassodes lapidosus* (Walckenaer)		
68	*Scotophaeus blackwalli* (Thorell)		

Clubionidae

69	*Clubiona corticalis* (Walckenaer)	MT	
70	*Clubiona pallidula* (Clerck)	MT	
71	*Clubiona phragmitis* C.L. Koch	MT	
72	*Clubiona terrestris* Westring	MT	
73	*Clubiona lutescens* Westring	MT	
74	*Clubiona comta* C.L. Koch	MT	
75	*Clubiona brevipes* Blackwall	MT	

Thomisidae – crab spiders

76	*Misumena vatia* (Clerck)		
77	*Xysticus cristatus* (Clerck)	MT	
78	*Oxyptila praticola* (C.L. Koch)		P

Philodromidae

79	*Philodromus dispar* Walckenaer	MT	
80	*Philodromus aureolus* (Clerck)	MT	
81	*Tibellus oblongus* (Walckenaer)	MT	

Salticidae

82	*Salticus scenicus* (Clerck)	MT	
83	*Heliophanus flavipes* (Hahn)	MT	

Lycosidae

84	*Pardosa pullata* (Clerck)		P
85	*Pardosa prativaga* (L. Koch)		P
86	*Pardosa amentata* (Clerck)		P
87	*Alopecosa pulverulenta* (Clerck)		

Pisauridae

88	*Pisaura mirabilis* (Clerck)	MT	

Agelenidae

89	*Tegenaria gigantea* Chamberlin and Ive		P
90	*Tegenaria domestica* (Clerck)		P

Theridiidae

91	*Steatoda bipunctata* (L.)	MT	
92	*Anelosimus vittatus* (C.L. Koch)	MT	
93	*Theridion varians* Hahn		
94	*Theridion mystaceum* L. Koch	MT	
95	*Theridion tinctum* (Walckenaer)	MT	
96	*Neottiura bimaculatum* (L.)	MT	
97	*Enoplognatha ovata* (Clerck)	MT	

Tetragnathidae

98	*Tetragnatha extensa* (L.)	MT	
99	*Tetragnatha montana* Simon	MT	
100	*Metallina segmentata* (Clerck)	MT	

Araneidae

101	*Araneus diadematus* Clerck	MT	
102	*Nuctenea umbratica* Clerck	MT	
103	*Araniella cucurbitina* (Clerck)		
104	*Zygiella x-notata* (Clerck)	MT	

Linyphiidae

105	*Ceratinella brevis* (Wider)		P
106	*Entelecara acuminata* (Wider)	MT	
107	*Gongylidium rufipes* (L.)	MT	
108	*Dismodicus bifrons* (Blackwall)	MT	
109	*Hypomma cornutum* (Blackwall)	MT	
110	*Oedothorax fuscus* (Blackwall)	MT	
111	*Oedothorax retusus* (Westring)	MT	
112	*Silometopus reussi* (Thorell)	MT	
113	*Tiso vagans* (Blackwall)	MT	
114	*Micrargus herbigradus* (Blackwall)		P
115	*Micrargus subaequalis* (Westring)	MT	
116	*Diplocephalus latifrons* (O.P.-Cambridge)		
117	*Milleriana inerrans* (O.P.-Cambridge)	MT	
118	*Erigone dentipalpis* (Wider)	MT	
119	*Erigone atra* Blackwall	MT	
120	*Porrhomma microphthalmum* (O.P.-Cambridge)		
121	*Meioneta rurestris* (C.L. Koch)	MT	
122	*Meioneta saxatilis* (Blackwall)	MT	
123	*Microneta viaria* (Blackwall)		P
124	*Centromerus dilutus* (O.P.-Cambridge)		P
125	*Centromerita bicolor* (Blackwall)		
126	*Bathyphantes gracilis* (Blackwall)	MT	
127	*Diplostyla concolor* (Wider)	MT	P
128	*Stemonyphantes lineatus* (L.)	MT	P
129	*Megalepthyphantes nebulosus* (Sundevall)		
130	*Lepthyphantes leprosus* (Ohlert)		P
131	*Lepthyphantes minutus* (Blackwall)	MT	
132	*Lepthyphantes tenuis* (Blackwall)	MT	P
133	*Lepthyphantes mengei* Kulczynski	MT	P
134	*Lepthyphantes ericaeus* (Blackwall)	MT	
135	*Lepthyphantes pallidus* (O.P.-Cambridge)	MT	
136	*Linyphia montana* (Clerck)	MT	P
137	*Linyphia clathrata* (Sundevall)		P
138	*Microlinyphia pusilla* (Sundevall)	MT	

polypeptides, but fewer than 0.1% of the world's spiders are dangerously poisonous to man. Apart from *Scytodes thoracica*, which shoots ensnaring silk to disable its prey from a distance, hunting spiders bite and inject poison at the first strike, but most other spiders ensnare and wrap their prey in silk before biting it.

A crab spider, *Misumena vatia* (above), lies in wait on flowers for its prey, and is often extremely well camouflaged, being variable and changeable in colour – white, pale green, yellow, or even pink

The chelicerae with their fangs strike vertically in Atypidae and other Mygalomorpha, but work horizontally in other spiders. In most species there are cheliceral teeth that, together with the massive coxae of the pedipalps, macerate the prey beyond all recognition while digestive enzymes are poured onto it. The resulting nutritious liquid is then sucked in, for no spider can ingest solid food. Thomisidae and Theridiidae, however, have no cheliceral teeth and inflict only a small wound on their intact prey, from which they suck the contents liquefied by extra-oral digestion.

Courtship and reproduction

The pair of segmented pedipalps, between the chelicerae and the first pair of walking legs, have a special function in males. The last, bulbous segment is variously modified to receive sperm from a drop that the male extrudes onto a specially made little silken web. Males then use their charged pedipalps as intromittent organs when they copulate with females. Males are vagabonds after their final moult, wandering around in search of females, which are usually larger; in most species, courtship involves the male dissuading the female from treating him as food. In some spiders, such as Thomisidae, direct contact is required for courtship to ensue; in others, female pheromones stimulate the males, and some Araneidae, for instance, mate immediately after the female's final moult. In Salticidae, however, visual signals are the primary releaser of courtship. In nocturnal hunters there is little courtship, but in Salticidae and Lycosidae there is an elaborate visual display, often involving much semaphoring with the legs, and in many web-makers the approach is tactile, involving coded vibrations of the web.

Many males have stridulatory organs; in Theridiidae, for instance, males vibrate their abdomens so that spurs on the front of the abdomen rub against ridges on the back of the thorax. A few females attack and eat males after mating but many, such as *Tegenaria* species, coexist in the same web, and the male is only eaten later, when he dies. Male *Pisaura mirabilis* present the female with a silk-packaged prey item, and mate when she is occupied with it, and in other species, males copulate while the female is feeding. In *Linyphia triangularis*, however, the male meets no resistance, and he destroys most of the female's web during courtship so that she can be interrupted neither by feeding nor by the attentions of competitors (Foelix 1982).

A wolf spider, *Pardosa amentata* (below), with its prey. Wolf spiders rely heavily on sight when hunting, and have eight simple eyes with convex lenses

A nursery web spider, *Pisaura mirabilis*, with her egg sac. When the young are about to emerge, she hangs the sac on low vegetation and weaves a 'nursery' tent of silk strands around it

Eggs are laid in a silken sac containing in some cases, such as Oonopidae, only two, in others, such as *Dolomedes*, more than 2,000 eggs. The sac may be suspended in the web, lodged in the retreat, when it may be stalked as in Clubionidae, or is carried around by the female, attached either to her fangs, as in Pisauridae, or attached to her spinnerets, as in Lycosidae. Early larval stages, which lack silk and poison glands, subsist on yolk, but these moult into cannibalistic spiderlings, resembling adults, which soon leave the egg sac.

Female Theridiidae feed their young from their mouths or on specially caught and macerated prey, and Lycosidae carry their young around on their backs, but most spiderlings soon disperse, either to make tiny webs for prey capture, or to take to the air. 'Ballooning' is the usual dispersal method of small species and of young. The little spider climbs to an elevated position, stands on tiptoe with the abdomen raised, pays out silk, and when a breeze catches the threads, releases its foothold to sail away.

As spiders grow they moult their skins, sometimes, as in Salticidae, in a specially prepared moulting chamber. Lost limbs can usually be regenerated at a subsequent moult. Different species mature and mate at different seasons, so that mature and active spiders can be found in all months. *Araneus diadematus*, for instance, mate in August, and in September or October lay eggs which overwinter and hatch the following spring; it is usually the following summer, more than a year later, before they mature (Bristowe 1971). Large spiders often take two, three, or even four years to mature, and females of *Steatoda bipunctata* may live several years, particularly indoors, although the life cycle of most species occupies only a year.

A total of 64 species of garden spiders were identified by R.F. Owen in the 1970s using Locket & Millidge (1951); I.M. Evans confirmed the identification of a few of the more difficult species. From 1986 to 1993, C.J. Smith identified spiders caught in the Malaise trap and in pitfalls, and from 1995 to 2001, J. Daws identified Malaise trap captures. From time to time, J. Daws also collected and identified spiders from the garden. An annotated list of spiders in the garden

These spiderlings of *Araneus diadematus* have emerged from the carefully guarded egg sac of the female onto her web

prior to 1986 was published in Owen, J. (1991). However, some of these species were discounted by Crocker & Daws (1996), so have been deleted from the garden list, which now totals 80 species.

SUMMARY

The 138 species of invertebrates, other than insects, identified from the garden, by no means form a definitive list. Special techniques are needed to find some groups such as false-scorpions, diligent searching of vegetation at all levels is necessary to find harvestmen and spiders, and groups such as Crustacea in the pond and soil-dwelling mites have not been investigated at all. Nevertheless, the varied list gives a flavour of the diversity of the fauna in a suburban garden.

11 : Vertebrates

VERTEBRATES, UNLIKE INSECTS and some other invertebrate groups, were not trapped in the Leicester garden, but continued observation led to accumulation of many records, particularly of birds. They are conspicuous residents and visitors, since there is dense vegetation for shelter and little threat from cats, which have always been discouraged. Indeed, over the 30 years, an astonishing variety of birds visited the garden or flew over it. The relatively small size of the garden and its location on a busy corner make it unsuitable for most mammals, but the pond, small though it is, attracts amphibians. In all, 3 amphibian species, 54 bird species and 7 mammal species were recorded.

Goldfinch, *Carduelis carduelis* (opposite), on a seed head of teasel (*Dipsacus fullonum*). As well as teasel seeds, goldfinches also fed on those of perennial cornflower (*Centaurea montana*) and common ragwort (*Senecio jacobaea*) in the Leicester garden

AMPHIBIANS

Common toad, *Bufo bufo* (above), has successfully bred in the Leicester garden, although they usually prefer a larger and deeper pond

Five species of amphibian occur in the county of Leicestershire: common frog, toad, smooth newt, palmate newt and great crested newt. Common frog and smooth newt are generally distributed and common in many places, and toads are widespread, although not as universally distributed as frogs. Palmate newt occurs in a few localities on Charnwood Forest and in east Leicestershire, and the great crested newt is widely although patchily distributed.

In the ten year period before 1975, installation of garden ponds was common and widespread, causing a rapid expansion of urban standing water (a trend that continued to the end of the century), although most ponds are of less than $2m^3$ in capacity. Frogs and smooth newts successfully invaded garden ponds, although toads were slower to establish themselves and great crested newts have had little success in spreading to urban developments (Mathias 1975).

Garden ponds are evidently important for the maintenance of breeding populations of amphibians, which have suffered countrywide from the drainage, pollution or infilling of field ponds, and, in the 1940s and 1950s, from large-scale collection of frogs for dissection in school biology classes, when this was regarded as an essential part of everyone's education. Certainly, frogs and

AMPHIBIANS

1 **Toad** *Bufo bufo* (L.)
2 **Common frog** *Rana temporaria* L.
3 **Smooth newt** *Triturus vulgaris* (L.)

smooth newts are now common in the city of Leicester, although toads are less common, as they normally maintain larger breeding populations than the typical small garden pond can support. There are also many ponds in the city on public land, such as parks and golf courses, and although old field ponds have been lost to urban development, many new ponds have been created. Some of these ponds are excellent for wildlife, including amphibians, and even great crested newts breed in a few of the larger ponds in the city. It seems that all city ponds are eventually colonized by amphibians, particularly frogs, whereas only about half of ponds on agricultural land are occupied. This is almost certainly because of the terrestrial requirements of amphibians, as they are invertebrate-eaters and shelter beneath vegetation. Garden ponds are particularly important not only because there are so many of them, well distributed, but also because lush, diverse garden vegetation provides a good habitat for amphibians away from the water, as their invertebrate food is plentiful.

Common frog, *Rana temporaria* (top), surprised among fallen leaves in the author's garden, where it was probably searching for invertebrates to eat

Smooth newt, *Triturus vulgaris* (above), which regularly bred in the garden pond. They hibernated beneath nearby vegetation, where my son found this one

Amphibians in the Leicester garden

I have recorded common frog, smooth newt and toad in my garden. Before the pond was drained, in 1976, frogs were repeatedly seen both in the pond in spring and summer, and in all other parts of the garden in summer and autumn, although there was no evidence of breeding. When, however, the pond was re-established, in early 1984, they quickly moved in and bred successfully. They probably recolonized so quickly because they were already in the garden; two were found in the rubble filling the pond cavity, when it was cleared away prior to reinstating the pond. In October 1984, a small frog, presumably born that spring, was found in a pitfall trap. Subsequently, frogs bred in the pond every year, and were to be seen sitting on stones by the water's edge in summer, and were frequently disturbed around the garden.

Smooth newts, too, were present in the pond before 1976, although breeding was not proved. Individuals were found hibernating in the rockery beside the pond, even in years when the pond was dry. When it was re-established, smooth newts quickly recolonized and bred successfully. They have often been found, in summer and winter, beneath a spreading clump of *Iberis sempervirens* close to the pond. In September 1985, a newly metamorphosed individual was found in a pitfall trap.

The first record of a toad was of a semi-torpid individual crouched beneath dead leaves some metres away from the pond in 1986. Subsequently, a toad has been seen in most years, including in the pond, and in some years they have successfully bred, small though the pond is. The breeding of three species of amphibian in it demonstrates that even a small pond is a valuable wildlife asset to a garden.

BIRDS

Leicestershire, being very much an inland county, does not have a bird list as large as some coastal counties on migration routes, but it does have a variety of habitats, including the internationally important Rutland Water. In 1986, 194 species of birds were recorded in the county (including Rutland) by the Leicestershire and Rutland Ornithological Society (LROS 1987). In the period 1983–1987, 108 species were recorded within the city boundary, including a fair number of ducks and other water birds (Lomax 1987). In the 1950s, my father, F. A. Bak, trapped and ringed 55 species in his 1,000m^2 garden, less than 2km from Leicester city centre, in an area of spacious gardens with many mature trees. The diverse traps he used, including a large one enclosing soft fruit bushes and baited with dripping water, caught unexpected visitors to a suburban garden such as sedge warbler (*Acrocephalus schoenobaenus*), grasshopper warbler (*Locustella naevia*), nightingale (*Luscinia megarhynchos*), brambling (*Fringilla montifringilla*), and the first marsh warbler (*A. palustris*) to be recorded in the county. Many additional species passed over the garden, or, like the local house martins and swifts, were not amenable to trapping.

Robins, *Erithacus rubecula* (top), are to be found in most gardens, where they shadow the gardener to snatch worms and other invertebrates unearthed by trowel or spade

Blue tits, *Parus caeruleus* (below), are frequent visitors to bird tables where they eat seeds and peck at nuts. They readily nest in nest-boxes set up in gardens as natural nesting holes are usually in short supply

The trapping list from my father's garden well illustrates the feature of birds that leads to their doing so well in gardens, even in a city: birds are supreme opportunists. Since they investigate every feeding possibility and are relentlessly curious, they are readily caught by traps that provide food, whether it be soft fruit, grain or meal-worms, and the sound of dripping water is a magnet to warblers, in particular, during hot summer weather. My father even brought black-headed gulls down onto his garden lawn by provision of sprats.

Feeding

Blue tits are adroit at levering open the blisters on holly leaves caused by holly leaf miners to extract the larvae, and have turned this skill to acquiring another food, for they tear open cardboard milk-bottle tops and hammer holes in foil ones to drink the cream. The habit was first noticed in the 1920s, but spread rapidly and now occurs in most parts of the country. Although there was undoubtedly some copying, the milk-stealing habit appears to have been acquired independently many times, and species other than blue tits have been involved.

Greenfinches, too, quickly acquire new eating habits, learning in the early 19th century to eat *Daphne mezereum* seeds in mid-summer while they are still soft, a habit that then spread at a rate of about 3km a year (Murton 1971).

Dunnock (or hedge sparrow), *Prunella modularis* (right), a cautious little bird, mainly an insect-eater, that often picks up fragments dropped from bird tables by the clumsy feeding activities of house sparrows

A male bullfinch, *Pyrrhula pyrrhula* (far right), with its distinctive rosy breast. A male and a somewhat duller plumaged female were often glimpsed in the shrubbery to the east of the Leicester house

I have seen countless examples of opportunism by birds in my garden, such as a blue tit investigating a new nest-box within hours of it being fixed. Robins are also opportunist, landing close by to snatch worms whenever soil is disturbed, pecking at large toadstools, presumably for insects, searching the timbers of a dismantled old oak gate for spiders and insects, and drinking melt-water dripping from icicles on the roof when a thaw set in after a prolonged cold spell. When I pulled up a root of *Leucanthemum* × *superbum* enclosed in an ant nest, exposing to view not only the ant brood but also unpigmented aphids on the root, a song thrush quickly moved in to eat ants and aphids alike.

BIRDS

B indicates species that bred in the garden.

4	**Heron** *Ardea cinerea* L.	
5	**Mallard** *Anas platyrhynchos* L.	
6	**Sparrowhawk** *Accipiter nisus* (L.)	
7	**Kestrel** *Falco tinnunculus* L.	
8	**Red-legged partridge** *Alectoris rufa* (L.)	
9	**Lapwing** *Vanellus vanellus* (L.)	
10	**Black-headed gull** *Larus ridibundus* L.	
11	**Common gull** *Larus canus* L.	
12	**Wood pigeon** *Columba palumbus* L.	B
13	**Collared dove** *Streptopelia decaocto* (Frivaldsky)	B
14	**Turtle dove** *Streptopelia turtur* (L.)	
15	**Cuckoo** *Cuculus canorus* L.	
16	**Tawny owl** *Strix aluco* L.	
17	**Swift** *Apus apus* (L.)	
18	**Green woodpecker** *Picus viridis* (L.)	
19	**Lesser spotted woodpecker** *Dendrocopus minor* (L.)	
20	**Skylark** *Alauda arvensis* L.	
21	**Swallow** *Hirundo rustica* L.	
22	**House martin** *Delichon urbica* (L.)	B
23	**Sand martin** *Riparia riparia* (L.)	
24	**Grey wagtail** *Motacilla cinerea* Tunstall	
25	**Pied wagtail** *Motacilla alba* subsp. *yarrellii* Gould	
26	**Wren** *Troglodytes troglodytes* (L.)	
27	**Dunnock** *Prunella modularis* (L.)	
28	**Robin** *Erithacus rubecula* (L.)	
29	**Blackbird** *Turdus merula* L.	
30	**Fieldfare** *Turdus pilaris* L.	B
31	**Song thrush** *Turdus philomelos* Brehm	
32	**Redwing** *Turdus iliacus* L.	B
33	**Mistle thrush** *Turdus viscivorus* L.	
34	**Chiffchaff** *Phylloscopus collybita* (Vieillot)	
35	**Willow warbler** *Phylloscopus trochilus* (L.)	
36	**Blackcap** *Sylvia atricapilla* (L.)	
37	**Goldcrest** *Regulus regulus* (L.)	
38	**Long-tailed tit** *Aegithalos caudatus* (L.)	
39	**Marsh tit** *Parus palustris* L.	
40	**Coal tit** *Parus ater* L.	
41	**Blue tit** *Parus caeruleus* L.	B
42	**Great tit** *Parus major* L.	
43	**Nuthatch** *Sitta europea* Blyth	
44	**Treecreeper** *Certhia familiaris* L.	
45	**Jay** *Garrulus glandarius* (L.)	
46	**Magpie** *Pica pica* (L.)	
47	**Jackdaw** *Corvus monedula* L.	
48	**Rook** *Corvus frugilegus* L.	
49	**Carrion crow** *Corvus corone* subsp. *corone* L.	
59	**Starling** *Sturnus vulgaris* L.	B
51	**House sparrow** *Passer domesticus* (L.)	
52	**Chaffinch** *Fringilla coelobs* L.	
53	**Greenfinch** *Carduelis chloris* (L.)	B
54	**Goldfinch** *Carduelis carduelis* (L.)	
55	**Linnet** *Carduelis cannabina* (L.)	B
56	**Redpoll** *Carduelis flammea* (L.)	
57	**Bullfinch** *Pyrrhula pyrrhula* (L.)	

Greenfinch, *Carduelis chloris* (above left), a regular bird table visitor which occasionally nested in a dense tangle of mock orange, *Philadelphus coronarius*,in the Leicester garden

Collared dove, *Streptopelia decaocto* (above right). Since arriving in Britain in 1955, collared doves have become a familiar sight gleaning seeds and nesting in gardens

Dunnocks exploit the untidy feeding activities of other birds, regularly gleaning morsels beneath the bird-table when coal tits peck somewhat ineffectually at a bag of nuts. One winter's day, dunnocks picked up seeds dropped on the snow by a bullfinch clumsily feeding at the seed-heads of antirrhinums. Greenfinches pick seeds from newly spread compost, and in hard weather wood pigeons strip broccoli plants to the mid-ribs of the leaves. Black-headed gulls circle high in the air feeding on flying ants whenever nests of black ant, *Lasius niger,* produce swarms of reproductive adults. I also observed house sparrows picking up birch seeds dropped from the tree by the feeding activities of the more agile redpolls. House sparrows will optimistically have a go at anything that might be edible, and one chased and snapped, without success, at a large *Aeshna* dragonfly crossing the garden.

Perhaps, however, the most striking instance of opportunism in birds is the speed and success of the colonization of English suburbia by collared doves, suggesting that there were hitherto unexploited opportunities for a small seed-gleaning dove. They reached Britain in 1955, and in the space of 15 years numbers increased to between 15,000 and 25,000 breeding pairs. They were first recorded in Leicestershire in 1961, by 1964 were widespread, first bred in the city in 1975 (Hickling 1978), and by 1985 were designated as common resident breeders in the city (LROS 1986).

Song thrush, *Turdus philomelos* (above). Once a familiar part of the scene and dawn chorus in the Leicester garden, song thrushes have become much less common, a national trend

Birds that bred in the Leicester garden

The core of my garden bird list is species that have bred in the garden, some of them every year from 1972 to 2001, others occasionally: wood pigeon, collared dove, house martin, dunnock, robin, blackbird, song thrush, blue tit, starling, greenfinch and linnet. However, the strong impression of the second fifteen years, 1987–2001, is that birds as a whole became much scarcer in the garden. House sparrows, once ubiquitous, virtually disappeared, and song thrushes, once always in the garden, were rarely seen.

From 1989 onwards, starlings were always to be seen on the lawn or flying to and fro, as they promptly took over two nest-boxes fixed under the house eaves in the hope of attracting breeding swifts. The constant presence and behaviour during the nesting season of goldcrests and bullfinches until about 1990, and redpolls until the birch was pollarded and then felled in 1994,

Great tit, *Parus major* (right), a regular garden visitor, with its insistent call of 'teacher, teacher'

Chaffinch, *Fringilla coelobs* (far right), an attractive and regular garden visitor

suggested that they too were breeding, although no nests were found. It is tempting also to include on the breeding bird list the wren that built a large unoccupied nest behind the honeysuckle against a fence, and the mallard, presumably one from the nearby park, that laid an egg in the strawberry patch.

Birds that visited the Leicester garden

Many other species of bird visited the garden, hunted in its airspace or perched on the chimney-pots or television aerial. Regular visitors included swift, mistle thrush, willow warbler, long-tailed tit, coal tit, great tit, magpie, carrion crow, house sparrow, chaffinch and goldfinch. Mistle thrushes visited at all times of year, including family parties in June. Long-tailed tits, often in family parties, great tits and carrion crows (on the chimney-pots or in tall trees), were frequent all year round. Magpies became much more abundant in the second fifteen years, 1987–2001, while, conversely, numbers of house sparrows declined dramatically. Swifts nested nearby every year, and each summer were major predators of the aerial plankton rising from the garden.

In 1987, and again in 1988, a pair of house martins successfully nested on the house, but in general, as aerial visitors to the garden, they became much less common in 1987–2001. Willow warblers were active in most years, feeding and often singing, particularly in early and late summer months, and were presumed to have bred in neighbouring gardens and scrubland. Chaffinches and goldfinches, too, were mainly summer visitors, the latter taking seeds of teasel, Oxford ragwort and perennial cornflower (*Centaurea montana*) but also probably eating aphids on the still-green cornflower seed-heads. Coal tits, however, were mainly winter visitors, although they were seen at all times of year.

Irregular visitors to the Leicester garden

Sparrowhawk, kestrel, red-legged partridge, turtle dove, tawny owl, lesser spotted woodpecker, fieldfare, redwing, chiffchaff, blackcap, marsh tit, nuthatch, treecreeper and jackdaw were irregular visitors to the garden. Sparrowhawks were not seen until 1987, but thereafter were sighted in the garden on several occasions, often carrying carcasses of birds, and on one occasion a male was watched plucking a collared dove on the ground. Kestrels were repeatedly seen overhead, occasionally perching on the television aerial or garage roof, and on

one occasion, hunting low over the garden. Red-legged partridges were twice disturbed from vegetable patches where there had been previously unexplained damage to lettuces and cabbages. A single turtle dove was seen on two occasions in the late 1990s, on one occasion eating elderberries.

Tawny owls were sometimes heard calling from chimney-pots, particularly in the first fifteen years, 1972–1986, lesser spotted woodpeckers sometimes flew over the garden in September or October, and fieldfares and redwings quite often visited in winter to feed on holly berries and rootle in leaf litter. Single chiffchaffs visited in April, August and September in different years, a pair of blackcaps was seen in December 2000, marsh tits were occasionally seen at different times of year, a single nuthatch was spotted in January 2000, and tree creepers were seen on four occasions. Jackdaws occasionally perched on a chimney-pot, and in June 1984 a young one spent some time walking around the garden and inspecting the pond.

Other birds passed over the garden in their foraging, roosting or dispersal movements; these included heron, lapwing, black-headed gull, common gull, swallow (although these became less common), pied wagtail, jay and rook. Skylarks and grey wagtails sometimes flew over, evidently on migration, particularly in the first fifteen years, 1972–1986, and cuckoo, sand martin and green woodpecker were each seen once.

Nationally, numbers of rooks declined from the mid- to late-1970s, probably because of the loss to Dutch elm disease of mature elms, in which they traditionally nested, and also because large areas of rough pasture, where they once fed, have been used for growing cereal crops. During the 1980s, some rooks were recorded in the county using electricity pylons for nesting; two rookeries were established in the city in the early- to mid-1980s, but have since been abandoned.

In the 1970s and 1980s there was a large winter roost of pied wagtails in trees in a busy thoroughfare in the centre of Leicester, but then, with development, the trees were removed, although pied wagtails may still be seen feeding in many parts of the city.

The total garden bird list is thus 54 species, which could marginally be boosted to 59 if racing pigeons and escaped cage-birds (three species of parrot and a cockatiel) are included. Authorities for the scientific names are from

Redwings, *Turdus iliacus* (above left), visit the Leicester garden in autumn to feed on berries of holly (*Ilex aquifolium*) and hawthorn (*Crataegus monogyna*)

Red-legged partridge, *Alectoris rufa* (above right), were occasionally disturbed from vegetable patches in the Leicester garden where they were eating the young leaves of lettuce (*Latuca sativa*) and other plants

Hedgehogs, *Erinaceus europaeus* (above left), rarely appeared in the Leicester garden after 1980

Wood mouse, *Apodemus sylvaticus* (above right), is the most abundant mammal in the city of Leicester

Witherby *et al.* (1946), except for the collared dove, taken from Coombs *et al.* (1981). In the cases of pied wagtail and carrion crow, subspecies is recorded, as other, visibly different, subspecies are known in the county.

A larger garden with greater structural diversity, especially one near the coast or a large body of inland water, would have a far longer bird list, but the garden list is impressive for a relatively small garden. Although there are no trapping figures to support this, my strong impression is that there were fewer birds in and over the garden in 1987–2001 than there were in 1972–1986, despite the addition of new species for the garden.

MAMMALS

For reasons of size and locality, the Leicester garden is not an ideal habitat for mammals, and only seven were recorded, none of them frequently. Nevertheless, in the city as a whole, 20 species were recorded during the period 1983–1987 (Lomax 1987), so other species might have been expected occasionally in the garden.

Hedgehogs were frequently seen in the garden in the 1970s, including a female with young in 1972. From 1980 on, they were rarely seen, the latest record being of one squashed in the road outside in 1985, but then one was found alive in the road outside in 2001, and released in the garden. They are believed to be common, although patchily distributed, in rural parts of the county, but numbers declined steadily in the city from about 1980. Bats that were occasionally seen flying above the garden at dusk, particularly during the period 1972–1986, were assumed to be pipistrelles. This, the commonest bat species in the county and nationally, is reasonably common in the city, having taken over roof- and wall-spaces of many modern houses for roosting. Nine species of bats, three of them rare, were recorded in the county in the 1980s, and at least three – pipistrelle, noctule and Daubenton's – occurred in the city 1983–1987 (Lomax 1987).

Grey squirrels were irregular but quite frequent visitors to the garden, particularly later in the 30 year period, presumably having come from nearby

MAMMALS

- **58** **Hedgehog** *Erinaceus europaeus* L.
- **59** **Pipistrelle** *Pipistrellus pipistrellus* (Schreber)
- **60** **Grey squirrel** *Sciurus carolinensis* Gmelin
- **61** **Field vole** *Microtus agrestis* (L.)
- **62** **Wood mouse** *Apodemus sylvaticus* (L.)
- **63** **House mouse** *Mus musculus* L.
- **64** **Fox** *Vulpes vulpes* (L.)

parks, where they are abundant. This introduced species was first recorded in the county in the late 1920s but is now widespread, particularly in urban and suburban areas, and is doing well (some would say rather too well).

Domestic cats, derived from *Felis sylvestris*, wreak havoc with small mammals and birds in gardens

Field voles, though common and widespread in the county on grassland and roadside verges, are not likely to visit a suburban garden, but, in August 1981, one was found damaged and dead, perhaps having been dropped by a kestrel. Its exposed flesh was attacked by large numbers of wasps. A wood mouse was seen in the backyard and on the patio on numerous occasions, particularly in the late 1990s. This is the most abundant mammal in the city and county, being even commoner than the house mouse. A dead house mouse was found in the garage, three were disturbed eating stored wild-bird seed in an outhouse, and one was often seen on the patio. The house mouse is common and widespread in the county, although probably under-recorded.

Foxes were not seen in the garden until 1992, although they had often been smelled before then. Since then, one or two foxes have regularly been seen, and their mating calls are a regular feature of nights in January and February. They became established in many British towns and cities in the inter-war years when there was a boom in building low-density housing with medium-sized gardens (Harris & Rayner 1986). They are common and widespread in the county of Leicestershire, and since the 1980s have been regularly seen in the city, some garden-owners having resident foxes, which they supply with food.

Dogs rarely entered the garden, but cats were frequent, and constantly had to be chased away, as they posed a threat to the garden's birds. The 70 domestic cats of a Bedfordshire village of 173 houses were judged in 1981–1982 to be major predators, 1,090 prey items (22 species of birds and 15 of mammals) being taken in the one-year period, the chief items being wood mice (17%), house sparrows (16%) and field voles (14%) (Churcher & Lawton 1987).

Moles are quite common in the city, but in many gardens, the soil is disturbed too much by digging for them to feel at home. Muntjac, small non-native deer, are occasionally reported in the city, and one was seen in the Humberstone area nearby, but the position of my garden, with roads on two sides, makes it unlikely that one will visit.

The mammal list is short, and other suburban gardens, adjacent to golf courses or open countryside, have more species. This illustrates particularly well that it is a medium-sized, ordinary, unsecluded, suburban garden. Insects, other invertebrates, and birds were diverse and abundant, but the habitat is neither extensive enough nor sufficiently undisturbed for there to be many mammals.

12 : The garden habitat

IN THE THIRTY-YEAR PERIOD 1972–2001, an astonishing diversity of animal species was recorded in the Leicester garden. It does not accord with the usual idea of a wildlife garden, but instead contains the elements of a conventional garden and is in no sense a wilderness. Neither do I believe it to be unusual in the richness of its animal life. There is every reason to suppose that the majority of gardens have an equally varied fauna. In other words, the diverse and abundant animal life of my garden is that of gardens in general.

A sunny border in the Leicester garden in the late 1970s (opposite). Against a backdrop of lilac (*Syringa vulgaris*) and Warminster broom (*Cytisus × praecox*), peonies (*Paeonia officinalis*) and perennial cornflower (*Centaurea montana*) spill onto the lawn; dog roses (*Rosa canina*) grow against the fence

Why are gardens so rich? What is it about gardens that enables them to support so many different animals? The answer lies in the nature of the process of gardening, which creates a varied, bountiful habitat, yet one which is in a continual state of flux. The essence of gardening is the creation and maintenance of high plant and structural diversity, while maintaining the habitat in an early successional stage, susceptible to colonization and change.

CONTRIVED PLANT DIVERSITY

A gardener crowds together in one place a far greater diversity of plants than is ever found in one place in the wild. Even in tropical rainforest, the habitat generally agreed to have the highest plant diversity, an area equivalent to that of a typical garden would not contain so many different plant species. Four hundred and twelve plant species in total grew in the garden between 1975 and 2001, 311 that were cultivated, and 101 that came in of their own accord; in 1984, for instance, my garden contained 264 species of flowering plant, excluding grasses, 197 of them intentionally cultivated, in a planted area of 741m^2, a species diversity of 3,563 species per hectare.

It is difficult to compare this with species diversity in tropical rainforest, because figures given for the latter usually refer only to trees. A typical 2.5 acre (approximately 1 hectare) plot of West African rainforest includes 20–25 species of tree of 16-inch (40cm) minimum diameter, 45–60 species of 8-inch (20cm) minimum diameter and 50–70 species of 4-inch (10cm) diameter (Hopkins 1965). The number of tree species in Malaysian or Brazilian rainforest plots of equivalent area might be considerably greater, as there are many more species of flowering plant there than in West Africa. It is unlikely that the inclusion of all types of plant would bring the species diversity of West African rainforest up to 3,563 species per hectare, and certainly, I was not aware of anything like this degree of small-scale diversity in West African forests when I lived there.

Part of the mixed planting bed beside the lawn, the so-called 'lawn bed', in the Leicester garden, with dwarf convolvulus (*Convolvulus tricolor*), sweet tobacco (*Nicotiana alata*), pot marigolds (*Calendula officinalis*), sweet peas (*Lathyrus odoratus*), French marigolds (*Tagetes patula*) and clarkia (*Clarkia unguiculata*), growing with runner beans (*Phaseolus coccineus*), radish (*Raphanus sativa*), tomatoes (*Lycopersicon esculentum*) and cabbage (*Brassica oleracea*)

It could be argued that most gardeners grow more-or-less the same range of species, so that the floristic composition of one garden is repeated again and again, making the average species diversity per hectare smaller, the larger the area considered. True, the composition of the plant diversity of one garden may be similar to that of other gardens in the same area of suburbia, but not identical, for different gardeners grow different species and cultivars and manage their land in different ways. Collectively, the gardens of suburbia are an area of incredibly high plant diversity.

Origins of the garden flora

I have called this diversity 'contrived plant diversity' because it is achieved by the decision of most gardeners to grow a wide range of different plants. Alien plant species, originating in many different parts of the world, are cultivated side by side with native plants. For instance, in the 14.5m^2 mixed lawn bed in my garden, runner beans (*Phaseolus coccineus*) from South America and sweet peas (*Lathyrus odoratus*) from Sicily surround a medley of tomatoes (*Lycopersicon esculentum*) from South America, courgettes (*Cucurbitus pepo*) from North America, spinach (*Spinacia oleracea*) from southwest Asia and native cabbages (*Brassica oleracea*) interplanted with sweet tobacco (*Nicotiana alata*) from southern Brazil, native cornflowers (*Centaurea cyanus*), pot marigolds (*Calendula officinalis*) from southern Europe, asters (*Callistephus chinensis*) from China and Japan, African and French marigolds (*Tagetes erecta* and *T. patula*) from Mexico, dwarf convolvulus (*Convolvulus tricolor*) from southwest Europe, *Clarkia unguiculata* from California, viper's bugloss (*Echium plantagineum*) and borage (*Borago officinalis*) from southern Europe, snapdragons (*Antirrhinum majus*) from the Mediterranean and poppies (*Papaver*

alpinum) from the European Alps. Growing between them are native 'weeds', such as chickweed (*Stellaria media*), and a persistent alien weed from South America, *Oxalis debilis*. This extraordinary assemblage of species, diverse taxonomically as well as in the country of origin, is typical of the plant diversity that a gardener contrives in a small space.

A view of the lawn bed in a different year, with sweet peas, runner beans, sweet tobacco, African marigolds, pot marigolds and cabbages

Since green plants are the starting points for all terrestrial food chains, and many, perhaps a majority of, insect herbivores are confined to one foodplant species, or at least a group of related species, diversifying the plants generates many different food chains. Plant diversity has been shown to be linked to diversity of herbivorous insects (e.g. Murdoch, Evans & Peterson 1972), which in turn is almost certainly linked to diversity of predatory insects. Latitudinal gradients in animal diversity are now generally assumed to stem, at least in part, from latitudinal gradients in plant diversity.

Because tree diversity, and by inference plant diversity as a whole, is greatest in the tropics, resources for herbivores are greater, and hence so are resources and opportunities for predators. While the situation is not identical in the artificial confines of a garden in the temperate region, it seems clear that high plant diversity, even if contrived, should generate high animal diversity by increasing consumer food chains. The abundant opportunities for herbivores are illustrated by the 68 species of moth that bred in the garden during the period 1972–2001, and the diversity of top consumers by the 529 species of ichneumonid trapped in the three years 1972–1974.

French marigolds (*Tagetes patula*) growing alongside cabbages (*Brassica oleracea*), where their aromatic foliage deters potential pests such as cabbage white butterflies (*Pieris rapae* and *Pieris brassicae*)

PERMANENT SUCCESSION

The activity of gardening entails regular clearing, digging, working and watering the soil, which renders it continually receptive to the establishment of native and naturalized plant species as weeds. One hundred and one species recorded in the Leicester garden during the period 1972–2001 were plants that had come in of their own accord. It is clear that while land is managed as a garden, contrived plant diversity is augmented by continual spontaneous introductions from outside that fill spaces as soon as they are created. Most gardeners never let their gardens alone, but are always clearing and re-planting. The cleared land is cultivated, watered, and often treated with fertilizers of one sort or another, so that the new plantings have to compete with a range of spontaneous introductions (weeds).

Were my garden abandoned, in about 30 years it would almost certainly become a dense mixed scrub of birch (*Betula pendula*), holly (*Ilex aquifolium*), ash (*Fraxinus excelsior*), elder (*Sambucus nigra*) and hawthorn (*Crataegus monogyna*), with brambles (*Rubus fruticosus*), bittersweet (*Solanum dulcamara*), nettles (*Urtica dioica*), several sorts of grass, and perhaps thistles (*Cirsium* species), smooth hawk's-beard (*Crepis capillaris*) and yarrow (*Achillea millefolium*) in the more open places. Eventually, the natural processes of plant succession would convert it to mixed deciduous woodland, the characteristic climax community of central England.

As it is, maintaining the land as a garden continually interferes with the natural progress of plant succession, preventing the establishment of a climax community, and keeping the plot in an early successional stage, receptive to invasion from outside. Without my efforts as a gardener, the apparent stability of ordered but contrived plant diversity would soon be lost, and the floristic richness would be reduced to perhaps only a twentieth the number of species that are maintained or arrive at present.

A continuous state of change

An aspect of gardening that is as significant for the garden community as making introductions, is the removal of unwanted plants. Not only are plants that have come in of their own accord treated as weeds and destroyed, but in addition, annuals and perennials are uprooted as soon as they have served their purpose as flowers or vegetables, and perennials and woody plants are often removed as the gardener's tastes or preferences change. Continued introductions to a garden, whether by deliberate planting or invasion, are constantly countered by changes of policy and weeding. At the same time, both cultivated plants and weeds die out naturally for reasons beyond the gardener's control, either because conditions were unsuitable from the outset, or because conditions have changed.

Wood avens, *Geum urbanum*, appeared by chance in the garden in 1976 and was initially encouraged for its novelty, but then became an aggressive weed

In my garden, some alien plants grown as vegetables, such as runner beans, or as flowers, such as African marigolds, cannot maintain themselves and are replanted every year. Certain cultivated plants, however, such as native forget-me-not (*Myosotis arvensis*) and Canadian goldenrod (*Solidago canadensis*), seed themselves so freely and effectively that they have to be rigorously controlled.

Various trees, such as almond (*Prunus dulcis*) and walnut (*Juglans regia*), were grown from seed and introduced, but did not survive. Two herbaceous border perennials, sneezeweed (*Helenium autumnale*) and smooth ox-eye (*Heliopsis helianthoides*), present when the garden was taken over in 1971, eventually died; both were subsequently grown from seed and replanted, but both eventually died again. Of annual flowers, cornflowers were grown in 18 years, *Petunia* cultivar in 17, but *Salpiglossis sinuata* in only five; among vegetables, broad beans (*Vicia faba*) were grown in only one year, asparagus peas (*Lotus tetragonolobus*) in nine, but cabbages (*Brassica oleracea*) in every one of the 30 years.

Unplanned garden introductions, or weeds, similarly arrive and quickly disappear or persist. Hairy tare, *Vicia hirsuta*, appeared in only one year, scarlet pimpernel, *Anagallis arvensis*, in four, but wood avens, *Geum urbanum*, which appeared in 1976 and was encouraged as an interesting addition to the flora, has since become one of the most persistent and aggressive weeds. Colt's-foot, *Tussilago farfara*, initially well-established in a herbaceous border, died out in 1977, perhaps as accumulation of dead plant material and dense soil cover created conditions in which this early successional species was a poor competitor.

In 1975, stinking chamomile (*Anthemis cotula*) and scentless mayweed (*Tripleurospermum inodorum*) appeared in the newly created lawn bed, where the soil was sandy and lacking in humus. As the soil changed in texture and fertility with repeated cultivation, these species disappeared, *T. inodorum* after four years and *A. cotula* after only two, although in 1986, one plant of *A. cotula* appeared elsewhere in the garden, growing between paving slabs.

Two arctiid moths, the muslin, *Diaphora mendica* (above left), and ruby tiger, *Phragmatobia fuliginosa* (above right), were regularly captured at light in the early 1970s. However, the ruby tiger disappeared after 1975 and the muslin was only caught sporadically after that date

The gardener's role

The plant community of a garden is never allowed to stabilize; it is continually interfered with and remains receptive to new arrivals from outside. Dominants are suppressed and prevented from forming an unbroken canopy of shade, so the ground is consequently repeatedly invaded by early successional species. The outcome is that a garden is maintained in a state of permanent succession, and every step towards establishment of a climax community is countered. Indeed, a gardener's prowess may be assessed in terms of his/her ability to check the normal processes of plant succession.

This continued state of flux in a garden, and the constant arrival of their own accord of new species creates yet more feeding opportunities for herbivores and so generates more food chains. Abundance of food resources combined with instability of the habitat makes a garden highly susceptible to temporary colonization. Two arctiid moths, muslin (*Diaphora mendica*) and ruby tiger (*Phragmatobia fuliginosa*), although uncommon, were regularly captured at light in the early 1970s, but no *P. fuliginosa* were seen after 1975. After 1976, *D. mendica* was absent until 1991, when one was caught, then another in 1992, after which it disappeared again.

Cheilosia bergenstammi, a hoverfly, was caught in 1972, 1973,1974 and again in 1988, but not in later years, whereas *C. vernalis* was first caught in 1978, and then again in 1980, after which it was taken almost every year, sometimes commonly. A striking black and yellow hoverfly, *Xanthogramma pedissequum*, was taken once in 1972, and then not until 1992, after which it was caught in every year but one, sometimes several individuals. The creation of a state of permanent succession is thus a source of further enrichment of the fauna.

STRUCTURAL HETEROGENEITY

A typical garden is a complex mosaic of tall and low vegetation, of open spaces and shade, forming an intricately structured and extremely patchy environment. Contrived plant diversity and permanent succession are thus combined with extreme structural heterogeneity. Within a small area there are elements of meadow, woodland and other habitats, so closely juxtaposed that a garden can be considered as a system of ecotones (transition zones between habitats). There are edges everywhere, between herbaceous borders and lawn, shrubbery and path, vegetable patch and compost heap, and so on.

Where two habitats meet, for instance where the trees of woodland give way to the open grass of a meadow, the edge effect leads to an increase in abundance and variety of organisms. Transition zones accommodate some species characteristic of each community, others that belong to both, and also characteristic ecotone species. Gardens are like vastly extended woodland edge, which may explain, in part, the richness of their insect fauna.

Ecotones in the garden

The occurrence of hoverflies in my garden may be used to illustrate the concept of a garden as a system of ecotones. As described in Chapter 5, 33 of the species recorded in the period 1972–2001 are usually associated with woodland or woodland edge, 11 with more open places, and 50, including such abundant species as *Platycheirus albimanus*, range over both types of habitat. It is not so easy to distinguish ecotonal species, because adults of woodland species must regularly move to the woodland edge, where there are more flowers, in order to feed. In the abundance of their flowers, gardens resemble woodland edge, and in utilizing this resource, many garden hoverflies are ecotonal species.

Hoverflies as a group are particularly suited to exploiting the patchiness of the garden habitat because adult and larval requirements are so different, and because larval life styles are so diverse. Structural heterogeneity together with high plant diversity probably also account for the high diversity of Ichneumonidae (533 species), in that a garden encompasses a multitude of potential ecological niches, so that the diverse egg-laying requirements of females are met.

Foliage height diversity

It has often been shown that animal diversity is positively correlated with foliage height diversity (e.g. MacArthur & MacArthur 1961 for birds in deciduous forests; Murdoch *et al.* 1972 for Homoptera in old fields). Foliage height diversity is one of the aspects of structural heterogeneity in a garden, and forms another element of the garden habitat that leads to diversity of the fauna.

The extreme patchiness of the garden habitat is illustrated by patchiness in the occurrence of insects. This environmental heterogeneity is not always apparent to the observer, although the behaviour of insects confirms its reality. Some sprawling plants of spotted dead-nettle (*Lamium maculatum*) in my garden support caterpillars of angle shades moth, *Phlogophora meticulosa*, in all seasons, year after year, whereas others of the same species nearby are never eaten.

Larva of angle shades moth, *Phlogophora meticulosa*, one of the most polyphagous moth species in the Leicester garden (see p60); larvae may be green or brown

Four species of hibernating hoverfly larvae (*Platycheirus albimanus, Melanostoma mellinum, M. scalare* and *Syrphus ribesii*) were found one winter in a patch of a *Saxifraga* cultivar, while another patch less than 3m away seemed not to shelter any. The size and composition of the hoverfly catches of two Malaise traps operated in separate parts of the garden in 1978 were quite different (see Owen, J. 1991).

Rosebay willowherb, *Chamerion angustifolium*, supports a complex foodweb, including 2-spot ladybirds, *Adalia bipunctata*; the adults and larvae feeding on aphids

Foodwebs

Clumps of rosebay willowherb (*Chamerion angustifolium*) in direct summer sunlight have been found to support complex foodwebs including aphids on the leaves, adults and larvae of 2-spot ladybird, *Adalia bipunctata,* and larvae of a hoverfly, *Syrphus ribesii,* eating the aphids, diplazontine ichneumonids parasitizing the hoverfly larvae, and vespid wasps, bumblebees (*Bombus* species) and hoverflies, such as *S. ribesii*, feeding on aphid honeydew. Full development of this foodweb seems to depend on the plant being in a sunny situation.

In his study of the organization of a plant-arthropod association in simple and diverse habitats, Root (1973) found that the number of herbivore species was greater in the diverse habitats, and suggested that the complexity of a community arises from the juxtaposition of a large number of plant-based component communities. This is another way in which extreme patchiness, coupled with plant diversity, enriches the insect fauna of a garden. Insect diversity is high as a consequence of plant and structural diversity. At the same time, high plant diversity coupled with permanent succession make gardens extraordinarily bountiful yet markedly unstable environments.

Complexity and stability of the garden habitat

High animal diversity and narrow ecological niches are features of climax communities and were once generally assumed to be related to stability of a community (Odum 1971). MacArthur (1955), for instance, thought that the number of possible energy pathways through a foodweb were a good indication of its stability. However, more recent investigation of community structure, particularly by model-building, often using randomly assembled foodwebs, has tended to show that the stability of a community declines with the complexity of the inter-relationships (Begon, Harper & Townsend 1986).

What is clear is that the complex communities of stable environments, such as tropical rainforest, are particularly susceptible to man-made perturbation from outside, whereas the simpler communities of more variable environments in temperate regions are more robust and resist perturbation. The nature of the relationship between the stability and complexity of a community seems to vary in different environments and situations, but the consensus now is that there is a tendency for complex communities to be less stable than simpler ones.

The garden habitat accords with more recent analyses of the relationship between stability and complexity, in that it supports a complex yet unstable community. It is, however, a man-made and hence artificial situation, where both instability and complexity are inherent to the management of the habitat; permanent succession and high plant and structural diversity being intentionally produced by the gardener. The richness and variability of the animal component of the community derives from the way in which the plants are managed.

It is no accident that insects dominate the garden fauna, but rather a consequence of the attributes of insects and the structurally complex nature of the habitat. Southwood (1978) argued that small size, metamorphosis, seasonal activity, short generation times and rapid breeding, all characteristic of insects, contribute to habitat exploitation in differing ways by sympatric species and therefore to greater diversity.

Hutchinson & MacArthur (1959) suggested 50 years ago that small, diversified elements of an environmental mosaic permit a greater species diversity of small than of large animals. The garden habitat, unstable though it is, can thus support an astonishing diversity of insects. It is unusual in that it supports complex communities in what is an inherently unstable habitat.

THE SUBURBAN LAWN

The lawn is so different from the rest of a garden, and is usually so sharply delimited from it both physically and functionally, that it is perhaps best regarded as a separate ecosystem. Maintenance of a lawn by mowing is akin to grazing, the lawn-mower being the grazer, and the attributes of grasses that make for a good lawn are adaptations to being grazed.

Most good lawn grasses are tufted perennials, which spread by means of underground stems that produce a high proportion of leafy shoots. Grass leaves grow from basal meristems, and so quickly make good loss from cropping. Cutting the leaves prevents production of flowering stems, and stimulates both thickening of the tuft and development of buds into leafy shoots. The buds are at or below ground level, where they are protected by the bases of old leaf sheaths and the soil from grazing (mowing) and trampling. Repeated mowing thus improves grass quality as well as promoting establishment of thick, dense turf.

To understand grass and its reaction to mowing, it is necessary to acknowledge the interdependence, indeed co-evolution, of grasses and large grazing ungulates. Grazers keep grasslands at an early successional stage by preventing the establishment of other sorts of plants, a successional stage so well-defined that it is sometimes called a grazing climax. A lawnmower similarly maintains a lawn as a grazing climax. Some plants, however, can co-exist with lawnmowers,

A suburban lawn requires regular mowing to keep it looking attractive. The lawnmower is acting in the same way as a grazing animal to stimulate growth of the grass plants

because of low, compact growth and short flowering stems.

My garden lawn, while being a compact, resilient turf, is not a pure stand of grass, but is interspersed with such plants as leptinella (*Cotula squalida*), yarrow (*Achillea millefolium*), bird's-foot-trefoil (*Lotus corniculatus*), field wood-rush (*Luzula campestris*), white clover (*Trifolium repens*), hop trefoil (*Trifolium campestre*), dandelion (*Taraxacum* species), greater plantain (*Plantago major*), selfheal (*Prunella vulgaris*) and daisy (*Bellis perennis*).

Lawn maintenance

Conventional practice is to remove not only weeds, dead grass and fallen leaves from a lawn, but also to collect and remove the grass-cuttings. Although I remove fallen leaves and smothering weeds, such as dandelions, from my lawn, most so-called weeds are tolerated and cut grass is left *in situ*. Lawns are usually regarded as needing fertilizers, but this is to make good the losses caused by removal of grass-cuttings. Leaving the cuttings to decompose on the lawn obviates the need for fertilizers.

Falk (1976) studied a conventionally managed lawn in California. From the 110m^2 lawn, 63kg dry weight of living and dead vegetation was removed by raking and mowing in the course of a year. This contained the equivalent of 3,300g of nitrogen, 960g of phosphorus and 1,850g of potassium. Fertilizers and manure equivalent to 1,720g nitrogen, 430g phosphorus and 860g potassium had to be applied to maintain a healthy lawn.

Maintenance of the lawn also involved the expenditure of considerable quantities of energy in the form of manpower for mowing and other work., petrol for driving the mower, electricity to pump water for irrigation, energy used in manufacturing fertilizers and other chemicals, and the energy content of seeds sown. Keeping the lawn 'reasonably attractive' required 1,865 calories per square metre per year, more than twice the 715 cal/m^2/yr used to grow an edible grass, maize (Falk 1976).

Common lawn weeds: these species, all of which grew in the Leicester lawn, are so low-growing that they can co-exist with a lawnmower.

1 Leptinella, *Cotula squalida*

2 Yarrow, *Achillea millefolium*

3 Bird's-foot-trefoil, *Lotus corniculatus*

4 Field wood-rush, *Luzula campestris*

5 White clover, *Trifolium repens*

6 Hop trefoil, *Trifolium campestre*

7 Dandelion, *Taraxacum* species

8 Greater plantain, *Plantago major*

9 Selfheal, *Prunella vulgaris*

10 Daisy, *Bellis perennis*

A song thrush, *Turdus philomelos*, collecting earthworms from a lawn where it can feed freely because it would see a potential predator before it came dangerously close

The lawn as a food source

Many animals utilize a lawn as a source of food in one form or another, and so transfer and recycle energy and nutrients, although the Californian study identified man as by far the dominant herbivore and scavenger in an intensively managed lawn community (Falk 1976). Homoptera, moth caterpillars and other animals feed from grass blades, and predatory ants, spiders and beetles run across the surface of the ground.

Most of the activity, however, is within the litter layer and in the soil beneath, where leatherjackets (crane fly larvae) and moth caterpillars, such as those of large yellow underwing, *Noctua pronuba*, feed on grass roots. Earthworms, nematode worms, various sorts of fly larvae, springtails, bacteria and fungi contribute to the breakdown of dead leaves and roots in the litter layer and the soil.

Birds, such as starlings and song thrushes, do particularly well for food on lawns, where they mainly extract lepidoptera larvae, leatherjackets and earthworms. For birds, a lawn represents concentrated food resources, in a site where they can feed freely right up to the edge but out in the open, able to see potential danger. Falk (1976) found that food utilization per unit area of lawn by suburban birds, at 46 cal/m^2/yr, greatly exceeded food utilization by birds on natural grassland (1.01–2.33 cal/m^2/yr).

The flocking habit of such birds as starlings probably maximizes food utilization from a lawn, because they respond to the downward swoop of one to a good food source by following, and the cumulative alertness of a flock means that on average each individual can spend more time feeding rather than looking around to check for predators.

FOOD SUPPLY

As is to be expected in a habitat where primary production (production of plant material) is high, green leaves are utilized as food by a wide variety of abundant insects and other animals in the Leicester garden. The most important chewers of leaves are moth caterpillars and sawfly larvae, but leaves are also consumed by other herbivorous insects, such as butterfly larvae and flea beetles, and the abundant slugs and snails rasp away considerable quantities of green plant tissue. Heteropta, and particularly Homoptera, suck nutrients from leaves and green stems, and presumably have an appreciable impact on the growth of plants.

A variety of insects, including moth caterpillars, aphids, leatherjackets and wireworms (elaterid beetle larvae) feed in different ways on plant roots, as do abundant, though minute, Symphyla. Butterflies, moths, bees, wasps and such flies as Syrphidae, take nectar from flowers, and bees, *Meligethes* beetles and many hoverflies utilize pollen. The flowers themselves (as opposed to their

Red admiral butterfly, *Vanessa atalanta*, one of many insects that feed on rotting fruit

products) are consumed by moth caterpillars, some beetles and earwigs.

Fruit is eaten whole by birds, and nibbled or sucked by moth caterpillars, adult moths and butterflies, some wasps and flies of various sorts. Seeds are eaten by birds and also nibbled by beetles, moth caterpillars (e.g. the grubs in pea pods) and other insects. Once it is dead, vegetation becomes much more uniform in quality, and animals that feed on it (primary decomposers) are less specific in what they eat. Slugs, earthworms, woodlice, millipedes, Symphyla, larvae of some hoverflies and other flies, staphylinid beetles, springtails and a range of other small animals eat decomposing plant material.

Predation

Diverse and abundant predators attack the herbivores, and each other. As in all habitats, most of the species are predatory or parasitic, whereas most of the individuals are herbivores. Adult and larval ladybirds, lacewing larvae and many hoverfly larvae eat aphids; vespid wasps feed to their larvae any soft-bodied insect they can catch; solitary wasps catch a range of insects and spiders as food for their larvae. Carabid and staphylinid beetles and centipedes patrol the soil surface and climb plants after prey; predatory bugs hunt over flowers and vegetation, and carabid larvae and centipedes hunt within the soil. Ubiquitous ants attack anything they can overwhelm; spiders set snares all over the garden, or hunt over the soil and vegetation, and long-legged harvestmen hunt on the ground and high on plants. Birds search the vegetation, the ground and the air for prey – and so on. Nothing is safe from predation. Sometimes it is devious rather than direct, as in the case of ichneumonids, whose larvae develop in and on other insects and spiders, slowly killing them.

Many individual insects and other animals do, however, escape predation, but there is no accumulation of dead animals for, as in all habitats, scavengers abound. These are known as secondary or higher order decomposers, depending on whether they eat dead herbivores or dead predators. Many animals are omnivorous, and, although usually predatory, feed on dead animal material when it is available. Calliphorid fly larvae abound in vertebrate corpses, which are often buried by scavenging silphid beetles, and vespid wasps chew meat for their larvae from damaged corpses.

Woodlice, harvestmen, ants, calliphorid and sarcophagid flies, and a multitude of minute insects, mites and bacteria feed on dead animal material. Scavengers such as *Eristalis* hoverfly larvae ingest minutely fragmented animal matter in wet micro-habitats, such as rot-holes in trees or the bottom of the pond. Nothing is wasted but, whether alive or dead, is consumed by some organism. The ultimate consumption of dead and fragmented animal and plant material is rarely obvious, because it is achieved by microscopic bacteria and fungi in the soil, but is evident from the eventual disappearance of all dead animals.

Much of the decomposition and consumption of dead plant material in a garden is concentrated by the gardener in the compost heap, creating a nutrient-rich compost that can be utilized in the garden. Unmethodical, rather casual, compost heaps, such as that in my garden, tend to be the focus of much visible activity. Wasps and various sorts of flies buzz around it, and hoverflies, such as *Syritta pipiens*, and psychodid flies visit to lay their eggs. The top, drier layers of the heap abound with woodlice, and the moist lower regions are a writhing mass of brandling worms. These are the readily visible decomposers, but, on a smaller scale, the compost heap is alive with nematode worms, mites and springtails, and at a microscopic level, it teems with bacteria and fungi.

FOODWEBS

I have identified many food chains in my garden, and many more, and the way in which they interlock with others to form food webs, can be inferred. Rearing moth caterpillars, sawfly larvae and predatory hoverfly larvae found in the garden has led to recognition of plant-moth and plant-sawfly food links and plant-aphid-hoverfly food chains. Sixty-eight species of moth caterpillar were found on 115 species of plant, giving a total of 298 plant-moth links identified to species. In many instances, however, reared moth caterpillars of 11 species ended up producing not moths but parasitic ichneumonids, and in a number of these cases, the parasitized moth caterpillar could be identified to species, giving five accurate plant-moth-ichneumonid food chains (see Owen, J. 1991). In most cases, the host caterpillar could be identified to family, but no further; nevertheless, this still enabled recognition of a food chain to a fairly detailed level. More than 40% of the species of ichneumonid captured in the Malaise trap in 1972–1973, and more than 29% of the individuals, parasitize Lepidoptera, so the plant-moth-ichneumonid food chain must be acknowledged as one of the important nutrient and energy pathways in the garden.

Ten species of sawfly have been identified from larvae eating nine different species of plant (Chapter 7), establishing ten plant-sawfly feeding links; two species of ichneumonid have been reared from sawfly larvae on birch, giving two plant-sawfly-ichneumonid food chains. A total of 11.9% of the ichneumonid species trapped in the garden 1972–1973, and 6.4% of the individuals, parasitize tenthredinid sawflies.

A food chain identified in the Leicester garden, part of a larger foodweb.

Cabbage (*Brassica* cultivar) (1) is fed on by aphids (*Brevicoryne brassicae*) (2), the aphids are eaten by larvae of a hoverfly (*Syrphus ribesii*) (3), which may be caught and eaten by a spider, *Enoplognatha ovata* (4)

Aphids in food chains

Of the nine species of hoverfly bred from predatory larvae found in the garden (see Chapter 5), five were feeding on six known species of aphid, giving a total of nine accurate plant-aphid-hoverfly food chains. Eight species of ichneumonid were reared from hoverfly larvae, giving seven known plant-aphid-hoverfly-ichneumonid food chains, in five of which all the elements were identified to species. The plant-aphid-hoverfly-ichneumonid food chain is evidently another important nutrient and energy pathway in the garden, 9% of the species of ichneumonnid trapped in 1972–1973, and more than 27% of the individuals, being parasites of aphid predators (Owen, J. 1991).

Ichneumonids are by no means the only consumers of hoverflies with predatory larvae. For instance, a wasp, *Vespula vulgaris*, was seen to catch and eat *Syrphus ribesii*, a hoverfly, and one was also caught and eaten by a spider, *Enoplognatha ovata*, which had a scaffolding web on a cabbage leaf. This latter gives another accurate food chain: cabbage–*Brevicoryne brassicae* (aphid)–*Syrphus ribesii*–*Enoplognatha ovata*.

Other elements of the foodweb including *Syrphus ribesii* can be inferred because *Ectemnius* species (solitary wasps) catch and store hoverflies for their larvae to eat, and swifts catch hoverflies, including *S. ribesii,* in the air, both to eat and to feed to their young (Parmenter & Owen, D.F. 1954). Some hoverfly larvae are almost certainly eaten by marauding predators, such as carabid and staphylinid beetles, and hoverfly larvae, of course, are not the only predators of aphids.

Ladybirds were abundant in the garden, particularly in some years (see Chapter 8), and these and their larvae must make considerable inroads on the garden's aphids. Ladybirds, too, are eaten by swifts, and may fall foul of other predators, as when newly hatched 7-spot larvae were seen to fall into a hammock web of a common house spider, *Tegenaria domestica*. Other spiders, such as *Tibellus oblongus*, seen hunting among aphids and ladybird larvae on creeping thistle, *Cirsium arvense*, undoubtedly also eat the larvae.

Complex relationships

Ichneumonids consume many other sorts of garden insects, and they also parasitize spiders and spider eggs, when they are probably operating as fourth or higher order consumers. Indeed, it is evident by observation that both of the major food chains described, i.e. plant-moth-ichneumonid and plant-aphid-hoverfly-ichneumonid, are only parts of complex foodwebs. For instance, caterpillars of dot moth, *Melanchra persicariae*, and cabbage moth, *Mamestra brassicae*, have each been found feeding on 38 species of plant, and such aphids as *Aphis fabae* are also markedly polyphagous.

Spittlebugs, whitefly larvae, and many beetles and bugs, as well as sawfly larvae, also consume plants. Caterpillars are eaten by blue tits, blackbirds and other insectivorous birds, as well as being collected by social and solitary wasps to feed their larvae, and many probably also fall prey to predatory beetles. Scavenging *Volucella* hoverfly larvae eat the excrement of wasp larvae, and sometimes the larvae themselves.

On the other hand, such insects as hoverflies, wasps, bumblebees and ants lap aphid honeydew, and when winged reproductive ants rise into the air, they are eaten by house sparrows, starlings, house martins, swifts, black-headed gulls and other birds. Lacewing larvae, as well as predatory hoverfly larvae and ladybirds, eat aphids, which are often parasitized by minute chalcid wasps. As has been shown, many other animals consume hoverflies, and, conversely, many predators are catholic feeders, the main constraints being size of prey relative to their own size and ability to deal with prey.

Volatile trophic levels

So complex are the interconnections of garden foodwebs, that it is impossible to recognize trophic levels, other than to identify the instantaneous trophic level of a feeding consumer. For instance, when a swift eats flying aphids it is operating as a secondary consumer, but when it eats ladybirds it is a tertiary consumer. A further complication is provided by the important, although usually less evident, decomposer food chains. These cannot be separated from consumer foodwebs but form part of an intricate feeding nexus because, for instance, song thrushes and blackbirds consume earthworms, although the latter are decomposers.

Furthermore, after their death, even top predators become incorporated into decomposer food chains, and such animals as harvestmen eat living prey and corpses. It is evident that in the bountiful garden situation, food supply is good for a multitude of animals with different feeding habits. This is a major factor in promoting the richness of the garden fauna and the complexity of the feeding relationships.

A garden such as mine evidently accommodates many complicated foodwebs which may, indeed, interlock, forming one vast, intricate foodweb. The garden community is not stable, but is constantly changing in composition as species disappear and new ones arrive. It remains, however, a large and varied community, because contrived plant diversity, permanent succession and structural heterogeneity diversify feeding opportunities, generate many different food chains and contribute to a multitude of ecological niches. This bountiful, intricately structured, yet unstable habitat is created and maintained by the activities of gardening *per se*.

It might be asked why, when there are such abundant resources for herbivores in my garden, they do not regularly get out of hand as pests. Part of the answer is that the very complexity of feeding relationships in the garden means that herbivores are in large degree controlled by predators of one sort or another. Also, as explained in Chapter 2, mixed planting reduces damage from so-called pests and, together with maintenance of good ground cover, encourages predators. Finally, there is the question of what is a pest in a garden. For most people, it is whatever interferes adversely with their enjoyment of what they grow. However, pollen beetles can be shaken from cut flowers, the odd caterpillar or slug can be picked off a cabbage, and anyway, like most gardeners, I grow more than I can cut or eat. There are no pests in my garden, because everything in it is a source of interest and enjoyment.

13 : Gardens and conservation

An area of the Leicester garden (opposite) near to the large apple tree. Hollyhock (*Alcea rosea*), royal mallow (*Lavatera trimestris*), fennel (*Foeniculum vulgare*), tarragon (*Artemisia dracunculus*) and sage (*Salvia officinalis*), with flowers of Shasta daisy, *Leucanthemum* x *superbum* in the foreground

The rich fauna of my garden demands comparison with the fauna not only of other gardens but also with other types of habitat, including nature reserves. For many of the animal groups investigated, the garden fauna includes an astonishingly large proportion of the species recorded in the British Isles, and it is worth assessing what all these animals are doing in the garden. The diversity and abundance of the garden fauna is a consequence of the activities of gardening, and so it is highly probable that other gardens, perhaps gardens in general, are just as rich. That being the case, gardens would seem to be of considerable significance for conservation.

THE FAUNA OF THE LEICESTER GARDEN IN RELATION TO THAT OF THE BRITISH ISLES

The fauna of the British Isles has been extensively investigated over many years, and is not only well known but also well documented. It is a rare occurrence for a new species to be added to the British list, other than in three sorts of circumstance. Firstly, when a group is taxonomically difficult, individuals are relatively small, and few people have specialized in its study; secondly, consequent on new arrivals in the country, either naturally or as a result of man's activities; and thirdly, when taxonomists split what had hitherto been recognised as one species into two. Overall, however, the British fauna is so well known that it provides a firm background against which to view the study garden's fauna, and with which the latter can be compared.

A number of animal groups are known to be represented in the Leicester garden by an astonishingly large proportion of British species (Table 13.1), particularly bearing in mind that Leicestershire is relatively poor faunistically. These groups tend to be flying insects, which is only to be expected, partly because of their powers of dispersal and partly because trapping methods used in the garden are particularly suited to catch them. In addition, many of the well-represented groups are flower visitors, which are not only attracted to gardens, but fare particularly well there. There are, however, exceptions to these broad generalizations, harvestmen, for instance, being represented by nearly half of British species, although representation of spiders, at about a sixth of the British list, is probably more typical of non-insect invertebrates.

The groups known to be well-represented in the Leicester garden are among those that have been intensively investigated. I can say with confidence that they

Orange ladybird, *Halyzia sedecimguttata* (above), one of the 13 ladybird species recorded in the Leicester garden, a figure that amounts to more than half the British species

Micromus variegatus (above right), a species of brown lacewing. It was an occasional visitor to the Leicester garden where almost 42% of British species of lacewings have been recorded

are well-represented because they have been well-listed (studied in detail). Species listed for other groups cannot be used in the same way; they have not been well-listed and so representation in the garden of the British fauna cannot be assessed. For instance, many individual earwigs have been collected and identified, but all are the same species and I am confident that there is only one species of earwig in my garden; no effort has been made, however, to collect false-scorpions, and the garden list of one species means only that one was noticed and identified, not that this is the only species in the garden. Moreover, many groups, such as mites (Acari) and the majority of families of flies (Diptera) have not been investigated at all.

Among insects, Tenthredinidae, a large family of sawflies, is represented in the study garden by 17.4%, nearly a fifth, of the British species; ladybirds (Coccinellidae) by 54.2%, more than half; and Ichneumonidae, and the moth family Geometridae by 26.3% and 34.0%, respectively. Lacewings and their allies

Table 13.1. Numbers of species in the Leicester garden, and in the British Isles, of groups well-listed in the garden

	Species in garden	Species in British Isles	Garden as % of British Isles
Native flowering plants	170	c.1,500	11.3
Centipedes	7	46	15.2
Harvestmen	11	23	47.8
Grasshoppers & crickets	4	28	14.3
Earwigs	1	4	25.0
Lacewings & allies	23	55	41.8
Butterflies	23	62	37.1
Macro-moths	282	881	32.0
Hoverflies	94	256	36.7
Sawflies	91	481	18.9
Ichneumons	533	2,028	26.3
Ants	2	36	5.6
Wasps	62	297	20.8
Bees	59	252	23.4
Ground beetles (Carabidae)	33	342	9.6
Ladybirds (Coccinellidae in part)	13	24	54.2

Common carpet, *Epirrhoe alternata* (above left), one of the 34% of the British list of Geometridae that occur in the Leicester garden

Tree wasp, *Dolichovespula sylvestris* (above right). All but one of the seven British species of social wasp were found in the Leicester garden

make up 41.8%, well over a third of the British list, and butterflies, moths of the large family Noctuidae, and hoverflies (Syrphidae) 37.1%, 40.3% and 36.7%, respectively, a third or more.

Solitary bees represent 19.9%, almost a fifth of British species, but social bees (Apidae) 56.0%, more than half. Similarly with wasps, 19.0%, or almost a fifth of British solitary species have been recorded in the garden, but all but one, i.e. six out of seven of the social species (Vespidae) have been. Evidently, butterflies, moths, hoverflies, ladybirds and social bees and wasps are particularly well represented in the garden, and general observation would suggest that this is true of gardens in general.

Extrapolated estimates

Animal groups that have been well-listed in the garden are represented there by between 5.6% and 54.2% of species known for the British Isles, native plants by 11.3% (Table 13.1). Extrapolating these proportions for animals to the total terrestrial fauna of the British Isles, it is possible to estimate within upper and lower limits the total number of animal species in the garden. According to the Biological Records Centre, Institute of Terrestrial Ecology, there are about 30,000 species of terrestrial animal in the British Isles, and on this basis there are between 1,680 and 13,050 in my garden. I know, despite the selective nature of my investigation, that there are more than the lower figure, and so could anticipate vastly increasing the species list by looking at more groups. Insect groups are represented in the garden by up to 54.2% of species known for the British Isles, and as the total British list of insects is about 22,500, there may be more than 10,000 species of insect in the garden.

The record of the flora and fauna of the 156.8ha Monks Wood Nature Reserve in Huntingdonshire, produced ten years after the opening of Monks Wood Experimental Station (Steele & Welch 1973), lists by group the species recorded there. Numbers of species of woodlice, lacewings and macro-moths were more or less the same as were recorded in the garden, numbers of Ichneumonidae and of solitary and social bees and wasps were much smaller than in the garden,

A male silver-washed fritillary, *Argynnis paphia*, a rare visitor to the Leicester garden. It was seen only once, in the hot, dry summer of 1976, when many nectar-feeding insects strayed far from their usual haunts in search of food

and numbers of harvestmen, spiders, tenthredinid sawflies, coccinellid and other beetles, hoverflies, Geometridae, Noctuidae and other macro-moths were greater, often twice as great.

Many of these larger species lists at Monks Wood are to be expected in view of the much larger area involved, and a group such as Coleoptera is bound to be better represented where there is woodland. The smaller species lists of Ichneumonidae and bees and wasps are more interesting. In the case of Ichneumonidae, the discrepancy clearly illustrates the unpopularity of the group and the difficulties inherent to its study. With bees and wasps, however, the explanation is almost certainly that a garden is a far more favourable habitat for them, particularly in view of the abundance of flowers there.

Relatively rare, or unexpected, species have turned up in the garden in most groups, such as silver-washed fritillary among butterflies (Chapter 3), a pyralid moth, *Sceliodes laisalis* (Chapter 4), and hoverflies *Ferdinandea ruficornis* and *Sericomyia silentis* (Chapter 5). It is, however, in the little-investigated groups, such as small parasitic wasps, that the real rarities have been recorded. Five species of Serphidae and 15 species of Ichneumonidae trapped or reared in my garden are believed to be additional to the British list, and four further ichneumonids are hitherto undescribed species (Chapter 9). This clearly illustrates the advantages, in terms of discoveries, of looking at a relatively little-known group.

Studies of comparable areas

The size of the species list for the garden, and the inclusion in it of unexpected species, demonstrates how much is to be gained by looking in detail, over a long period, at a small area of land. Gilbert White, the doyen of British naturalists, realized this as long ago as 1768, when he wrote to Thomas Pennant 'all nature is so full, that that district produces the greatest variety which is the most examined' (White 1947). The few published long-term studies comparable to that of the garden are of much greater areas, such as the investigation by Francis Evans and collaborators of the natural history of an abandoned field in Michigan,

USA (Evans 1975), and the ongoing study of Wytham Wood, near Oxford (see Elton 1966). They have certainly borne out Gilbert White's dictum, as indeed did Frank Lutz when he investigated the insect fauna of his suburban New York garden (Lutz 1941).

The fauna of the majority of nature reserves is far less well known than that of my garden, usually for lack of specialists with time to collect and identify species. Indeed, evaluations of areas threatened by development or proposed as nature reserves have to be conducted on far less information. Complete faunal lists for nature reserves or equivalent areas are hard to come by, even if they exist, and, indeed, Sites of Special Scientific Interest (SSSIs) have largely been designated on the basis of their floras.

The fauna of Buckingham Palace garden was investigated as far as possible in the 1960s (Bradley & Mere 1966, McClintock *et al.* 1964) and surveyed in more detail in the 1990s (Plant 1999, 2001). In the Palace garden, birds were represented in the 1990s by 58 species, but there was only one species of amphibian, although there were six of mammals. One species of false-scorpion was recorded along with four woodlice, six earthworms, six harvestmen, eight slugs, eight snails, 16 myriapods and 112 species of spider. Among insects, Lepidoptera were intensively investigated, yielding 640 species over the 40-year period, but the lists for other insect groups are less complete.

In terms of number of other insect species, Buckingham Palace garden was recorded as having one Thysanura, 42 springtails, two Ephemeroptera, two Odonata, two Orthoptera, one Dictyoptera, one Dermaptera, three Psocoptera, 11 Thysanoptera, 176 Hemiptera, 22 Neuroptera, 20 Trichoptera, 287 Coleoptera, 142 Hymenoptera (including 16 ichneumonids), and 827 Diptera (including 39 syrphids). The large number of springtails, Thysanoptera and Diptera, compared with numbers for the Leicester garden, show that specialists in these groups participated. On the other hand, the numbers of ichneumonids, in particular, in my garden list are high because a specialist in the group looked at trap samples. This illustrates the bias introduced into species lists by the interests and expertise of those involved in the investigation.

Often an area has been particularly well surveyed, using several collecting methods, for one particular taxonomic group, as at Malham Tarn Estate Nature Reserve in North Yorkshire, which is one of the best-surveyed sites in the country for Diptera. This survey was initiated in 1954–1958, and by the end of 1983, the Diptera list included more than 1,000 species (largely due to the diligence of Henry Disney, who also participated in the investigation at Buckingham Palace) and was believed still to be incomplete (Disney *et al.* 1982, Disney 1986). This ongoing survey has involved a range of collecting methods, the expertise of many specialists, and a lot of time and effort.

The much smaller number of Diptera species recorded in my garden (145) is partly a consequence of a much smaller area, but also results from the incompleteness of the garden survey. One family of Diptera (Syrphidae) has been looked at in detail in the garden, others have been partially investigated, but many are totally unknown. That the difference stems from a difference in effort is borne out by the sizes of the species lists of Syrphidae (a group on which a lot of effort has been expended in the garden) for the two sites – 94 for the garden, 79 for the far greater area of the Malham Tarn Estate Nature Reserve.

Animal activities: The author's photographs of some of the wide range of animal behaviour she observed in her garden

1 A mating pair of green-veined whites, *Pieris napae*

2 Pupae of a parasitic braconid wasp emerging from a stricken caterpillar of poplar hawk moth, *Laothoe populi*

3 A blackbird, *Turdus merula*, panting to cool himself on a hot day

4 Seven-spot ladybird, *Coccinella septempunctata* , laying batches of eggs on an iris leaf

5 Aphids (*Aphis sambuci*) feeding on elder, *Sambucus nigra*

6 White-tailed bumblebee, *Bombus lucorum*, feeding at a *Tagetes patula* flower

WHAT ARE SO MANY ANIMAL SPECIES DOING IN THE GARDEN?

A brief survey of the garden fauna and its activities would perhaps give the impression that relatively few of the animals are residents and that most are temporary visitors for food. This might be true of flying insects and vertebrates, but it cannot be true for terrestrial invertebrates such as snails, slugs, earthworms, woodlice, centipedes, millipedes, harvestmen and spiders. These are evidently feeding and breeding in the garden, and are undoubted residents.

As for the vertebrates, all three amphibians have bred in the pond and evidently feed in the garden, 11 species of bird have definitely bred and another three possibly so, and among mammals, house mice have bred in an outhouse and in the garage. All these must be classed as garden residents. Another 18 species of bird, and probably four of mammals, regularly feed in or over the garden, although I have no record of their breeding there.

Herbivorous insects

The situation is, however, more difficult to assess for flying insects, especially as so many records come from Malaise trap captures rather than observation. The picture is clearer for herbivores than it is for carnivores, because they are, on the whole, more sedentary and easier to observe. Four species of butterfly definitely bred in the garden (and the meadow brown would have done so had the lawn not been cut); by definition, their caterpillars feed in the garden, as did the adults of these and the other 18 recorded species (Chapter 3). At least four of the garden butterflies must thus be classed as residents.

The feeding situation is harder to assess with moths, for some may have been drawn in to the mercury-vapour light without ever settling at flowers. However, at least 20% of moths recorded in the garden have bred there and can thus be classed as residents.

Like butterflies and moths, adult sawflies are mobile, and may be caught in the Malaise trap on temporary visits to the garden, but at least ten species are known to have bred (Chapter 7), so nearly 11% of the recorded sawflies are residents. Similarly, 12% of Homoptera have been recorded feeding and breeding on garden plants (Chapter 7), and this is undoubtedly a minimum figure, as many small Auchenorrhyncha are high on trees out of view. The general impression is that at least 20% of those insects recorded in the garden that are herbivorous at some stage of their lives are residents.

Carnivorous insects

A large proportion, if not all, of the carnivorous adult insects trapped in the garden probably feed there, predators tending to take food as and when it is available. Such insects as hoverflies with predatory larvae are flower-feeders as adults, when they certainly feed in the garden. Residence, as with herbivores, can be recognized on the basis of breeding, but there are relatively fewer records of breeding carnivores, largely because their presence and activities are difficult to monitor. Probably the majority of the 93 species of Heteroptera trapped (not all carnivores, however), and of the 32 species of carabid beetle, bred in the garden, because they are not particularly prone to flight.

Of the 57 species of hoverfly with predatory larvae recorded in the garden, nine (16%) have been found breeding, and about 27% of hoverflies in general are assumed to do so (Chapter 5). Twenty species of Ichneumonidae, which as parasitoids are specialized predators, have been bred from parasitized larvae of moths, sawflies and hoverflies found in the garden, and ichneumonid larvae have been found on spiders (see Owen, J., Townes & Townes 1981). This is only 4% of the number of ichneumonid species recorded in the garden, but the proportion that breed is certainly larger, because I have reared in captivity only a few types of insects.

The overall figure for carnivorous insects breeding in the garden is probably at least 20% and possibly higher. That it is erroneous to assume that insects trapped rarely in the garden and rare in county or country are chance visitors, is demonstrated by *Hyposoter tricolor*, an ichneumonid. When two individuals were caught in the Malaise trap 1972–74 it was a new record for Britain, but subsequently two were bred from geometrid larvae feeding in the garden (Owen, J., Townes & Townes 1981).

The answer to the question 'What are so many animals doing in the garden?' would thus appear to be that a large proportion, if not the majority, are feeding, and from 20% to all, depending on taxonomic group and mobility, are breeding and hence resident. The many species of flying insect caught only once in the Malaise trap (about 25% in total, ranging from 13.8% in hoverflies, through 17.9% in Sphecidae, solitary wasps, to 29.3% in Heteroptera) are regarded as vagrants, chance visitors that happen to get trapped in the course of their wanderings.

GARDENING FOR CONSERVATION

I manage my garden in such a way as to enhance it as a habitat. Every effort is made to grow plants with flowers that are attractive to insects, or fruits attractive to birds. Furthermore, flowering is prolonged, where possible, by dead-heading. High plant diversity is maintained, and cultivation of plants of different heights and growth form contribute to structural heterogeneity. Pruning and clearing are kept to a minimum, and delayed until winter whenever possible. Good ground cover is maintained, both by cultivated plants, and failing that, by weeds. Above all, and of paramount importance in a wildlife garden, no pesticides or other noxious chemicals are used, other than the occasional application of slug pellets. The result is a lush garden, alive with animals, although one that still furnishes produce including flowers, herbs, salads, vegetables and fruit, because mixed planting keeps so-called pests down to manageable proportions.

As has been stressed earlier, the study garden with its rich wildlife is not a wilderness, but a neat, productive suburban garden. There is no reason to believe that it differs fundamentally from other gardens, and every reason to suppose that gardens in general are good habitats, rich in animal life. This is particularly so with the prevalence, since about 1975, of so-called wildlife gardening. A 1980s book on creating a wildlife garden (Gibbons & Gibbons 1988) describes such gardens as 'oases of diverse, semi-natural, and controlled environments, where nature can find some sanctuary from pesticides and habitat loss.'

Wildlife gardens are seen as one of the answers to a dilemma that increasingly perplexes Western man – how, in this modern civilization, to preserve the plants, animals and habitats that continue to be destroyed by the process of civilization? Transport, housing, amenities for all, and generation of the wealth that pays for emancipation from poverty, disease and ignorance, have eroded the natural world and threaten its very existence. Towns have spread, to the detriment of the countryside, and modern agriculture has made many wild plants and animals rare. The same processes of civilization have allowed for the development of new feelings, sensibilities and values about the world that is being destroyed.

Conservation *versus* progress

Nineteenth- and twentieth-century legislation on conservation issues has its roots in the desire of amateur natural historians of the eighteenth century to protect wild plants and animals that for centuries had been despised and systematically destroyed. An attempt to resolve the dilemma of how to retain elements of the past, while living in the present, has led to a vigorous conservation movement and the establishment of nature reserves and other conservation areas, although these have been described as 'fantasies which enshrine the values by which society as a whole cannot afford to live' (Thomas 1983). Nevertheless, nature reserves of one sort or another, that conserve animals and plants, species and communities, and habitats, satisfy intellectual curiosity and give aesthetic pleasure, as well as providing interest and recreation.

The most-cited criteria for evaluation of a wildlife conservation area are diversity of habitats and/or species, naturalness, rarity of habitats and/or species, and area (Usher 1986). Gardens would appear to meet the criterion of diversity of species, some of which may be rare, and, collectively, the 8,000km^2 of gardens in Britain (Loram *et.al.* 2007) meet the criterion of area. Gardens are by definition unnatural, and the medley of micro-habitats that they contain are created. On diversity of species alone, however, a garden such as mine would rate highly for conservation. Describing the total garden area in terms of satisfying the area criterion is no doubt misleading, because the garden area is fragmented into millions of individual gardens, each under different management. This may, however, be no bad thing.

Large *versus* small

The value given to a large area for conservation purposes stems from island biogeography theory, which states that larger areas hold more species (MacArthur & Wilson 1967). Nature reserves have been likened to habitat islands in an inhospitable sea of environment, and so island biogeography theory is thought to apply to them. However, a number of small nature reserves may have more species than a single large one of equivalent area (Higgs & Usher 1980), and this almost certainly arises from habitat differences between reserves (Gilpin & Diamond 1980). Indeed, it is almost certainly increased habitat diversity that makes a larger island richer in species than a small one. It seems that it is quality and diversity of habitat that make for a good nature reserve, because it can serve as a refuge for more species, and that the island biogeography theory should not be applied uncritically to nature reserves (Simberloff 1986).

The vast number of small gardens in Britain, no two exactly alike, are (and with the popularity of wildlife gardening, are increasingly becoming) a complex of small nature reserves. The collective species list of these gardens must be far larger than would exist on the land under one management policy, or perhaps even naturally The 1,700 reserves held in 1987 by the 46 county naturalists' trusts affiliated to the Royal Society for Nature Conservation collectively accommodated 17,000ha of woodland and 7,600ha of meadows and pastures (Perring 1987). This area is even larger now, and is a major holding in terms of conservation, and valuable because the reserves are natural areas, but the area is small compared with the garden area.

The reserves held by the county wildlife trusts are in many cases last-ditch attempts to conserve a fragment of a habitat that is under threat locally or nationally. They are a bastion against the destruction of communities and habitats by development. But the garden habitat as such is in no sense threatened, despite 'garden grabbing' to build more houses in cities. On the contrary, the encroachment of towns and cities on the countryside means the creation of more gardens, and with the movement towards gardening for wildlife, and widespread anxieties about use of pesticides, gardens are becoming more and more hospitable as habitats and refuges. It is perhaps even more relevant now than it was in 1975 (Owen, J. & Owen, D.F. 1975) to ask 'Are suburban gardens England's most important nature reserve?'

14 : The garden fauna: changes and trends

Small tortoiseshell butterflies, *Aglais urticae*, feeding on nectar of Michaelmas daisy (opposite)

In my garden as, almost certainly, in many others, the very process of gardening and caring for my environment has created a rich habitat that abounds in plants and animals of many different kinds. The total numbers of animals and plants in different taxonomic groups recorded in the study garden in the 30 years 1972–2001 are shown in Tab. 14.1. This list cannot be considered as exhaustive because, for many practical reasons, only selected groups have been identified; the soil fauna, for instance, is unknown, apart from the larger invertebrates. The total of 2,673 species is a minimum figure anyway, because some beetles only identified to genus probably include more than one species. The 2,673 species comprise 474 plants, 1,997 insects, 138 other invertebrates, and 64 vertebrates.

Table 14.1. Numbers of species of plants and animals recorded in the Leicester garden, 1972–2001.

Flowering plants (including grasses)	436
Flowerless plants	38
Flatworms (Turbellaria)	3
Snails and slugs (Gastropoda)	17
Earthworms (Oligochaeta)	5
Leeches (Hirudinea)	1
Woodlice (Isopoda)	8
Centipedes (Chilopoda)	7
Millipedes (Diplopoda)	5
False-scorpions (Pseudoscorpiones)	1
Harvestmen (Opiliones)	11
Spiders (Araneae)	80
Dragonflies (Odonata)	7
Grasshoppers and crickets (Orthoptera)	4
Earwigs (Dermaptera)	1
Psocids (Psocoptera)	18
Bugs (Hemiptera)	183
Lacewings and allies (Neuroptera)	23
Butterflies (Lepidoptera)	23
Macro-moths (Lepidoptera)	282
Micro-moths (Lepidoptera)	93
Caddisflies (Trichoptera)	5
Hoverflies (Syrphidae-Diptera)	94
Other flies (Diptera)	51
Sawflies (Symphyta)	91
Ichneumons (Ichneumonidae-Hymenoptera)	533
Other parasitic wasps (Serphidae & Heloridae)	24
Ants (Formicidae-Hymenoptera)	2
Wasps (Hymenoptera)	62
Bees (Hymenoptera)	59
Beetles (Coleoptera)	442+
Amphibians	3
Birds	54
Mammals	7
GRAND TOTAL	**2,673+**

Changes and trends

Five groups of insect were trapped and identified for the full thirty-year period: butterflies, macro-moths, hoverflies, social Hymenoptera and ladybirds. Interestingly, although very different sorts of insects, the 30-year changes in all these groups tell much the same story; that they were not so abundant in the second half of the trapping period, 1987–2001, as they had been in 1972–1986 (Figs. 3.1, 4.1, 5.1, 6.1, 6.2, 6.3 and 8.4). This bears out the findings of the UK-based Centre for Ecology and Hydrology, which in 2004 published figures showing an alarming decline in butterflies, birds and plants (Thomas *et al.* 2004).

Within this general pattern of decline, butterflies, moths, hoverflies, social Hymenoptera and ladybirds differed in when they had a good year. Butterflies and moths did particularly well in the period 1973–1976; this was never repeated, though butterflies had a good year in 1983, and moths in 1985 and 1996. The best hoverfly years, by far, were 1975 and 1978, although they also did well in 1982, 1985 and 1997. Ladybirds were abundant in 1975, 1976 and 1977 (1976 being 'the ladybird year') and nothing approaching these numbers was ever trapped again.

Bumblebees, *Bombus* species, were trapped and identified from 1972 to 2001, but with a four-year gap (1978–81) during that time. Numbers of *Bombus* were high 1972–1977, and were still high in 1982, but declined thereafter. In general terms, the period 1972–1986 was good, but later years, 1987–2001, were relatively poor apart from a good catch in 1993.

Numbers of two social wasp species, *Vespula vulgaris* and *V. germanica,* were high in 1972–1986, 1979 being the peak year; thereafter, numbers of *V. vulgaris* remained quite good, but after 1979, numbers of *V. germanica* dropped off and never recovered.

In other words, the social bee and wasp figures accord with those for butterflies, moths, hoverflies and ladybirds, suggesting that garden insects were nothing like as numerous after 1986. Different groups had good years during this overall slump, but the actual year of this differed for the five groups, implying that garden insects were in decline, although particular weather conditions boosted figures in some years.

It is not clear why particular weather conditions should have led to good years. 1993, when *Bombus* numbers were relatively high, had a cool, wet summer with only average amounts of sunshine, although it followed a warm, wet summer in 1992, with average amounts of sunshine, when queen production may have been good.

The good years for *Vespula vulgaris* (1996) and *V. germanica* (1998) probably have more to do with the cyclical effect of their population dynamics, although 1996, when moths also had a good year, was a hot, dry, bright summer, which evidently also suited wasps. 1997, when hoverflies did well, was a warm, wet, dull summer, during which, however, there was a mass immigration of *Episyrphus balteatus*, which made up more than 45% of the hoverfly catch.

Solitary bees and wasps were sorted from Malaise trap samples and identified from 1975 to 2001, covering all but the first three years of the 30-year period (Figs. 6.4 and 6.5). Numbers of solitary bees were high from 1975 to 1978, with 1978 the peak year, despite a cool, damp, dull summer. Their numbers generally remained reasonably good, despite some poor years, although the totals of 1975–1978 were never matched. Solitary wasps, on the other hand, had more good years in the 1990s, with peaks in 1993 and 1997, despite these not being particularly good summers, both being wet. The overall impression is that numbers of solitary wasps improved in the later part of the 30 years, thus differing from those for most other insect groups.

Lacewings were identified from 1978 onwards, although the samples for 1997 and 1998 were lost. The peak year was 1987, a cool, damp, dull summer, and numbers usually fell far short of this peak. Numbers declined dramatically after 1994, when the large birch tree was pollarded and subsequently felled, and this is unsurprising as the lacewing catch was made up largely of *Wesmaelius subnebulosus*, a species associated with birch.

Beetles, other than ladybirds, were identified from Malaise trap samples from 1981 to 2001 (Fig. 8.2). Numbers were much higher from 1993 onwards, particularly in 1995, 1999 and 2001, so beetles followed a pattern unlike most

Much nearby farmland has been built on since the 30-year study began, destroying many habitats. However, over time, the gardens of the new houses should mature and attract insects and other animals

garden insects, in that they evidently became much more abundant towards the end of the 30-year period of the garden study. Beetles caught by pitfall trapping were only identified from 1979 to 1990, but during this 12-year period, they tended to become more common, although numbers were down in 1989 and, particularly, 1990 (Fig. 8.1).

Sawflies, too, were identified from Malaise trap samples from 1981 onwards, although the 1993 sample was lost. Their numbers were high usually until a peak year in 1992; numbers were generally lower thereafter, although the period from 1997 to 2000 was good, suggesting that sawfly numbers were at least being maintained, and certainly not declining (Fig. 7.1).

Psocoptera were identified from 1983 onwards, annual numbers fluctuating widely, with many poor years, and peaks in 1984, 1998 and 1999 (Fig. 7.2). This group, also, did not show the downward trend most groups showed in the later years of the study, although 2001 was worse than the previous four years.

Heteroptera, identified from 1984 to 2001, tended to show a downward trend from 1994 on, the peaks being widely spaced in 1984, 1988 and 1994 (Fig. 7.3). 1984 was a hot, dry, bright summer, and 1994 was hot and bright, although damp, but 1988 was cool, damp and dull.

Explaining the trends

Thus, the overall impression from Malaise trapping is that insect numbers declined during the 30-year period 1972–2001. However, four groups (beetles, sawflies, solitary wasps and Psocoptera) did not accord with this general pattern, and beetles and solitary wasps actually became more abundant. Perhaps the slight but sustained warming of the weather suited them, although I would have expected this to apply to other groups, such as butterflies and hoverflies.

It is much more likely that the explanation for the general downward trend is that habitats at large, beyond the garden, have become unsuitable for many insect groups. For instance, many hectares of nearby farmland have been developed and built on since the early years of the study. Development in

In the wider countryside there has been a trend towards the intensification of agriculture. This has led to the loss of hedgerows and headlands that supported much wildlife.

general, and changes in the remaining countryside, as farming becomes more mechanized and chemical-dependent, have led to reductions in insect life.

That leaves the question why this should have had no adverse effect on psocids, sawflies, beetles and solitary wasps (the converse in the case of the last two groups). Solitary wasps, being predatory, can exploit what prey is available, within certain limits, and may not have been affected by the decline in some of that prey. Beetles are collectively so diverse a group, that they may have benefited from the decline of other groups leaving niches unexploited. Psocoptera, grazing indiscriminately on organic debris on vegetation, will have suffered no declines in food and habitat. Sawflies have larval stages that are ecologically very like caterpillars, but the adults are very different from butterflies and moths, suggesting that the decline in the latter two groups stems from adverse effects on the adult stages.

Overall, however, the picture is not good, and insect life in the garden shows no sign of returning to the boom years of the 1970s. Prolonged investigation of the garden's fauna has, however, shown that the picture can change over 30 years, and it is as well to remain optimistic, particularly with increasing temperatures and the changes proposed in management of the countryside, particularly with regard to wildlife and habitat maintenance.

My overriding impression of the garden fauna, despite disturbing recent trends, is one of great richness and interest, so that Gilbert White's remark still holds true, that 'all nature is so full, that that district produces the greatest variety which is the most examined' (White 1947). Certainly, there is always something going on in a garden to attract the attention of the gardener and naturalist, which is why I have always found gardening such an enjoyable, rewarding and absorbing activity.

References

Alford, D.V. (1973) *Bumblebee distribution maps scheme: guide to the British species.* Reprinted from *Entomologist's Gazette* 21, 22 and 23. Bee Research Association, London.

Archer, M.E. (1985) Population dynamics of the social wasps *Vespula vulgaris* and *Vespula germanica* in England. *Journal of Animal Ecology* 54: 473–485.

Archer, M.E. (1990a) The solitary aculeate wasps and bees (Hymenoptera:Aculeata) of an English suburban garden. *Entomologist's Gazette* 41: 129–142.

Archer, M.E. (1990b) The aculeate solitary wasps and bees (Hymenoptera: Aculeata) of Leicestershire. *Transactions of the Leicester Literary and Philosophical Society* 84: 9–25.

Archer, M.E. (2001) Changes in abundance of *Vespula germanica* and *V. vulgaris* in England. *Ecological Entomology* 26: 1–7.

Arvill, R. (1973) *Man and Environment.* Penguin, Middlesex.

Azmeh, S., Owen, J., Sorensen, K., Grewcock, D. & Gilbert, F. (1998) Mimicry profiles are affected by human-induced habitat changes. *Proceedings of the Royal Society of London B* 265: 2285–2290.

Baines, C. (1984) *Chris Baines' Wildlife Garden Notebook.* The Oxford Illustrated Press, Somerset.

Baines, C, (1985) *How to Make a Wildlife Garden.* Elm Tree Books, London.

Barnard, P.C. (1978) A check-list of the British Neuroptera with taxonomic notes. *Entomologist's Gazette* 29: 110–124.

Begon, M., Harper, J.L. & Townsend, C.R. (1986) *Ecology.* Blackwell Scientific Publications, Oxford.

Betts, C. (1986) Symphyta. In: Betts, C. (ed.) *The Hymenopterist's Handbook*, pp. 21–23. The Amateur Entomologists' Society, London.

Birch, M.C. (1970) Structure and function of the pheromone-producing brush-organs in males of *Phlogophora meticulosa* (L.) Lepidoptera: Noctuidae) *Transactions of the Royal Entomological Society of London* 122: 277–292.

Bradley, J.D. & Mere, R.M. (1966) Natural history of the garden of Buckingham Palace. Further records and observations 1964–65. Lepidoptera. *Proceedings of the South London Entomological and Natural History Society 1966*: 15–17.

Brickell, C. (ed.) (1989) *Gardeners' Encyclopedia of Plants and Flowers.* Dorling Kindersley, London.

Bristowe, W.S. (1971) *The World of Spiders.* Collins, London.

Chandler, P. (ed.) (1998) *Checklists of Insects of the British Isles (New Series). Part 1: Diptera.* Royal Entomological Society of London.

Chinery, M. (1976) *A Field Guide to the Insects of Britain and Northern Europe.* Collins, London.

Chinery, M. (1977) *The Natural History of the Garden.* Collins, London.

Chinery, M. (1986) *The Living Garden.* Dorling Kindersley, London.

Churcher, P.B. & Lawton, J.H. (1987) Predation by domestic cats in an English village. *Journal of Zoology, London* 212: 439–455.

Clark, J.T. (1979) Psocoptera activity as monitored by a Malaise trap. *Entomologist's Monthly Magazine* 114: 94–95.

Cloudesley-Thomson, J.L. (1968) *Spiders, Scorpions, Centipedes and Mites.* Pergamon Press, Oxford.

Coe, R.L. (1953) *Syrphidae. Handbooks for the Identification of British Insects, X (1).* Royal Entomological Society of London, London.
Colebourn, P. & Gibbons, B. (1987) *Britain's Natural Heritage: Reading our Countryside's Past.* Blandford Press, London.
Colyer, C.N. & Hammond, C.O. (1968) *Flies of the British Isles.* Frederick Warne & Co. Ltd, London.
Conn, D.L.T. (1972a) The genetics of the bee-like patterns of *Merodon equestris. Heredity* 28: 379–386.
Conn, D.L.T. (1972b) The genetics of mimetic colour polymorphism in the large narcissus bulb fly, *Merodon equestris* Fab. (Diptera: Syrphidae). *Philosophical Transactions of the Royal Society of London* 264: 353–402.
Conrad, K.F., Woiwod, I.P., Parsons, M. & Fox, R. (2004) Long-term population trends in widespread British moths. *Journal of Insect Conservation* 8: 119–136.
Conrad, K.F., Warren, M.S., Fox, R., Parsons, M.S. & Woiwod, I.P., (2006) Rapid declines of common widespread British moths provide evidence of an insect biodiversity crisis. *Biological Conservation* 132 (2006): 279–291.
Coombs, C.F.B., Isaacson, A.J., Murton, R.K., Thearle, R.J.P. & Westwood, N.J. (1981) Collared doves (*Streptopelia decaocto*) in urban habitats. *Journal of Applied Ecology* 18: 41–62.
Crocker, J. & Daws, J. (1996) *Spiders of Leicestershire and Rutland.* Loughborough Naturalists' Club and Kairos Press, Leicestershire.
Crocker, J. & Daws, J. (2001) *Spiders of Leicestershire and Rutland: Millenium Atlas.* Loughborough Naturalists' Club and Kairos Press, Leicestershire.
Crowson, R.A. (1981) *The Biology of the Coleoptera.* Academic Press, London.
Danks, H.V. (1971) Biology of some stem-nesting aculeate Hymenoptera. *Transactions of the Royal Entomological Society of London* 122: 323–399.
Davis, B.N.K. (1978) Urbanization and the diversity of insects. *Symposia of the Royal Entomological Society of London* 9: 126–138.
Disney, R.H.L. (1986) Assessments using invertebrates: posing the problem. In: Usher, M.B. (ed.) *Wildlife Conservation Evaluation*, pp. 271–293. Chapman & Hall, London.
Disney, R.H.L., Erzinclioglu, Y.Z., Henshaw, D.J. de C., Howse, D., Unwin, D.M., Withers, P. & Woods, A. (1982) Collecting methods and the adequacy of attempted fauna surveys, with reference to the Diptera. *Field Studies* 5: 607–621.
Dixon, A.F.G. (1985) *Aphid Ecology*. Blackie, Glasgow.
Dony, J.G., Jury, S.L. & Perring, F.H. (1986) *English Names of Wild Flowers.* Botanical Society of the British Isles, Reading.
Eason, E.H. (1964) *Centipedes of the British Isles.* Frederick Warne & Co. Ltd., London.
Elton, C.S. (1966) *The Pattern of Animal Communities.* Methuen, London.
Emmet, A.M. (ed.) (1988) *A Fieldguide to the Smaller British Lepidoptera*. British Entomological and Natural History Society, London.
Emmet, A. & Heath, J. (1991) *The Moths and Butterflies of Great Britain and Ireland. Vol. 7, Part 2.* Harley Books, Essex.
Evans, F.C. (1975) The natural history of a Michigan field. In: Wali, M.K. (ed.) *Prairie: A Multiple View*, pp. 27–51. University of North Dakota Press, Grand Forks.
Falk, J.H. (1976) Energetics of a suburban lawn ecosystem. *Ecology* 57: 141–150.
Fitter, R.S.R. (1945) *London's Natural History*. Collins, London.
Fitter, R., Fitter, A. & Blamey, M. (1974) *The Wild Flowers of Britain and Northern Europe.* Collins, London.
Fitton, M.G., Graham, W.W.R., Boucek, Z.R.J., Ferguson. N.D.M., Huddleston, T., Quinlan, J. & Richards, O.W. (1978) *Handbooks for the Identification of British Insects,* 11 (4). Royal Entomological Society of London, London.
Foelix, R.F. (1982. *Biology of Spiders*. Harvard University Press, Harvard.
Forsythe, T.G. (1987) *Common Ground Beetles.* Naturalists' Handbooks 8. Richmond Publishing Co. Ltd., Berkshire.
Fox, R., Conrad, K.F., Parsons, M.S., Warren, M.S. & Woiwod, I.P. (2006) *The State of Britain's Larger Moths.* Butterfly Conservation and Rothamsted Research, Dorset.
Frank, K.D. (1988) Impact of outdoor lighting

on moths: an assessment. *Journal of the Lepidopterists' Society* 42: 63–93.
Frankie, G.W. & Ehler, L.E. (1978) Ecology of insects in urban environments. *Annual Review of Entomology* 23: 367–387.
Free, J.B. & Butler, C.G. (1959) *Bumblebees*. Collins, London.
Gault, S.M. (1976) *The Dictionary of Shrubs in Colour*. Ebury Press and Michael Joseph, London.
Gibbons, B. & Gibbons, L. (1988) *Creating a Wildlife Garden*. Hamlyn, London.
Gilbert, F.S. (1981) Foraging ecology of hoverflies: morphology of the mouthparts in relation to feeding on nectar and pollen in some common urban species. *Ecological Entomology* 6: 245–262.
Gilbert, F.S. (1986) *Hoverflies*. Naturalists' Handbooks 5. Cambridge University Press, Cambridge.
Gilbert, F.S. & Owen, J. (1990) Size, shape, competition, and community structure in hoverflies (Diptera: Syrphidae). *Journal of Animal Ecology* 59: 21–39.
Gill, D. & Bonnet, P. (1973) *Nature in the Urban Landscape: A Study of City Ecosystems*. York Press, Baltimore.
Gilpin, M.E. & Diamond, J.M. (1980) Subdivision of nature reserves and the maintenance of species diversity. *Nature* 285: 567–568.
Greenslade, P.J.M. (1964) Pitfall trapping as a method for studying populations of Carabidae (Coleoptera). *Journal of Animal Ecology* 33: 301–310.
Hammond, C.O. (1977) *The Dragonflies of Great Britain and Ireland*. Curwen Books, London.
Harding, P.T. & Sutton, S.L. (1985) *Woodlice in Britain and Ireland: Distribution and Habitat*. Institute of Terrestrial Ecology, Monks Wood Experimental Station, Huntingdon.
Harris, S. & Rayner, J.M.V. (1986) A discriminant analysis of the current distribution of urban foxes (*Vulpes vulpes*) in Britain. *Journal of Animal Ecology* 55: 605–611.
Hay, R. & Synge, P.M. (1971) *The Dictionary of Garden Plants*. Ebury Press and Michael Joseph, London.
Heath, J., Pollard, E. & Thomas, J. (1984) *Atlas of Butterflies in Britain and Ireland*. Viking, Middlesex.
Hickling, R. (1978) *Birds of Leicestershire and Rutland*. Leicestershire and Rutland Ornithological Society, Leicester.
Higgs, A.J. & Usher, M.B. (1980) Should nature reserves be large or small? *Nature* 285: 568–569.
Hopkins, B. (1965) *Forest and Savanna: An Introduction to Tropical Plant Ecology with Special Reference to West Africa*. Heinemann, London.
Hoskins, W.G. (1957) *Leicestershire: An Illustrated History of the Landscape*. Hodder & Stoughton, London.
Hubbard, C.E. (1974) *Grasses*. Penguin, Middlesex.
Hutchinson, G.E. & MacArthur, R.H. (1959) A theoretical ecological model of size distribution among species of animals. *American Naturalist* 93: 117–125.
Imms, A.D. (1964) *A General Textbook of Entomology*. Methuen & Co. Ltd., London.
Jones, D. (1983) *The Country Life Guide to Spiders of Britain and Northern Europe*. Country Life Books, Middlesex.
Keble Martin, W. (1967) *The Concise British Flora in Colour*. Ebury Press and Michael Joseph, London.
Kennedy, C.E.J. & Southwood, T.R.E. (1984) The number of species of insects associated with British trees: a re-analysis. *Journal of Animal Ecology* 53: 455–478.
Kerney, M.P. & Cameron, R.A.D. (1979) *A Field Guide to the Land Snails of Britain and North-West Europe*. Collins, London.
Killington, F.J. (1936) *A Monograph of the British Neuroptera*. Ray Society, London.
Kloet, G.S. & Hincks, W.D. (1964) *A Checklist of British Insects*. Handbooks for the Identification of British Insects, XI (1). Royal Entomological Society of London, London.
Leather, S.R. (1986) Insect species richness of the British *Rosaceae*: the importance of host range, plant architecture, age of establishment, taxonomic isolation and species area relationships. *Journal of Animal Ecology* 55: 841–860.
Le Quesne, W.J. & Payne, K.R. (1981) *Cicadellidae (Typhlocybinae) with a Checklist of the British Auchenorrhyncha (Hemiptera, Homoptera)*. Handbooks for the Identification

of British Insects, II (2C). Royal Entomological Society of London, London.
Lindroth, C.H. (1974) *Coleoptera: Carabidae.* Handbooks for the Identification of British Insects, IV (2). Royal Entomological Society of London, London.
Locket, G.H. & Millidge, A.F. (1951) *British Spiders. Vols. I & II.* Ray Society, London.
Lomax, P. (1987) *Leicester Habitat Survey*. City Wildlife Project, Leicester.
Loram, A., Tratalos, J., Warren, P.H. & Gaston, K.J. (2007) Urban Domestic Gardens (X): the extent and structure of the resource in five major cities. *Landscape Ecology* 22: 601–615.
Lott, D. & Owen, J. (2003) *Bruchidius varius* (Olivier) (Bruchidae) in Leicestershire. *The Coleopterist* 12 (3): 113.
LROS (1974–1987) *The Birds of Leicestershire and Rutland*. Annual reports, 1972–1986 of the Leicestershire and Rutland Ornithological Society, Leicester.
Luff, M.L. (1968) Formalin as an attractant. *Entomologist's Monthly Magazine* 104: 115–116.
Luff, M.E. (1975) Some features influencing the efficiency of pitfall traps. *Oecologia* 19: 345–357.
Lutz, F.E. (1941) *A Lot of Insects: Entomology in a Suburban Garden*. Putnam, New York.
MacArthur, R. (1955) Fluctuations of animal populations, and a measure of community stability. *Ecology* 36: 533–536.
MacArthur, R.H. & MacArthur, J.W. (1961) On bird species diversity. *Ecology* 42: 394–398.
MacArthur, R.H. & Wilson, E.O. (1967) *The Theory of Island Biogeography*. Princeton University Press, New Jersey.
Majerus, M. & Kearns, P. (1989) *Ladybirds*. Naturalists' Handbook 10. Richmond Publishing Co. Ltd., Berkshire
Mathias, J.H. (1975) A survey of amphibians in Leicestershire gardens. *Transactions of Leicester Literary and Philosophical Society* 69: 28–41.
May, R.M. (1987) How many species are there? *Nature* 324: 514.
McClintock, D. *et al.* (1964) Natural history of the garden of Buckingham Palace. *Proceedings of the South London Entomological and Natural History Society 1963*. London.
Miess, M. (1979) The climate of cities. In: Laurie, I.C. (ed.) *Nature in Cities*, pp. 91–114. John Wiley, New York.
Mitchell, A. (1974) *A Field Guide to the Trees of Britain and Northern Europe*. Collins, London.
Murdoch, W.W., Evans, F.C. & Peterson, C.H. (1972) Diversity and pattern in plants and insects. *Ecology* 53: 819–829.
Murton, R.K. (1971) *Man and Birds*. Collins, London.
New, T.R. (1971) An introduction to the natural history of the British Psocoptera. *Entomologist* 104: 59–97.
New, T.R. (1974) *Psocoptera*. Handbooks for the identification of British Insects, I(7). Royal Entomological Society of London, London.
Nicholson, B.E., Wallis, M., Anderson, E.B., Balfour, A.P., Fish, M. & Finnis, V. (1972) *The Oxford Book of Garden Flowers*. Oxford University Press, Oxford.
Odum, E.P. (1971) *Fundamentals of Ecology*. W.P. Saunders, Philadelphia.
Owen, D.F. (1978) The effect of a consumer, *Phytomyza ilicis*, on seasonal leaf-fall in holly, *Ilex aquifolium*. *Oikos* 31: 268–271.
Owen, D.F. (1982/83) Fluctuations in abundance of Coccinellidae. *Entomologist's Record* 94: 225–228, 95: 29–31.
Owen, D.F. (1986) Do native plants support a richer lepidopteran fauna than alien plants? *Environmental Conservation* 13: 359–362.
Owen, J. (1981) Trophic variety and abundance of hoverflies (Diptera: Syrphidae) in an English suburban garden. *Holarctic Ecology* 4: 221–228.
Owen, J. (1983) Effects of contrived plant diversity and permanent succession on insects in English suburban gardens. In: Frankie, G.W. & Koehler, C.S. (eds.) *Urban Entomology: Interdisciplinary Perspectives*, pp. 395–422. Praeger, New York.
Owen, J. (1991) *The Ecology of a Garden: The First Fifteen Years*. Cambridge University Press, Cambridge.
Owen, J. & Gilbert, F.S. (1989) On the abundance of hoverflies (Syrphidae). *Oikos* 55: 183–193.
Owen, J. & Owen, D.F. (1975) Suburban gardens: England's most important nature reserve? *Environmental Conservation* 2: 53–59.
Owen, J., Townes, H. & Townes, M. (1981)

Species diversity in Ichneumonidae and Serphidae (Hymenoptera) in an English suburban garden. *Biological Journal of the Linnean Society* 16: 315–336.

Parmenter, L. & Owen, D.F. (1954) The swift, *Apus apus* L., as a predator of flies. *Journal of the Society for British Entomology* 5: 27–33.

Perring, F. (1987) A last word. *Natural World* 19: 5.

Phillips, R. (1981) *Mushrooms and Other Fungi of Britain and Europe.* Pan Original, London

Plant, C.W. (1994) *Provisional Atlas of the Neuroptera, Megaloptera, Raphidioptera and Mecoptera of the British Isles.* Institute of Terrestrial Ecology, Huntingdon.

Plant, C.W. (ed.) (1999) The Natural History of Buckingham Palace Garden, London. Part 1. *London Naturalist* 78 (Suppl.):1–108

Plant, C.W. (ed.) (2001) The Natural History of Buckingham Palace Garden, London. Part 2. *London Naturalist* 80 (Suppl.):109–330.

Pollard, E. (1971) Hedges VI. Habitat diversity and crop pests: a study of *Brevicoryne brassicae* and its syrphid predators. *Journal of Applied Ecology* 8: 751–780.

Pollard, E., Hall, M.L. & Bibby, T.J. (1986) *Monitoring the Abundance of Butterflies.* Research and Survey in Nature Conservation Series, 2. Nature Conservancy Council, Peterborough.

Pope, R.D. (1977) *Coleoptera and Strepsiptera.* Handbooks for the Identification of British Insects, XI (3). Royal Entomological Society of London, London.

Primavesi, A.L. & Evans, P.A. (eds.) (1988) *Flora of Leicestershire.* Leicestershire Museums Service, Leicester.

Pye, N. (1972a) The regional setting. In: Pye, N (ed.) *Leicester and its Region*, pp. 1–15. Leicester University Press for the British Association for the Advancement of Science, Leicester.

Pye, N. (1972b) Weather and climate. In: Pye , N. (ed.) *Leicester and its Region*, , pp. 84–113. Leicester University Press for the British Association for the Advancement of Science, Leicester.

Root, R.B. (1973) Organization of a plant-arthropod association in simple and diverse habitats: the fauna of collards (*Brassica oleracea*). *Ecological Monographs* 43: 95–124.

Rose, F. (1981) *The Wild Flower Key.* Frederick Warne & Co. Ltd., London.

Salisbury, E.J. (1935) *The Living Garden.* G. Bell & Sons Ltd, London.

Sankey, J.H.P. (1988) *Provisional Atlas of the Harvest-spiders (Arachnida: Opiliones) of the British Isles.* Institute of Terrestrial Ecology, Natural Environment Research Council, Huntingdon.

Satchell, G.H. (1947) The ecology of the British species of Psychoda. *Annals of Applied Biology* 34: 611–621.

Simberloff, D. (1986) Design of nature reserves. In: Usher, M.B. (ed.) *Wildlife Conservation Evaluation*, pp. 315–37. Chapman & Hall, London.

Skinner, B. (1998) *Colour Identification Guide to Moths of the British Isles.* Viking, London.

Smith, K.G.V. (1969) *Diptera: Conopidae.* Handbooks for the Identification of British Insects, X (3a). Royal Entomological Society of London, London.

Smith, R.M., Gaston, K.J., Warren, P.H., & Thompson, K. (2006) Urban Domestic Gardens (VIII): environmental correlates of invertebrate abundance. *Biodiversity and Conservation* (2006) 15: 2515–2545.

Southwood, T.R.E. (1961) The number of species of insect associated with various trees. *Journal of Animal Ecology* 30: 1–8.

Southwood, T.R.E. (1978) The components of diversity. *Symposia of the Royal Entomological Society of London* 9: 19–40.

Southwood, T.R.E. & Leston, D. (1959) *Land and Water Bugs of the British Isles.* Frederick Warne & Co. Ltd., London.

Speight, M.C.D. (1978) The genus *Paragus* (Dipt.: Syrphidae) in the British Isles, including a key to known and possible British Isles species. *Entomologist's Record and Journal of Variation* 90: 100–107.

Stace, C. (1997) *New Flora of the British Isles*, 2nd edition. Cambridge University Press, Cambridge.

Steele, R.C. & Welch, R.C. (1973) Monks Wood: A Nature Reserve Record. The Nature Conservancy: Natural Environment Research Council.

Stellemann, P. (1978) The possible role of insect visits in pollination of reputedly anemophilous plants, exemplified by *Plantago lanceolata*, and syrphid flies. *Linnean Society Symposium* 6: 41–46.

Stokoe, W.J. & Stovin, G.H.T. (1948) *The Caterpillars of British Moths*. Frederick Warne & Co. Ltd., London.

Stork, N.E. (1988) Insect diversity: facts, fiction and speculation. *Biological Journal of the Linnean Society* 35: 321–337.

Stubbs, A. & Chandler, P. (eds.) (1978) *A Dipterist's Handbook*. The Amateur Entomologists' Society, London

Stubbs, A.E. & Falk, S.J. (1983) *British Hoverflies: An Illustrated Identification Guide*. British Entomological and Natural History Society, London.

Sutton, S. (1972) *Woodlice*. Ginn & Company Ltd., London.

Tahvanainen, J.O. & Root, R.B. (1972) The influence of vegetational diversity on the population ecology of a specialized herbivore, *Phyllotreta cruciferae* (Coleoptera: Chrysomelidae). *Oecologia* 10: 321-346.

Thomas, J. A., Telfer, M. G., Roy, D. B., Preston, C. D., Greenwood, J. J. D., Asher, J., Fox, R., Clarke, R.T., Lawton, J. H. (2004) Comparative losses of British butterflies, birds and plants and the global extinction crisis. *Science* 303: 1879–1881

Thomas, K. (1983) *Man and the Natural World: Changing Attitudes in England 1500–1800*. Allen Lane, London.

Thompson, F.C., Vockeroth, J.R. & Speight, M.C.D. (1982) The Linnean species of flower flies (Diptera: Syrphidae). *Memoirs of the Entomological Society of Washington* 10: 150–165.

Thompson, K. (2006) *No Nettles Required: The Truth About Wildlife Gardens*. Eden Project Books, London.

Todd, V. (1948) The habits and ecology of the British harvestmen (Arachnida, Opiliones), with special reference to those of the Oxford district. *Journal of Animal Ecology* 18: 209–229.

Torp, E. (1984) *De Danske Svirrefluer (Diptera: Syrphidae)* [The Danish Hoverflies] Danmarks Dyreliv, Vol. 1. Fauna Boger, Copenhagen (in Danish).

Townes, H. (1972) A light-weight Malaise trap. *Entomological News* 83: 239–247.

Townes, H. & Townes, M. (1981) A revision of the Serphidae (Hymenoptera). *Memoirs of the American Entomological Institute* 32.

Usher, M.B. (1986) Wildlife conservation evaluation: attributes, criteria and values. In: Usher, M.B. (ed.) *Wildlife Conservation Evaluation*, pp. 3–44. Chapman & Hall, London.

Weaver, C.R. & King, D.R. (1954) Meadow Spittlebug *Philaenus leucophthalmus* (L.). Research Bulletin 741. Ohio Agricultural Experiment Station, Ohio.

White, G. (1947) Letter XX to Thomas Pennant. *The Natural History of Selborne*. J. Fisher (ed.). The Crescent Press, London.

Wiegert, R.G. & Owen, D.F. (1971) Trophic structure, available resources and population density in terrestrial vs. aquatic ecosystems. *Journal of Theoretical Biology* 30: 69–81.

Williams, G. (1963) Seasonal and diurnal activity of harvestmen (Phalangida) and spiders (Araneida) in contrasted habitats. *Journal of Animal Ecology* 31: 23–42.

Williams, P.H. (1982) The distribution and decline of British bumblebees (*Bombus*). *Journal of Apicultural Research* 21: 236–245.

Witherby, H.F., Jourdain, F.C.R., Ticehurst, N.F. & Tucker, B.W. (1946) *The Handbook of British Birds*. H.F. & G. Witherby Ltd, London.

Yeo, P.F. & Corbet, S.A. (1983) *Solitary Wasps*. Naturalists' Handbooks 3. Cambridge University Press, Cambridge.

Index

Page numbers in **bold** refer to illustrations

A

D

E

F

N

T

U

V

W

Photographic credits

Michael Ballard
5; 13; 153; 191a; 193c

BioImages
120; 162b; 163b; 165b; 180;
213d; 219

Joe Botting
108; 132; 165c; 218b

Simon Garbutt
3; 8; 10a & b; 11a & b; 44; 47b; 74d & h; 100; 173 a & b; 197; 209c, i & j; front cover; back cover (main image)

Mike Grant
29a; 47a; 209b, e, f, g & h; 211; 226; 229; 230

Leicester Mercury Media Group
41

Jonathan Michaelson
80a; 85; 90; 94; 96; 97a & b; 107; 109; 118b & d; 123; 126; 127; 129a & d; 133; 135a & b; 136; 143; 154a & b, 155; 161b; 164a; 165a; 167; 178; 179; 181; 185a & b; 186; 187; 218a; 219a

RHS
20b RHS / Barry Phillips;
53b RHS / Barry Phillips;
72a RHS / Andrew Halstead;
73c RHS / John Maynard;
74a RHS / Wendy Wesley;
74b RHS / Tracy Leverett;
74c RHS / Carol Sheppard;
74e RHS / Graham Titchmarsh;
74f RHS / Barry Phillips;
74g RHS;
74i RHS / Tim Sandall;
74j RHS / Carol Sheppard;
74k RHS / Rebecca Ross;
74l RHS;
75b RHS;
75c RHS;
80b RHS / John Maynard;
98b RHS;
104 RHS / Carol Sheppard;
105a & b RHS / Andrew Halstead;
12 RHS / P. Becker;
131b RHS / Keith Harris;
141 RHS / Keith Harris;
151 RHS / Andrew Halstead;
161a RHS / John Maynard;
203 RHS;
208 RHS / Tim Sandall;
209d RHS Advisory;
213a RHS / Rebecca Ross

Richard Revels
xvi; 45; 46a & b; 52a, b & c; 53a; 54; 56; 67a & b; 68a, b, c & d; 72b; 73a & b; 77; 78; 101; 102; 103; 110; 114; 118a, c & d; 122; 125; 129b & c; 142; 156; 158; 160; 170; 175; 176; 188; 189; 191b; 192a & b; 193a & b; 194a & b; 195a & b; 196a & b; 204a & b; 205; 206c; 210; 213b; 220

RSPB
162a; 174; 213c

Jane Sebire
98a; back cover (author portrait)

Brian Valentine
89; 115a & b; 159; 163a; 166; 168; 169; 183

All other images by **Jennifer Owen**

Artwork

Simon Garbutt
Inset map 7 (relief details © Digital Wisdom 1993 / Mountain High Maps ®);
garden plan 15, 16 & 17

PCGraphics (UK) Ltd
Maps 7, 40 (contain Ordnance Survey data © Crown copyright and database right 2010); relief details (© Digital Wisdom 1993 / Mountain High Maps ®)

Publisher's acknowledgements

The publisher would like to thank the following for their help: Jay Gearing, Andrew Halstead, Michelle Housden, Diana Levy, Rebecca Matthews Joyce, Andrew Salisbury, Carol Sheppard, Ken Thompson, and Cambridge University Press